Signal Transduction and Protein Phosphorylation

NATO ASI Series

Advanced Science Institutes Series

A series presenting the results of activities sponsored by the NATO Science Committee, which aims at the dissemination of advanced scientific and technological knowledge, with a view to strengthening links between scientific communities.

The series is published by an international board of publishers in conjunction with the NATO Scientific Affairs Division

A	**Life Sciences**	Plenum Publishing Corporation
B	**Physics**	New York and London
C	**Mathematical and Physical Sciences**	D. Reidel Publishing Company Dordrecht, Boston, and Lancaster
D	**Behavioral and Social Sciences**	Martinus Nijhoff Publishers
E	**Engineering and Materials Sciences**	The Hague, Boston, Dordrecht, and Lancaster
F	**Computer and Systems Sciences**	Springer-Verlag
G	**Ecological Sciences**	Berlin, Heidelberg, New York, London,
H	**Cell Biology**	Paris, and Tokyo

Recent Volumes in this Series

Series A: Life Sciences

Signal Transduction and Protein Phosphorylation

Edited by

L. M. G. Heilmeyer

Institute for Physiological Chemistry I
Ruhr University—Bochum
Bochum, Federal Republic of Germany

Plenum Press
New York and London
Published in cooperation with NATO Scientific Affairs Division

Proceedings of a NATO/FEBS Summer School on
Signal Transduction and Protein Phosphorylation,
held September 14-26, 1986,
at the Korgialenios School, on the Island of Spetsai, Greece

Library of Congress Cataloging in Publication Data

NATO/FEBS Summer School on Signal Transduction and Protein Phosphorylation (1986: Korgialenios School)
    Signal transduction and protein phosphorylation.

    (NATO ASI series. Series A, Life sciences; vol. 135)
    "Proceedings of a NATO/FEBS Summer School on Signal Transduction and
Protein Phosphorylation, held September 14-26, 1986, at the Korgialenios
School, on the Island of Spetsai, Greece"—T.p. verso.
    "Published in cooperation with NATO Scientific Affairs Division."
    Includes bibliographies and index.
    1. Phosphoproteins—Metabolism—Congresses. 2. Phosphorylation—Con-
gresses. 3. Proteins—Metabolism—Congresses. 4. Cellular control mechanism
—Congresses. I. Heilmeyer, L. M. G. (Ludwig M. G.), 1937-   . II. North Atlantic
Treaty Organization. Scientific Affairs Division. III. Federation of European
Biochemical Societies. IV. Title. V. Series: NATO ASI series. Series A, Life
Sciences; v. 135. [DNLM: 1. Chemistry, Organic—congresses. 2. Phosphotrans-
ferases—metabolism—congresses. 3. Proteins—metabolism—congresses. QU
55 N2857s 1986]
QP552.P5N38   1986   599'.01'852   87-15398
ISBN 978-1-4757-0168-5   ISBN 978-1-4757-0166-1 (eBook)
DOI 10.1007/978-1-4757-0166-1

© 1987 Plenum Press, New York
Softcover reprint of the hardcover 1st edition 1987
A Division of Plenum Publishing Corporation
233 Spring Street, New York, N.Y. 10013

PREFACE

A NATO Advanced Study Institute on "Signal Transduction and Protein Phosphorylation" was held to overview recent developments in this area. The participants in the Institute dealt with protein phosphorylation as the most prevalent mode of regulation of cellular processes.

First, methods needed to analyze the complex cascade systems involved were reviewed, including protein sequencing, crystallography, characterization and isolation of membrane proteins, use of monoclonal or polyclonal antibodies and application of fluorescent probes. In great detail the x ray crystallographic structure of glycogen phosphorylase was presented. This enzyme is located at the end of a signal cascade triggered by the hormonal activation of the membrane-bound adenylate cyclase. The interaction of the hormone/receptor with the catalytic subunit of the adenylate cyclase involves GTP-binding proteins. The function of these recently detected intermembrane coupling factors were reviewed, as well as the structure and properties of various protein kinases.

Major emphasis was placed on Ca^{2+} as a second messenger, its metabolism, mechanism of release and uptake from intracellular stores and its role on cell motility and muscle contraction.

Two classes of protein phosphatases were discussed. They differ in their subunit structure and substrate specificity and are subject of a highly complex regulatory mechanism as yet not fully understood.

The general principles of regulation by signal transduction and protein phosphorylation/dephosphorylation were presented in the context of specific cellular processes. These included control of protein synthesis at the translational level and the mechanism of action of interferon. The discussion included the role of protein tyrosine kinases which are structurally related to some oncogene products and, therefore, implicated in various aspects of cell development and transformation.

This text presents the content of the major lectures and important posters displayed and discussed during the Institut's program. It is the hope that inclusion of recent results discussed in the poster sessions presents the reader an impression on the forefront of research in this area. Initiating this book the editor hopes to convey the proceedings of the NATO Advanced Study Institute on "Signal Transduction and Protein Phosphorylation" to a larger audience and to offer a comprehensive account of those developments in an area which is growing very fast.

Ludwig Heilmeyer

Bochum, February 1987

CONTENTS

II. REGULATION OF MUSCLE CONTRACTION

III. STRUCTURE FUNCTION RELATIONSHIP OF CASCADE ENZYMES

IV. CONTROL OF CELLULAR PROCESSES

I. REGULATORY PROPERTIES OF SIGNAL CASCADES

PROTEIN PHOSPHORYLATION: A HISTORICAL OVERVIEW

Edmond H. Fischer

Department of Biochemistry
University of Washington
Seattle, Washington 98195

This first lecture will offer a rapid overview of one of the most prevalent mechanisms organisms have devised to regulate their cellular function, that is, phosphorylation-dephosphorylation of proteins.

Prior to 1950, a few phosphoproteins were known. Their main function was thought to be almost exclusively related to food sources for the embryo or the young. This was the case for instance for phosvitin of the egg yolk, casein from the milk, etc. But there was no suggestion that protein phosphorylation could be part of a regulatory mechanism. In fact, at that time, very little was known as to how metabolic processes could be regulated, except that one knew that cellular events could not be controlled solely by adjusting the rate at which proteins or enzymes were synthesized. Cells had to have ways of modulating the activity of their enzymes once these had been produced. They had to have the capability of controlling their economy, rapidly and at all time, in response to changing internal or external conditions, to adapt to their environment or satisfy their metabolic needs.

To describe the origin of the covalent control of enzymes by phosphorylation-dephosphorylation, one must retrace the early history of glycogen phosphorylase with which it is intimately linked. When first discovered in the mid-30's by Parnas in Poland and Carl and Gertie Cori in this country, this enzyme was known to have an absolute requirement for AMP for activity. In 1939, Kiessling in Germany reported that he had obtained a form of the enzyme that was active even in the absence of AMP. The Cori's were so incredulous that the report was largely discounted until three years later when Arden Green in the Cori's own laboratories crystallized phosphorylase in a form that, indeed, did not require AMP for activity. They called this active form phosphorylase a, and assumed that it must be the native enzyme since, in the crude extract, it was rapidly converted to the earlier type, that required AMP, which they called phosphorylase b. They made the very logical assumption that in native phosphorylase a, AMP was covalently bound to the protein as a prosthetic group. Conversion of a to b would then be brought about by a "prosthetic group-removing enzyme". But this is when all the trouble started: if this hypothesis were correct, AMP would have to be released during the reaction. They found none. In fact, using the most sensitive analytical procedure available at that time, they could detect no AMP, adenine or ribose, either attached to the native enzyme or liberated during the reaction. The Cori's knew that the enzyme existed in two forms, but did not know how these two forms differed and they actually dropped the problem for the next ten years.

Clarification of what really took place occurred ten years later when, with Ed Krebs, we attempted to introduce new procedures for the purification of the enzyme. The classical Cori procedure called for an early filtration of the muscle extract through a battery of filter papers. The operation was cumbersome: as soon as a filter paper became clogged up, one would switch the gunky mess to the next, and so on. So that step was replaced by a centrifugation. But, no matter how carefully or how fast we worked, the preparations always failed: we could never obtain the so-called native, active phosphorylase a, only the degraded b form. Until one day, in desperation, we decided to repeat the Cori preparation to the letter, paper filtration and all, checking every step, analyzing every fraction from beginning to end. And to our amazement, we found out that the very first extract actually contained the inactive enzyme - not the active form. Yet, when the same extract was passed through filter paper, the solution that emerged was active. A conversion of phosphorylase b to a had taken place.

Filtration through paper as a means of activating the enzyme was really unexpected. Actually, it was not as bizarre as originally thought: it so happened that, at that time, all filter papers were contaminated by calcium ions. These were picked up by the extract while passing through the paper, and it was really the calcium ions that did the trick. If the filters were pre-washed with acid or EDTA, no conversion occurred; if calcium ions were then added directly to the extract, activation resulted. It was rapidly found that calcium did not work alone: ATP always present in fresh muscle extract was also required. If the extracts were aged for a while and ATP was hydrolyzed, no conversion occurred until ATP was readded. It then became clear that ATP was involved in a phosphorylation reaction; this was confirmed by the use of radioactive ATP. The original form is totally inactive in the absence of AMP, the phosphorylated species is fully active whether AMP is present or not. Phosphorylation causes a change in conformation of the protein such that the active site which is masked in the b form becomes catalytically operative. The reaction was obviously enzymatic: it had to be catalyzed by a phosphorylase kinase while the reverse reaction had to involve the removal of the phosphate group by a specific phosphatase. Through the action of these two enzymes, phosphorylase could be shuttled back and forth between these two forms in response to cellular demands.

Calcium was essential for the reaction and yet, when the system was finally purified, it was clear that calcium did not participate directly in the b to a conversion, only Mg^{++} and ATP. This meant that calcium had to act at an earlier step, i.e., the activation of phosphorylase kinase, implying that this enzyme itself also existed in an inactive and active state and that calcium was absolutely required for the activation process. This indicated that a sort of cascade of two or three successive reactions was taking place: activation of phosphorylase kinase by calcium ions; the active kinase then activating phosphorylase, which would finally initiate the degradation of glycogen.

It was not known at that time whether this was a rare event restricted to the control of carbohydrate metabolism, or a more widespread type of regulatory mechanism that would apply to other cellular processes. One thing, however, was already clear: the physiologists had known for many years that muscle contraction is triggered when calcium ions are released in muscle cells in response to a nerve impulse. To maintain contraction, one needs energy in the form of ATP. The finding that both processes, muscle contraction and glycogen degradation were triggered simultaneously by calcium ions, explained how these two physiological events could be regulated in concert.

The next observation was that when Ca^{2+}-activated phosphorylase kinase was preincubated with ATP, its activity increased many-fold. The reaction was slow but greatly accelerated by the addition of cAMP. Most probably, it involved a phosphorylation of phosphorylase kinase. Did cAMP simply increase the rate of an autophosphorylation or did it act through a second kinase (it was called "kinase kinase" at first) that might be present as a contaminant? It was only when this enzyme was purified by Ed Krebs and Don Walsh that its nature as a cAMP-dependent protein kinase of broad specificity could be firmly established. At that time, Sutherland had demonstrated that cAMP is produced as a second messenger to turn on a number of metabolic processes (including the activation of phosphorylase) following the binding of a circulating hormone such as adrenaline to a specific receptor in the cell membrane. Today, we know that the hormonal control of glycogen degradation follows this well-known cascade of successive reactions.

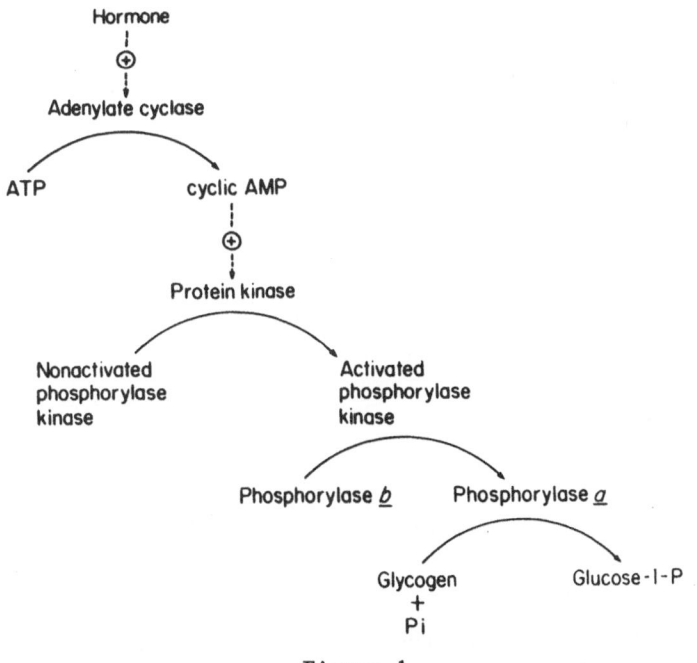

Figure 1.

Everyone of these enzymes is subjected to both allosteric and covalent modification and one might wonder why an organism found it necessary to utilize these two types of regulation? Or again, why have two systems of control evolved simply to convert the inactive form of an enzyme into its active conformation? The

5

need for allosteric control is obvious: it will primarily reflect intracellular conditions or respond to intracellular needs. For instance, when a muscle contracts, ATP is consumed and glycogen or glucose must be metabolized to resynthesize the ATP needed to maintain contraction. Therefore, many glycolytic or Krebs cycle enzymes would be expected to be activated by AMP/ADP and inhibited by ATP. However, if one had to rely solely on allosteric activation, one would affect simultaneously all the enzymes susceptible to these effectors unless strict intracellular compartmentation existed. In contrast, covalent control is mediated by regulatory enzymes which are generally highly specific. It provides the possibility of affecting a single step without necessarily having to touch any other. Furthermore, covalent modification "freezes" an enzyme in a given conformation, often rendering it insensitive to modulation by the usual allosteric effectors.

In contrast, covalent control by phosphorylation-dephosphorylation will respond primarily to external signals.

CONTROL OF CELLULAR PROCESSES BY PHOSPHORYLATION

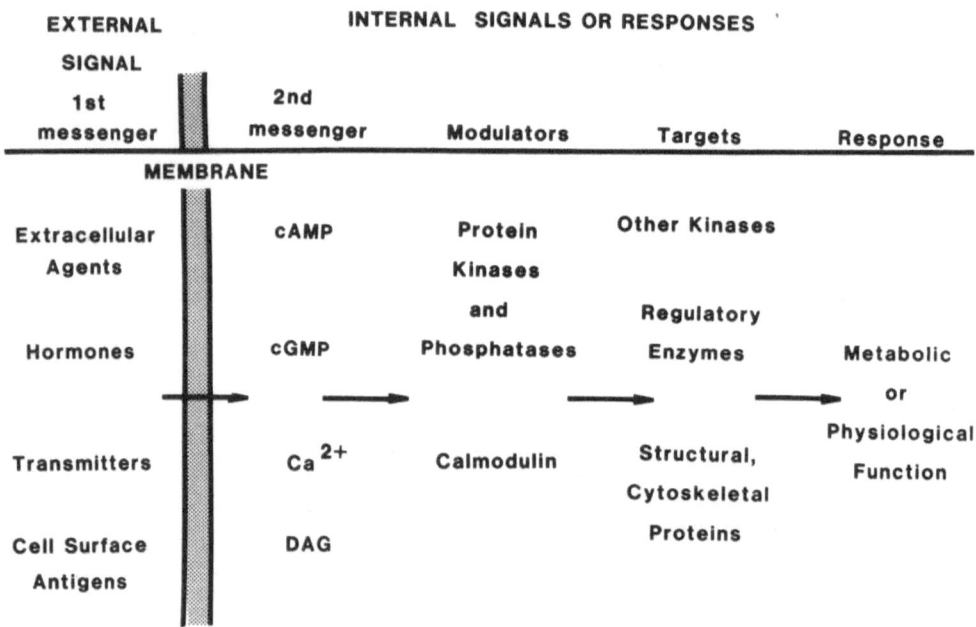

Figure 2.

Consider for example, the control of carbohydrate metabolism in the liver. The prime purpose for regulating glycogenolysis in the liver is to maintain blood glucose levels for the benefit of other organs such as the brain or the erythrocytes, particularly during fasting. Therefore, the sensory mechanism will be external, and the signals sent to the liver to maintain glucose homeostasis will come from the outside -- principally in the form of hormones released from the adrenals and the pancreas. These external signals will act on the plasma membrane, and their effects will be transmitted inside the cell by various second messengers, such as cyclic nucleotides, calcium, diacylglycerol or other compounds yet to be recognized. Each of these will interact with specific modulator proteins which, in turn, might phosphorylate another enzyme, and ultimately elicit some kind of metabolic or physiological response. One already sees here the elements of a cascade system of control.

What other advantages might such cascade reactions hold? The most obvious is that by having an enzyme acting on an enzyme acting on yet another enzyme, one will have a considerable amplification of an original signal -- something on the order of a million-fold or more. This is why extremely minute amounts of a hormone, or any other compound serving as a metabolic signal, can bring about the mobilization of large quantities of a reserve polysaccharide such as glycogen within a very short time. Stadtman and Chock at NIH have considered this problem from a quantitative point of view and made a theoretical analysis of simulated mono- or polycyclic cascade models. They showed that the process was extraordinarily sensitive to even small variations in the concentrations of the various effectors to such an extend that a ten-billion-fold amplification could be theoretically achieved in a four-cycle cascade process in which each parameter varies only by a factor of two.

Equally important, if not more so, is that cascade reactions have pleiotropic functions.

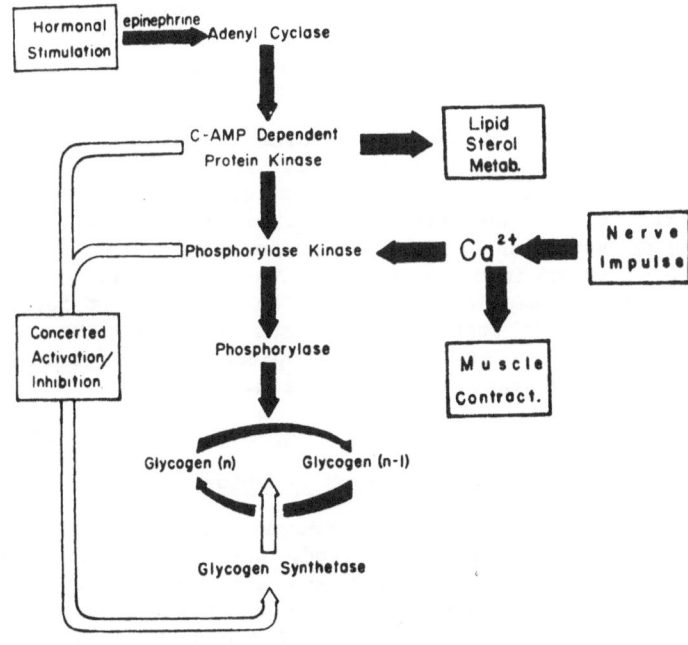

LINKAGE OF PHYSIOLOGICAL PROCESSES
THROUGH CONTROL ENZYMES

Figure 3.

That is, each of the regulatory enzymes involved can affect a number of proteins from other metabolic pathways, in effect linking different physiological processes to one another. As already mentioned, adenylate cyclase serves as the link between the endocrine system and many intracellular pathways, including those of carbohydrate metabolism. Phosphorylase kinase links the same process to muscle contraction: following neural stimulation, calcium released from the sarcoplasmic reticulum will trigger contraction by binding to the troponin complex on the thin filament, or to myosin light chain kinase through its modulator protein, calmodulin. By the same token, it will trigger glycogen breakdown by activating phosphorylase kinase. The third member of this cascade, the cAMP-dependent protein kinase, plays a pleiotropic role par excellence, since it can link carbohydrate metabolism to a number of other pathways such as lipid metabolism (by acting on the hormone-sensitive lipase), cholesterol metabolism (through cholesterol esterase), steroid metabolism (through HMG CoA reductase) and so on. Both cAMP-dependent protein kinase and phosphorylase kinase will also act on glycogen synthase; however, while protein phosphorylation activates phosphorylase, it inhibits glycogen synthase. This affords a synchronous but opposite control of the activity of the two enzymes. It renders the system exquisitely sensitive to slight variations in the concentration of a number of effectors acting in a reciprocal manner. It also prevents the establishment of a futile cycle which would be engendered by the simultaneous synthesis and breakdown of glycogen, thus providing a safeguard against a useless utilization of ATP.

There is perhaps another advantage of relying on the covalent modification of proteins. All such regulatory enzymes have oligomeric structures, and there is no reason why the modification reaction should proceed in an all or none fashion. In the case of phosphorylase, if partially phosphorylated intermediates were produced,it is possible that the properties of the enzyme would be neither those of phosphorylase b nor a. These intermediates which could be active in the presence of substrates (P_i or G1P) would be strongly inhibited by physiological concentration of G6P. Therefore, small variations in the relative proportion of these sugar esters might afford a sensitive and effective means of controlling glycogen utilization.

Finally, it would be quite impossible to pack all the information needed for such multiple interactions within a single protein. Therefore, all the excess of information required for calcium regulation, hormone recognition or interaction with other metabolic pathways can only be stored in annex molecules, just as one would store excess information in a second disc on a computer when the first one has become saturated. In fact, it is for this reason that most of these regulatory enzymes are themselves made up of different subunits, each with the complement of information it needs to carry out its characteristic function.

Cellular regulation by enzyme phosphorylation

When control of enzyme activity by reversible phosphorylation was first uncovered, one didn't know whether this would be a unique occurrence or a more general phenomenon which might apply to a number of other systems. Would protein phosphorylation be restricted to enzymes involved in carbohydrate metabolism? Would enzymes involved in nitrogen metabolism be covalently modified by, let's say, amidation-deamidation, or those of lipid metabolism by acetylation or methylation, and so forth? While other types of covalent modifications have been described (such as the adenylylation and uridylylation controlling bacterial glutamine synthetase; the ADP-ribosylation of elongation factor 2 and the G-protein of adenylate cyclase or tranducin, catalyzed by diphtheria or cholera toxin respectively; the methylation of the chemotactic receptor in bacteria, etc.), it is clear that phosphorylation reactions play a fundamental role in cellular processes. Just as an example, the following figure illustrates a few of the systems which are modified or regulated by phosphorylation in a schematized liver cell.

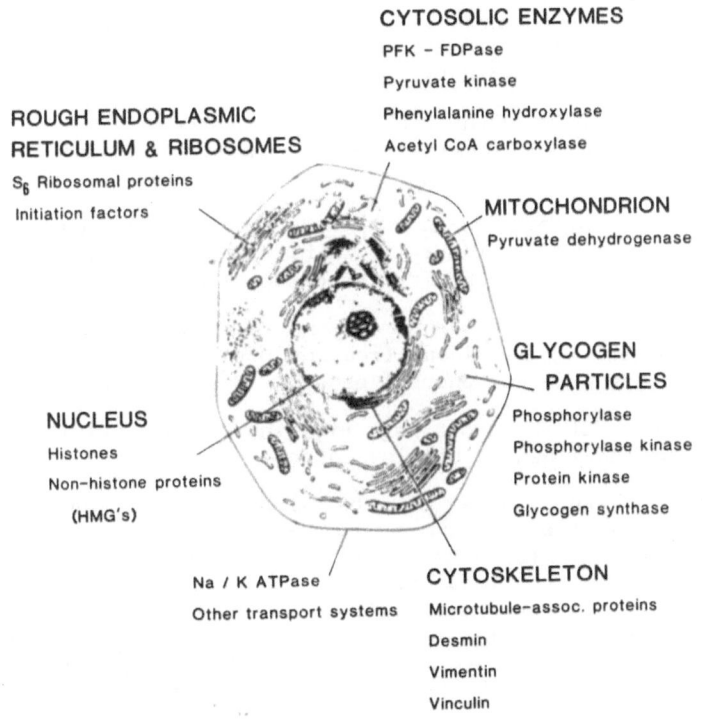

CYTOSOLIC ENZYMES
PFK – FDPase
Pyruvate kinase
Phenylalanine hydroxylase
Acetyl CoA carboxylase

ROUGH ENDOPLASMIC RETICULUM & RIBOSOMES
S_6 Ribosomal proteins
Initiation factors

MITOCHONDRION
Pyruvate dehydrogenase

GLYCOGEN PARTICLES
Phosphorylase
Phosphorylase kinase
Protein kinase
Glycogen synthase

NUCLEUS
Histones
Non-histone proteins
(HMG's)

Na / K ATPase
Other transport systems

CYTOSKELETON
Microtubule-assoc. proteins
Desmin
Vimentin
Vinculin

Figure 4.

It includes a great many cytosolic enzymes such as those implicated in carbohydrate, amino acid, lipid and sterol metabolism, as well as the mitochondrial pyruvate dehydrogenase, various ribosomal proteins or initiation factors involved in protein synthesis, histones and non-histone proteins associated with chromatin and found in nucleosomes, a number of cytoskeletal elements and membrane proteins involved in ion transport or cell-cell interaction. In neurons, tyrosine and tryptophan hydroxylase are phosphorylated, as well as myelin basic protein and the two synapsins described by Walter and Greengard. In muscle, troponin-I and troponin-T are phosphorylated, together with one of the light chains of myosin and certain components of the sarcoplasmic reticulum implicated in calcium transport. Essentially all hormone receptors, including the acetylcholine receptor at the neuromuscular junction, transport ATPases, etc. are phosphorylated, and this list increases almost daily.

Some of the highlights in the development of the field of phosphorylation/dephosphorylation are listed below:

1955 - Interconversion of phosphorylase. Fischer, Krebs, Sutherland.

1957 - cAMP as a second messenger. Sutherland.

1968 - cAMP-dependent protein kinase: the way cAMP expresses its presence in eukaryotic cells. Krebs and Walsh.

1969 - Phosphorylation of the pyruvate dehydrogenase complex: first example of regulation of mitochondrial enzyme not directly involved in carbohydrate metabolism. Reed.

1970-1971 - Calmodulin as a mediator of Ca^{2+} effects. Ca^{2+}/calmodulin is involved in the activation of Ca^{2+}-dependent protein kinases. Hartshorn.

1978 - The product of the src gene from oncogenic Rous sarcoma virus is a protein kinase. This kinase phosphorylates tyrosyl residues in proteins. Collet, Erickson; Tony Hunter.

1979-1983 - Diacylglycerol, Ca^{2+}/phospholipid- activated protein kinase C, and identification of this enzyme as the receptor for the tumor promoter phorbol esters. Nishizucka.

1980 - EGF receptors have tyrosine kinase activity triggered by binding of the ligand. Stanley Cohen.

Perhaps one of the most exciting developments has resulted from the finding a few years ago that certain oncogenic retroviruses bring about cell transformation by triggering phosphorylation reactions and that the kinases involved catalyze the phosphorylation, not of seryl or threonyl residues as had been observed until then, but of tyrosine side chains. Furthermore, such protein tyrosine kinases are common to all cells, and are induced in certain receptors upon binding of their specific hormone or growth factor.

So you see, protein phosphorylation has come a long way since its early discovery. A process that started as a simple, isolated event in the control of glycogen metabolism has found its way into just about every phase of the life of a cell. Many of its more significant aspects will be covered in detail in the course of this conference by some of the people who have contributed the most to the development of the field. I am sure we can all look forward to a most exciting meeting.

REFERENCES

Fischer, E.H., Brautigan, D.L., (1982) TIBS, Vol. 7, No. 1, 0. 3-4

Fischer, E.H., (1983) Bulletin de L'Institut Pasteur 81, p. 7-31.

Krebs, E.G., (1981) Cukrr, topics in Cell. Reg., 18, 401-419.

Cohen, P. (1982) Nature 296, 613-620.

SIGNAL TRANSDUCTION THROUGH cAMP and cGMP

Jackie D. Corbin, Stephen J. Beebe, Charles E. Cobb,
Sharron H. Francis, Jack N. Wells*, Stanley L. Keeley[+],
Thomas W. Gettys, Peter F. Blackmore, Lynn Wolfe,
and Leslie R. Landiss

Howard Hughes Medical Institute Lab., Dept. of Molecular
Physiology and Biophysics, and *Dept. of Pharmacology,
Vanderbilt Univ., Nashville, TN 37232, USA

[+]Present address: Cardiovascular Research, Bristol-Myers Co.
2404, West Pennsylvania, Evansville, IN 47712

SIGNAL TRANSDUCTION THROUGH CASCADE SYSTEMS

Cells respond to hormones, neurotransmitters and other agents in two opposing ways. These are referred to as amplification (enhancement) and adaptation (diminution) (1). Amplification enables an organism or cell to respond to a very faint signal such as a low blood hormone concentration. Adaptation prevents constant background stimulation, or excessive stimulation, of a pathway. The cAMP cascade system illustrates two different kinds of amplification: the first is magnitude amplification, which is an increase in output molecules in greater numbers than input molecules; and the second is sensitivity amplification, which is a greater percentage increase in ouput than the percentage increase in input. An example of magnitude amplification would be the production of 100 cAMP molecules from 1 active molecule of adenylate cyclase, and sensitivity amplification could be an increase in cAMP-dependent protein kinase activity of 200º/o by 100º/o increase in cAMP. Both kinds of amplification can occur at each step of a cascade, although the overall magnification in a cascade can also be calculated. For example, the overall magnitude amplification for glucagon stimulation of glycogen breakdown in the liver is represented by the number of glucose molecules produced divided by the number of glucagon molecules added, and can be greater than 10,000.

One way that sensitivity amplification can be achieved is through stimulation of one enzyme of a cyclic system and inhibition of the opposing enzyme of the cycle. This would occur if a protein kinase were stimulated and a phosphoprotein phosphatase inhibited by the same regulatory agent. Sensitivity amplification can also be achieved through cooperative stimulation of an allosteric enzyme by its activator. The cAMP-dependent protein kinase is such an allosteric enzyme as indicated by its relatively high Hill constant of 1.6 - 1.8 (2). Other steps of the cAMP cascade may also exhibit cooperative effects of activators, and the sum of the cooperative effects on all of the enzymes of the cascade could theoretically result in a very high sensitivity amplification for the whole pathway. However, as seen in Fig. 1 for epinephrine regulation

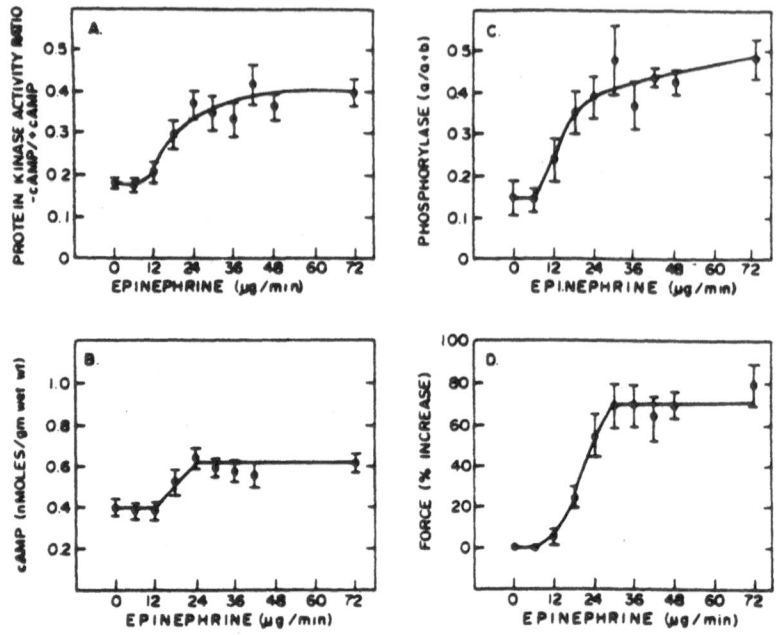

Figure 1. Effect of epinephrine concentration on various steps of the cAMP cascade in the perfused rat heart. Incubation time = 2 min. [Data from S.L. Keely and J.D. Corbin, Am. J. Physiol. 233(2): H269-H275(1977)].

of phosphorylase in the perfused rat heart, a doubling in the epinephrine concentration (from 12 to 24 µg/min) results in only slightly more than doubling of the phosphorylase activity (from activity ratio of 0.15 to 0.40) and only a 50°/o increase in the contractile force. Thus, the overall sensitivity amplification is marginal or absent. There is also no apparent sensitivity amplification for increases in cAMP although the increase in protein kinase activity exhibits some sensitivity amplification.

It appears that the reason for apparent lack of sensitivity amplification at particular steps of the cAMP cascade, or in the overall cascade, is not that it is absent, but that it is masked by adaptation. The cAMP cascade system illustrates several types of adaptation. Biological systems have evolved "leak" steps such as cyclic nucleotide phosphodiesterases, phosphoprotein phosphatases, or negative feedback regulation in order to prevent excessive amplification. When the cascade is activated by hormone elevation, the substrates (cAMP, phosphoproteins, etc.) of the respective enzymes are elevated, thus by mass action causing increased activities of these enzymes. The same enzymes may be activated further by feedback regulation. The "low Km" or "hormone-sensitive" phosphodiesterase is activated by cAMP-dependent protein kinase stimulation in several mammalian tissues (3,4). This presumably occurs by catalytic subunit-catalyzed phosphorylation either of an intermediate protein or of the phosphodiesterase itself. The level of cellular cAMP is therefore determined by the balance between the state of activation of

adenylate cyclase and activation of phosphodiesterase as illustrated in
Fig. 2. The phosphodiesterase activation thus compromises any
sensitivity amplification that might occur at the cAMP step of the
cascade.

cAMP RECEPTORS IN PRO- AND EUKARYOTES

cAMP serves a similar physiological function in bacteria and
mammals: when sugar availability is low, cAMP is elevated, thus causing
increased availability of sugar. The two types of cAMP receptors which
mediate these effects are homologous proteins (5), the bacterial protein
(6) being termed catabolite gene activator protein (CAP) and the
mammalian protein being named regulatory subunit of cAMP-dependent
protein kinase (2). By binding directly to DNA the CAP protein mediates
cAMP activation of expression of genes for proteins involved in transport
and metabolism of certain sugars, while the regulatory subunit is
responsible for cAMP modulation of enzyme activities through protein
phosphorylation. CAP does not possess, nor is it associated with,
protein kinase activity; the regulatory subunit lacks the DNA binding
domain exhibited by CAP, although it is known that cAMP-dependent protein

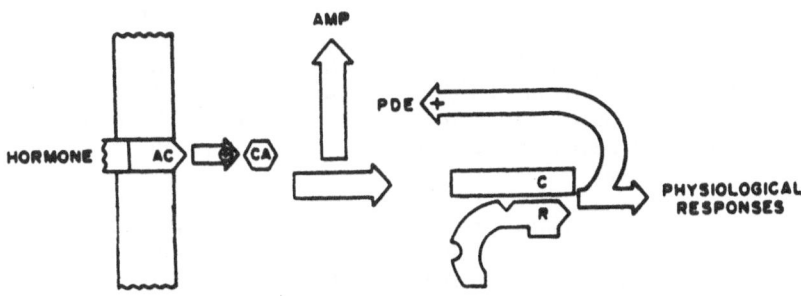

Figure 2.   Adaptation of the cAMP cascade by feedback activation of
phosphodiesterase. For simplicity the cAMP-dependent protein kinase is
presented as the dimer rather than the tetramer.

kinase is responsible for cAMP effects on gene expression in mammals
(7). All effects of cAMP in mammals that have been characterized to date
are mediated by cAMP-dependent protein kinase. This enzyme has been
studied in detail with regard to isozymes, structure, mechanism of
action, and physiological roles. The enzyme is a tetramer which contains
two regulatory and two catalytic subunits. The regulatory
subunit has two cAMP binding sites on each subunit, which are referred to

as sites 1 and 2 (8). The binding of cAMP at these two sites is
cooperative and the cooperativity is mainly due to intrasubunit
interactions. Although it is known that during cAMP activation of the
enzyme the catalytic subunit dissociates from the regulatory subunit, the
precise roles of multiple cAMP binding sites on the regulatory subunit in
the activation process are not clear. Preliminary results suggest that
two of the four binding sites can be occupied by cAMP without causing
significant activation. This binding is distributed equally between
sites 1 and 2. At present it is not certain whether the two cAMP
molecules are bound on the same subunit, on different subunits, or
whether they occupy both of the site 1 or site 2 domains in a population
of half-saturated holoenzyme molecules. It is possible that the inactive
ternary complex is a "primed" enzyme in basal tissues. It exhibits a
higher Hill constant and binds cAMP at lower concentrations in vitro than
does the cAMP-free form of the holoenzyme. The formation of an inactive
ternary complex by cAMP binding to the holoenzyme and activation of the
enzyme by further cAMP elevation can be illustrated by the equation:

$$R_2C_2 + 2 \text{ cAMP} \rightleftharpoons R_2(\text{cAMP})_2 \cdot C_2 \xrightarrow{2 \text{ cAMP}} R_2(\text{cAMP})_4 + 2 C$$
$$\text{inactive} \qquad\qquad\qquad \text{inactive} \qquad\qquad\qquad \text{active}$$

MECHANISM AND ROLE OF cGMP-DEPENDENT PROTEIN KINASE

cGMP- and cAMP-dependent protein kinase are evolutionarily related
proteins (9). They have similar, but not identical, substrate
specificities in vitro. Each has a regulatory and catalytic component,
although these two components of the cGMP kinase are covalently linked by
peptide bond. The regulatory component of each enzyme has two cyclic
nucleotide binding domains on a single subunit (10,11). These binding
sites have different cyclic nucleotide analog specificity and are
positively cooperative. With the cGMP enzyme, cGMP is bound only by site
1 at low cGMP concentrations, while binding at site 2 occurs at higher
cGMP concentrations. About 50⁰/o of total enzyme activation occurs
after occupation of both site 1 domains. At higher cGMP concentrations
both site 2 domains become occupied, causing activation of the remaining
50⁰/o of the activity.

cGMP and cGMP-dependent protein kinase have been known to be present
in tissues for many years, but the biological function(s) have remained
obscure. The cGMP enzyme has a more restricted distribution than does
the cAMP enzyme. The resolution of two peaks of cGMP-dependent activity
by DEAE-cellulose chromatography of extracts of certain tissues such as
the pig coronary artery suggests the presence of multiple isozymes as is
the case for cAMP-dependent protein kinase. Recently, it has been
suggested that certain effects of the newly discovered hormone,
atriopeptin, are mediated by cGMP-dependent protein kinase. Of
particular interest has been that atriopeptin-induced smooth muscle
relaxation in artery walls is possibly brought about by activation of
this enzyme. This possibility has been tested by comparing the potencies
of cGMP and cAMP analogs in activating the purified cyclic nucleotide
kinases in vitro with the potencies of these analogs in relaxation of the
intact pig coronary artery strip. The results suggest a role of the
cGMP-dependent protein kinase in regulation of smooth muscle relaxation.

REFERENCES

1. Koshland, D.E., Jr., Goldbeter, A., and Stock, J.R. (1982) Science
   217:220-225.
2. Flockhart, D.A. and Corbin, J.D. (1982) CRC Crit. Rev. Biochem.
   12:133-186.

3. Corbin, J.D., Beebe, S.J., and Blackmore, P.F. (1985) <u>J. Biol. Chem.</u> 260:8731-8735.
4. Gettys, T.W., Blackmore, P.F., Redmon, J.B., Beebe, S.J., and Corbin, J.D. (1987) J. Biol. Chem. 262, (In press).
5. Weber, I.T., Takio, K., Titani, K., and Steitz, T.A. (1982) <u>PNAS</u> 79:7679-7683.
6. de Crombrugghe, B., Busby, S., and Buc, H. (1984) <u>Science</u> 224:831-838.
7. Beebe, S.J., Koch, S.R., Granner, D.K., and Corbin, J.D. (1985) 13th International Congress of Biochemistry (Abstracts) Amsterdam, p. 649.
8. Rannels, S.R. and Corbin, J.D. (1981) J. Biol. Chem. 256:7871-7876.
9. Lincoln, T.M. and Corbin, J.D. (1983) <u>Adv. Cyclic Nucleotide Res.</u> 15:139-192.
10. Corbin, J.D. and Doskeland, S.O. (1983) <u>J. Biol. Chem.</u> 258:11391-11397.
11. Corbin, J.D. Ogreid, D., Miller, J.P., Suva, R.H., Jastorff, B., and Doskeland, S.O. (1986) <u>J. Biol. Chem.</u>, 261:1208-1214.

SIGNAL CASCADES IN REGULATION OF GLYCOGENOLYSIS

Ludwig M. G. Heilmeyer, Jr.

Institut für Physiologische Chemie
Abt. für Biochemie Supramolekularer Systeme
Ruhr-Universität Bochum, Universitätsstr
4630 Bochum, West-Germany

Glycogen phosphorylase was the first enzyme recognized some thirty years ago to be regulated by phosphorylation/dephosphorylation. Over the years many fundamental observations allowed to formulate cascades by which extracellular signals could be coupled to the physiological response - glycogenolysis (1). Before trying to analyze this "glycogenolytic cascade" on the basis of accumulated physiological and biochemical data it seems worthwhile to summarize some general principles inherent in intracellular signal pathways. Extracellular agonists like hormones or neurotransmitters are pleiotropic, however, they provoke a cell specific physiological response. As an initiating event agonists combine with specific receptors on the cell surface. Plasma membrane-localized signaling systems, transmit and amplify the signal; they release intracellularly a second messenger. All known second messengers again are pleiotropic and potentially can influence simultaneously more than one process. An advantage of this pleiotropism is the possibility to coordinate cellular processes e.g. cell motility with energy metabolism or, for example, to prevent futile cycling by shutting down glycogen synthesis when glycogenolysis is activated. In analogy to the extracellular event second messengers interact with specific receptors; their saturation constitutes the corresponding intracellular signal. Somewhere in the cascade the pleiotropic signal must be translated into a specific signal which is carried out by a "specificator". Activation or inactivation of a specificator is ultimately responsible for a specific cellular response triggered by the extracellular signal. The principle steps in a signal cascade are outlined in Fig. 1. Generally, cascades allow signal amplification. Even more important might be that the steady state level of an interconvertible enzyme, located in a cascade, can change in response to a signal with very high sensitivity, a phenomenon called 'signal sensitisation' (2). This signal sensitisation is an important inherent property of signal cascades (2). Often, the specificator is a point at which two signal pathways merge. Therefore, the specificator can exert two functions, namely signal translation and signal selection. In principle, this

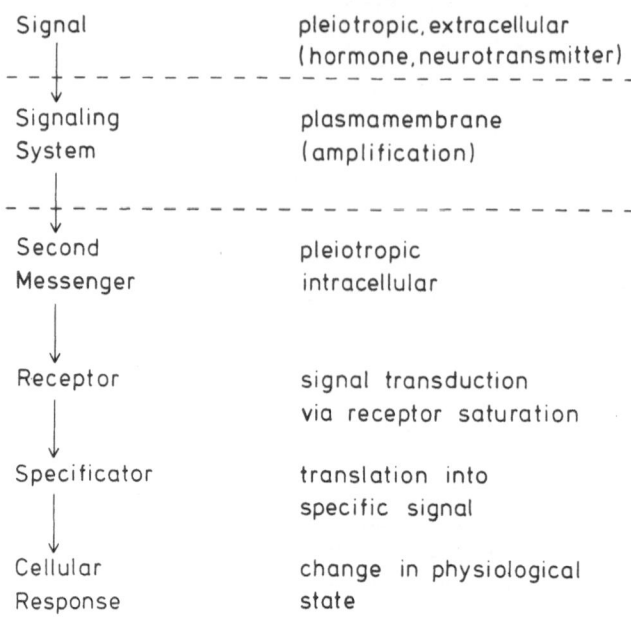

| Signal | pleiotropic,extracellular (hormone,neurotransmitter) |

Signaling System — plasmamembrane (amplification)

Second Messenger — pleiotropic intracellular

Receptor — signal transduction via receptor saturation

Specificator — translation into specific signal

Cellular Response — change in physiological state

Fig. 1. Principal Components of |Signal Pathways

selection may result in simply additive or synergistic or antagonistic responses. Moreover, there are many interconnections of signal pathways. Components of one signal pathway modulate the signal transduction in the other pathway and vice versa. Such a signal network allows fine tuning of cellular responses in relation to two or more extracellular signals (for a diagrammatic representation see Fig. 2).

Two extracellular signals, the hormone epinephrine and plasma membrane depolarization upon acetylcholine release at the neuro muscular junction are well known to trigger glycogenolysis in muscle (Fig. 3) (3). Epinephrine initiates a complex reaction sequence in the plasma membrane involving G-proteins which finally results in adenyl cyclase activation. It causes an increase in the concentration of the intracellular pleiotropic signaling molecule, cyclic AMP. This nucleotide releases the catalytic subunit from the cyclic AMP dependent protein kinase which again constitutes a pleiotropic signal since the free catalytic subunit is able to phosphorylate quite a variety of proteins. The first well characterized substrate for this protein kinase was phosphorylase kinase. In the signal cascade for hormone activation of glycogenolysis phoshorylase kinase represents the specificator. Phosphorylation of phosphorylase kinase enhances its enzyme activity which then convertes inactive glycogen phosphorylase b to the active a form thereby triggering glycogen degradation. The product glucose 1-phosphate enters glycolysis to fullfill energy requirements of the cell (Fig. 3). Alternatively, glycogenolysis is stimulated when muscle contracts which occurs upon depolarization of the sarcolemn. Generally, it is accepted that depolarisation causes Ca^{2+} release from the terminal cysternae of the sarcoplasmic reticulum resulting in an increase of ca. one order of magnitude in sarcoplasmic free Ca^{2+}. Phosphorylase kinase again is the specificator which converts the pleiotropic signal Ca^{2+} into the specific signal i.e.

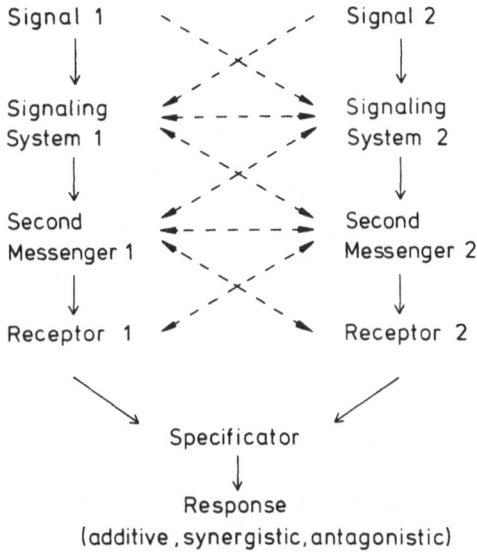

Fig. 2. Signal Selection from Signal Networks

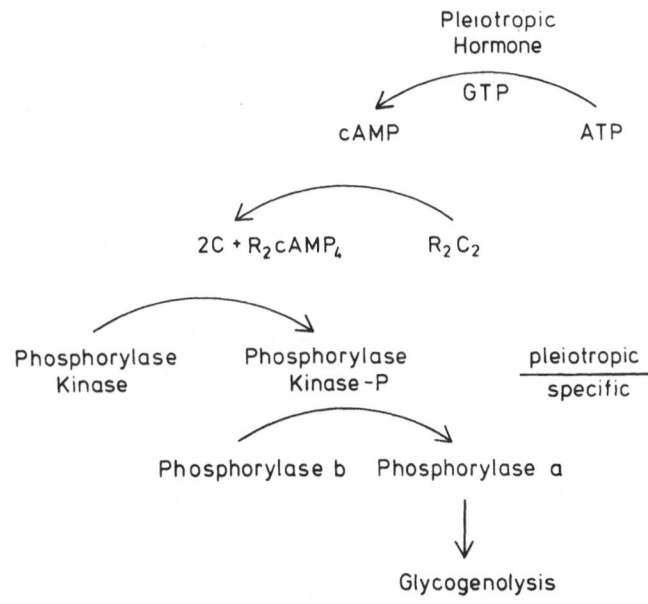

Fig. 3. Signal Transduction in Hormonal Stimulation of Glycogenolysis

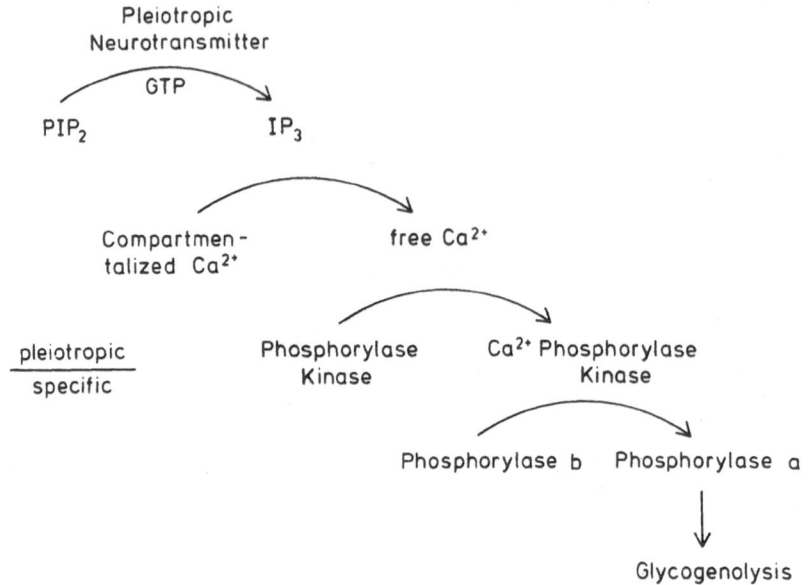

Fig. 4. Signal Transduction in Neuronal Stimulation of Glycogenolysis

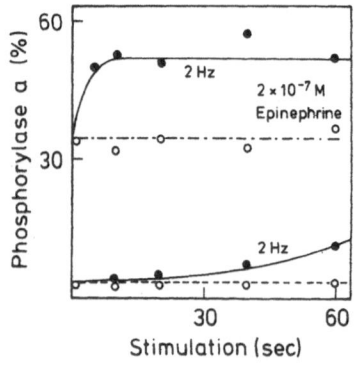

Fig. 5. Submaximal Electrical and/or Hormonal Stimulation of Phosphorylase
a Formation in Frog Skeletal Muscle

phosphorylase a formation. Ca^{2+} binds to its receptor, calmodulin, which is an integral part of phosphorylase kinase. The reaction of calmodulin with Ca^{2+} activates the enzyme which then triggers phosphorylase b to a conversion and thereby also glycogen degradation (Fig. 4).

It has been amply documented that in intact muscle phosphorylase a formation can be triggered along these two signal pathways, separately. Either electrical stimulation or ß-agonists e.g. isoproterenol can stimulate phosphorylase a formation up to 100 % conversion (for review see 4) Many carefull studies, however, indicate a more complex interrelationship between extracellular signaling and intracellular response i.e. phosphorylase b to a conversion. An example is shown in Fig. 5. Repetitive direct electrical stimulation at 2 Hz of frog skeletal muscle leads to a slow accumulation of phosphorylase a. The system behaves like an integrator: phosphorylase a accumulates with time proportional to the amount of single twitches. Epinephrine, at the choosen concentration, augments the phosphorylase a level to ca 30 % in resting muscle. More importantly, the hormone accelerates the phosphorylase a accumulation upon repetitive stimulation at 2 Hz i.e. a synergistic effect is clearly demonstrable (see Fig. 5). Detailed analysis of solely hormonal activation of glycogenolysis in intact mammalian muscle allows to differentiate three phases. 1. phosphorylase a formation can occur without a change in cAMP or conversion of phosphorylase kinase to the activated form; 2. the phosphorylase a level increases proportional to the rise of the cAMP level and conversion of non activated phosphorylase kinase to the activated form; 3. phosphorylase a accumulates further but cAMP increases to a greater amount which is not reflected in a greater formation of activated phosphorylase kinase.

These few examples out of an enormous amount of data available in literature shall demonstrate that there is no direct correlation between the extent of extracellular stimulation and intracellular response. Maximal response upon either neuronal or hormonal stimulation can be easily explained on the basis of current schemes of the glycogenolytic cascades (Fig. 3 and 4). However, at submaximal stimulation which might represent the more physiological event new phenomena like an integrator like function upon repetitive neuronal stimulation or synergistic effects upon simultanueous submaximal neuronal and hormonal stimulation or independence from the second messenger concentration are detectable. These more complex phenomena are not yet understood and might become understandable when each component of the signal network is known on a molecular level. The knowledge of the molecular details of phosphorylase kinase is certainly of crucial importance due to its location as specificator at the merging point of signal pathways in the glycogenolytic cascades (Fig. 6). Simultaneous phosphorylation and Ca^{2+} saturation migth result in synergism as observable in intact muscle (see Fig. 5).

There is a large amount of information available on the regulation of phosphorylase kinase activity by Ca^{2+} or phosphorylation; in comparison there is much less structural information available especially on the two high molecular weight subunits α and β (see below). In the following structural aspects of each subunit are summarized separately and these informations are combined with functional aspects.

The specific and selective functions of phosphorylase kinase in the glycogenolytic cascades are underlined by the fact that it is a very complex enzyme. It was first isolated over ten years ago and shown to be a multimeric enzyme composed of four kinds of subunits called α, β, γ and δ (Fig. 7). Each subunit is present in the complex approximately 4 fold yielding an enzyme of a native molecular weigth of ca. 1.25 million. Cohen assigned a structure of $α_4$, $β_4$, $γ_4$, $δ_4$ (5) whereas Krebs and coworkers found a higher relative amount of C, a subunit which is equivalent to the γ - subunit (6). Most workers in the field found varying amounts of this γ-subunit in the isolated holocomplex. Electron microscopic images reveal "butterfly" or "chalice" like forms (Fig. 7) (8, 9). The dimension of one butterfly wing or of the cup of the chalice are similar, 18-20 nm x 10-11 nm. Minimally, the isolated enzyme contains 8 moles phosphate/mole of protein. Isolation of the enzyme in presence of protein phosphatase inhibitors increases the phosphate content up to 20 moles phosphate/mole protein. ^{31}P NMR spectroscopy identifies all of these phosphates as phosphoserines which seem all to be tightly complexed with other amino acid residues of the molecule (10, 11). Recently, ADP has been characterized as an allosteric effector influencing the activity of the enzyme. Maximally 8 moles ADP/mole of protein can be bound (12). Saturation of the enzyme with Ca^{2+} also results in binding of 8 moles Ca^{2+}/mole of protein in absence of Mg^{2+}. Millimolar concentrations of Mg^{2+} induce additional Ca^{2+} binding sites so that 16 moles Ca^{2+}/mole protein can be bound (13). Indirectly, it was concluded that induction of these 8 additional Ca^{2+} binding sites requires the binding of 8 moles Mg^{2+}/mole of protein (7). Optimal activity of this enzyme is expressed at rather unphysiological conditions like high pH values of about 8.6 or extremely high concentrations of Mg^{2+} (ca. 50 mM) in combination with high concentrations of free Ca^{2+} (ca. 100 µM) (7). Thus, the α β γ δ -tetramer represents a kind of inactive storage form in the cell, from which by different activation processes species are generated which can express activity under physiological conditions. Characterization of the enzymatic activity as function of the free Ca^{2+} concentration allows to differentiate three kinds of activities called A_0, A_1 and A_2. A_0 is Ca^{2+} independent and is expressed at nanomolar free Ca^{2+}. It represents only less than 1 % of the optimal activity. A more than 30fold enhancement of the A_0 activity is observed when calmodulin i.e. the δ-subunit is forced to dissociate from the holoenzyme. Therefore, it is concluded that calmodulin plays a dual role in phosphorylase kinase: on the one hand it represses the A_0 activity of the Ca^{2+} free form and - like in other Ca^{2+}/Calmodulin dependent systems - activates the Ca^{2+} dependent activities A_1 and A_2 (7, 14, 15).

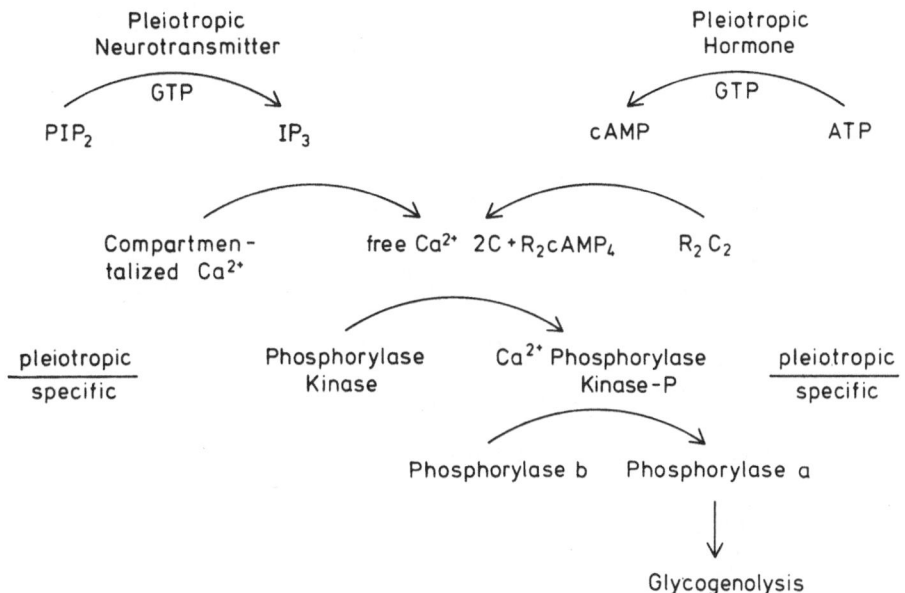

Fig. 6. Simultaneous Hormonal and Neuronal Stimulation of Glycogenolysis

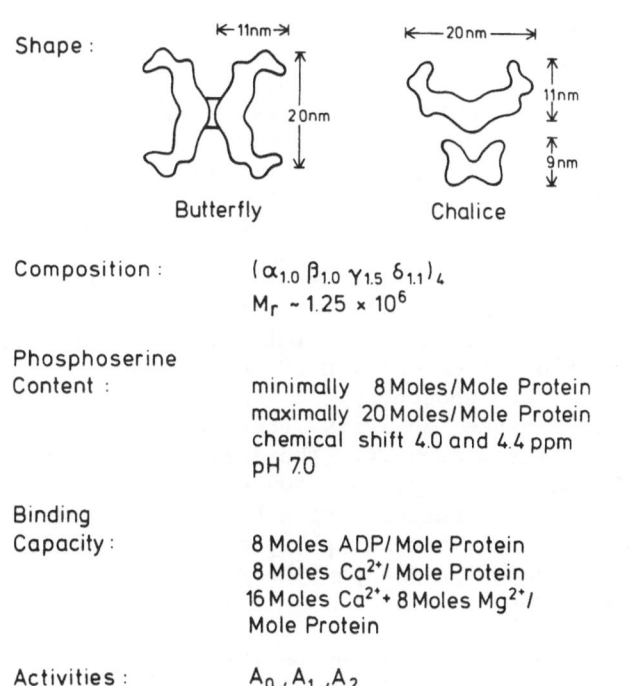

Fig. 7. Molecular Description Holophosphorylase Kinase

Mr   ~140,000
Mr α'-Isoform: ~138,000

N-Terminus
α + α' :   M R S R S N S G V R L D S Y A R L

ATP/ADP   K M Q D G Y F G G A R
Binding Site :   FTC

Phosphoryla-
table Sites   P
cAMP Kinase :   G V E F R R L S I S T E S Q P P D G G H S

                                   P
Auto₁ :   T G I M Q L K S E I K

                                P
Auto₂ :   E F G V E R S V R P T D S N —

                                P
Auto₃ :   — V S P A I S I H E I G A V G (A T K)

Phosphoserine
Content :   1.8 ± 0.2 Moles /Mole Protein

Fig. 8. Structural Information on the α–Subunit of Phosphorylase Kinase

The α subunit is the largest subunit of the enzyme. Its molecular weight is in the range of 140,000 (Fig. 8) as determined by sodium dodecyl sulfate gel electrophoresis. The sequence of the 16 N-terminal amino acid residues shows some relationship to the $pp60^{vsrc}$ tyrosine kinase (11). Very specifically and exclusively this subunit can be labelled with fluoresceineisothiocyanate (16) and ATP protects the protein from beeing labelled (17). From this modified protein a labelled peptide was isolated which shows a cluster of glycine residues as observed in other adenine nucleotide binding domains. Therefore, it is concluded that the α subunit contains a nucleotide binding domain (17); to identify it as a part of a catalytic domain a higher amount of primary sequence information is required. Enhancement of the enzyme activity can occur by phosphorylation of a specific serine residue catalyzed by the cAMP dependent protein kinase. According to the specificity requirement of the cAMP dependent protein kinase this serine residue is located two amino acid residues downstream of an arginine (Fig. 8). Self phosphorylation of phosphorylase kinase which is coupled to enhancement of enzyme activity (see below) also incorporates phosphate into the α subunit. The sequences around these serine residues are shown in Fig. 8. Non of these phosphoserines are located two residues downstream of a basic amino acid (18). The α subunit contains ca. 1.8 moles phosphate/mole protein. Non of these phosphates is located in the phosphorylation sites which have been determined until now.

Phosphorylation catalyzed either by the cyclicAMP dependent protein kinase or by selfphosphorylation enhances the A_1 activity. This activity appears to be stimulated half maximally at about 1 μM free Ca^{2+}. Therefore, in intact muscle A_1 will be expressed during muscle contraction which is triggered also by a rise of the intracellular free Ca^{2+} to about this level.

24

The β subunit (M_r 127,000), the second largest subunit of the holocomplex has a blocked N-terminus (Fig. 9). As isolated it contains 1.3 mole phosphate/mole of protein. According to the ^{31}P NMR spectra (see above) this phosphate is also present as phosphoserine (10). Several serine residues can be phosphorylated in vitro. The sequence around a serine residue phosphorylatable by the cAMP dependent protein kinase shows the consensus sequence characteristic for sites which can be phosphorylated by this kinase i.e. it is located two amino acid residues downstream of a basic residue (Fig. 9) (19). This same serine is phosphorylated during selfphosphorylation of the enzyme. It is not yet clear if this phosphorylation is catalyzed by phosphorylase kinase itself or by contamination of the enzyme preparation with cAMP dependent protein kinase. Similarly to the observation on the α - subunit a second serine which is phosphorylated during selfphosphorylation shows no homologies in its surrounding to those sites phosphorylated by the cAMP dependent protein kinase (Fig. 9) (18).

Phosphorylation of the β subunit by the cAMP dependent protein kinase causes an increase of the low affinity Ca^{2+} dependent activity A_2. In the non phosphorylated enzyme expression of A_2 activity requires about 30 μM free Ca^{2+} for half maximal activation and either alkaline pH values or high Mg^{2+} concentrations. Mg^{2+} has been characterized as an inducer of Ca^{2+} binding. A_2 activity can be expressed at neutral pH and low Mg^{2+} (ca. 1 mM) following β -subunit phosphorylation catalyzed by the cyclic AMP dependent protein kinase. It is concluded that β -subunit phosphorylation controls the affinity for Mg^{2+} which inturn regulates Ca^{2+} activation of the A_2 activity (20). The β-subunit certainly also contains a nucleotide binding domain. Many ATP analogs are labelling preferentially the β-subunit which in some instances can be correlated with a decrease of enzymatic activity (21). If such a nucleotide binding domain belongs to a catalytic center is yet unclear. However, a catalytically active proteolytic fragment of phosphorylase kinase thought to be derived from the β subunit has been isolated. The structure of this domain is yet unknown (22).

M_r :	~127,000
N-Terminus :	Acetylated
ATP/ADP Binding Site :	positive, not identified
Phosphorylatable Sites cAMP Kinase :	A R T K R S G $\overset{P}{S}$ I Y E P L K
Auto$_1$:	A V L F $\overset{P}{S}$ L A E D D Y K
Auto$_2$:	R S G $\overset{P}{S}$ I Y E P L K
Phosphoserine Content :	1.3 ± 0.1 Moles/Mole Protein

Fig. 9. Structural Information on the β-Subunit of Phosphorylase Kinase

M_r :   44,673 known from primary
                          sequence

ATP
Binding Domain :   L G R G V S S,D F G,
                        C G T P

Catalytic Site :   Y A V K I I D

Calmodulin
Binding Domain :   C - Terminus

Catalytic
Activity : A_0, active without
                        calmodulin

Fig. 10. Structural Information on the γ-Subunit of Phosphorylase Kinase

Evidences are accumulating that the γ-subunit (M_r 45,000) is catalytically active without Ca^{2+}/calmodulin, i.e. this subunit expresses the Ca^{2+} independent activity A_0 (Fig. 10) (23, 24). Calmodulin binds to the C terminial region of the molecule which can be jugded from the primary sequence of the γ-subunit that has been established recently by Reimann et al. (25). A glycine rich sequence is found in the γ-subunit which is a well known feature for a nucleotide binding domain. Ca. 11 residues downstream a lysine residue surrounded by hydrophobic amino acids is located which has been characterized in several protein kinases as part of the catalytic center.

The δ subunit (M_r 17,000) is almost identical to calmodulin (Fig. 11). Its primary sequence is known (26) and its three dimensional structure has been determined (27). The molecule is doubly headed; each head consists of two Ca^{2+} binding domains connected by a central helix. One head houses the two high affinity Ca^{2+}/Mg^{2+}, the other the two low affinity Ca^{2+} specific binding sites. Moreover, calmodulin binds additionally 2 moles Mg^{2+} specifically which have not yet been located in the molecule. Calmodulin which is an integral subunit of holophosphorylase kinase exerts two functions in this enzyme (15). On the one hand it suppresses in its Ca^{2+} free form the Ca^{2+} independent activity A_0; on the other hand it stimulates the activity A_1 when the Ca^{2+}/Mg^{2+} high affinity binding sites are saturated with Ca^{2+} and it stimulates the activity A_2 upon saturation of the Ca^{2+} specific binding domains with Ca^{2+}. As mentioned above a prerequisite for Ca^{2+} to be able to bind to these specific sites is the presence of Mg^{2+} on the Mg^{2+} specific sites which act as inducer for these Ca^{2+} specific sites. The affinity for Mg^{2+} is regulated by ß-subunit phosphorylation (20).

26

M_r :   16,680 known from primary
                  sequence

Three Dimensional
Structure :   two heads containing each
                  two Ca^{2+} binding domains,
                  central α helix

Metal Binding
Sites :   2 Ca^{2+}/Mg^{2+} high affinity
                  2 Ca^{2+} specific, low affinity
                  2 Mg^{2+} specific

Function :   suppression of Ca^{2+}
                  independent activity A_0
                  activation of A_1 due to
                  Ca^{2+} binding to high affinity
                  sites
                  activation of A_2 due to
                  Ca^{2+} binding to low affinity
                  and Mg^{2+} binding to Mg^{2+}
                  specific sites

Fig. 11. Structural Information on the δ-Subunit (Calmodulin)
of Phosphorylase Kinase

This overview demonstrates that several species of phosphorylase kinase exist e.g. forms phosphorylated on different subunits, dissociated species etc. It could be visualized that signal selection which has been characterized to be a primary function of phosphorylase kinase is based on the formation of different phosphorylase kinase species through which intracellular signals could be transmitted to phosphorylase. Therefore, it shall be tried to name specific species of phosphorylase kinase which could serve as signal transmitters i.e. which would express catalytic activity in accordance with the extracellular signal and its intracellular event, i.e. second messenger enhancement.

The correlation of phosphorylase a formation with muscle contraction can be most easily be explained with activation of A_1 activity since both events are triggered by approximately the same free Ca^{2+} concentration. Phosphorylation of the α subunit by the cAMP dependent protein kinase produces a species which can express this A_1 activity. Therefore, it might be that the flow of information from the nerve impuls to phosphorylase involves an subunit phosphorylated phosphorylase kinase' i.e. the extent of the phosphorylase a formation upon neuronal stimulation would depend on parallel, submaximal hormonal stimulation of the tissue (Fig. 12).

The integrator like function might involve a hysteresis loop of Ca^{2+} and Mg^{2+} induced conformational changes of calmodulin (28). It has been shown that Mg^{2+} is able to stabilize a Ca^{2+} independent "active state" of calmodulin. In analogy, first binding of Ca^{2+} and Mg^{2+} and subsequent only removal of Ca^{2+} produces an enzyme form which expresses activity without bound Ca^{2+} i.e. an A_0 like activity (29). Therefore, one might assume that during muscle contraction a species accumulates expressing the A_0 activity which could serve as a molecular integrator.

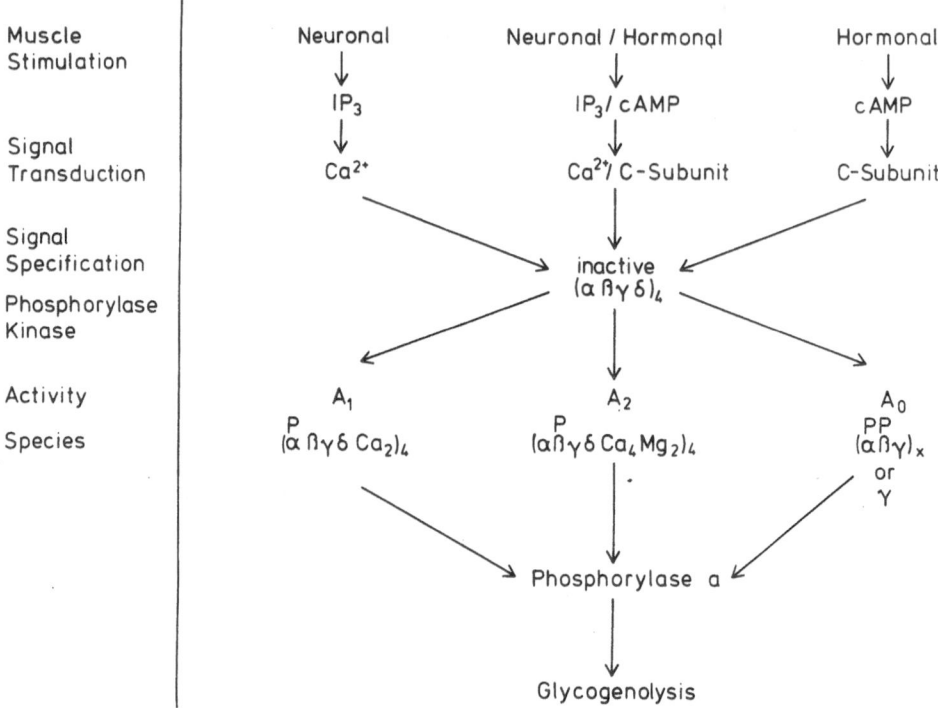

Fig. 12. Synopsis of Signal Specification on Phosphorylase Kinase

Upon hormonal stimulation β subunit phosphorylation has been demonstrated. The most dominant effect of this phosphorylation consists in activation of the A_2 activity. Since this activity requires 30 μM free Ca^{2+} it can only be active in the cell during contraction. Therefore, the synergistic effect exerted by epinephrine on contracting muscle could involve a β subunit phosphorylated phosphorylase kinase.

The involvement of A_0 in signal transduction of the hormone epinephrine in resting muscle seems reasonable (15). A_0 is the only activity which can be expressed in a quiescent muscle cell since the free Ca^{2+} concentration is very low (ca. 50 nM). Even the high affinity Ca^{2+} dependent activity A_1 requires a ca. 20 fold higher Ca^{2+} concentration for half maximal stimulation. A_0 activation occurs upon dissociation of calmodulin from the enzyme. However, until now it was not possible to demonstrate calmodulin dissociation upon phosphorylation of the enzyme. Certainly upon epinephrine stimulation of the muscle cell the free Ca^{2+} concentration does not increase substantially since the hormone does not trigger contraction i.e. the free Ca^{2+} concentration remains below the treshold. However, up to 100 % conversion of phosphorylase b to the a form can be induced by the hormone. The alternate hypothesis that the Ca^{2+} sensitivity of the enzyme is increased by phosphorylation in such a way that at Ca^{2+} concentrations even below the treshold for contraction phosphorylated phosphorylase kinase becomes Ca^{2+} saturated is much debated in literature. Data on the stimulation of A_1 and A_2 activities upon α - or β -subunit phosphorylation would rather indicate an increase in V_{max} than Ca^{2+} sensitivity.

Enhancement of the V_{max} of A_1 in comparison to the A_2 activity could be interpreted as an apparent increase in Ca^{2+} sensitivity since A_1 is activated at a ca. 10 fold lower free Ca^{2+} concentration than the A_2 activity. However, at the free Ca^{2+} concentration of the resting muscle cell A_1 would be active to a degree of maximally ca. 5 % which could not explain the complete conversion of phosphorylase to the a form even under maximal hormonal stimulation. Therefore, A_0 activation upon hormonal stimulation would still be a more plausible explanation which seems to be an attractive hypothesis for further studies on the specificator phosphorylase kinase.

References

(1) Fischer, E.H., this volume, Plenum Publ. Corp., pp. (1987).
(2) Goldbeter, A. and Koshland, D.E., Jr., Proc. Natl. Acad. Sci. USA, 78, 6840-6844 (1981).
(3) Pickett-Gies, C.A. and Walsh, D.A., The Enzymes, Vol XVII, 395-459 (1986).
(4) Malencik, D.A. and Fischer, E.H., Calcium and Cell Function (Cheung, W.Y., ed.), Vol. III, 161-188 (1982).
(5) Cohen, P., Eur. J. Biochem. 34, 1-14 (1973).
(6) Hayakawa, T., Perkins, J.P. and Krebs, E.G., Biochemistry 12, 574-580 (1973).
(7) Kilimann, M.W. and Heilmeyer, L.M.G., Jr., Biochemistry 21, 1727-1734 (1982).
(8) Cohen, P., Current Top. Cell. Regul. 14, 117-196 (1978).
(9) Schramm, H.J., and Jennissen, H.P., J. Mol. Biol. 181, 503-516 (1985).
(10) Kilimann, M.W., Schnackerz, K.D. and Heilmeyer, L.M.G., Jr., Biochemistry 23, 112-117 (1984).
(11) Crabb, J.W. and Heilmeyer, L.M.G., Jr., J. Biol. Chem. 259, 6346-6350 (1984).
(12) Cheng, A., Fitzgerald, T.J. and Carlson, G.M., J. Biol. Chem. 260, 2535-2542 (1985).
(13) Kilimann, M.W. and Heilmeyer, L.M.G., Jr., Eur. J. Biochem. 73, 191-197 (1977).
(14) Kilimann, M.W. and Heilmeyer, L.M.G., Jr., Biochemistry 21, 1735-1739 (1982).
(15) Hessova, Z., Varsanyi, M. and Heilmeyer, L.M.G., Jr., Eur. J. Biochem. 146, 107-115 (1985).
(16) Sotiroudis, T.G. and Nikolaropoulos, S., FEBS Lett. 176, 421-425 (1984).
(17) Sotiroudis, T.G., Crabb, J.W. and Heilmeyer, L.M.G., Jr., unpublished.
(18) Meyer, H. E. and Heilmeyer, L.M.G., Jr., unpublished.
(19) Cohen, P., Watson, D.C. and Dixon, G.H., Eur. J. Biochem. 51, 79-92 (1975).
(20) Heilmeyer, L.M.G., Jr., Jahnke, U., Kilimann, M.W., Kohse, K.P. and Sperling, J.E., Cold Spring Harbor Conferences on Cell Proliferation, Vol. 8, 321-329 (1981).
(21) Gulyaeva, N.B., Vulfson, P.L. and Severin, E.S., Biokhimiya 43, 373-381 (1977).
(22) Fischer, E.H., Alaba, J.O., Brautigan, D.L., Kerrick, W.G.D., Malencik, D.A., Moeschler, H.J., Picton, C. and Pocinwong, S., in: Versatility of Proteins, (C.H. Li, ed.), Academic Press, New York, 133-145 (1978).

(23) Skuster J.R., Chan, C.K.F and Graves, D.J., J. Biol. Chem. 255, 2203-2210 (1980).

(24) Chan, K.F.J. and Graves, D.J., J. Biol Chem. 257, 5948-5955 (1981).

(25) Reimann, E.M., Titani, K., Ericsson, L.H., Wade, R.D., Fischer, E.H. and Walsh, K.A., Biochemistry 23, 4185-4192 (1984).

(26) Grand, R.J.A., Shenolikar, S. and Cohen, P., Eur. J. Biochem. 113, 359-367 (1981).

(27) Sudhakar Babu, Y., Sack, J.S., Greenhough, T.J., Bugg, C.E., Means, A.R. and Cook, W.J., Naure 315, 37-40 (1985).

(28) Kohse, K.P. and Heilmeyer, L.M.G., Jr., Eur. J. Biochem. 117, 507-513 (1981).

(29) King, M.M. and Carlson, G.M., Arch. Biochem. Biophys. 209, 517-523 (1981).

THE ROLE OF ZERO-ORDER ULTRASENSITIVITY AMPLIFICATION IN THE REGULATION OF

THE GLYCOGEN PHOSPHORYLASE A - PHOSPHORYLASE B CYCLE

R. D. Edstrom, J. S. Bishop and M. H. Meinke

Departments of Biochemistry and Medicine

University of Minnesota, Minneapolis, Minnesota, 55455

Biological signaling systems generally contain amplification steps to offset attenuations of the signal. Diminution of signal may be caused by dilution of messenger molecules, desensitization of receptors and by the activities of specific antagonistic processes. Two general classifications of biochemical amplification have been described by Koshland et al. (1). Magnitude amplification depends on the ability of a single enzyme molecule to catalyze the conversion of many molecules of substrate. Several steps of magnitude amplification may be linked in a cascade such as found in blood clotting or hormonally induced glycogenolysis. Sensitivity amplification is a measure of the fractional change in response compared to the fractional change in signal. In the case of sensitivity amplification, the signal may be a small change in a relatively large concentration of stimulatory species with the response being a relatively larger change in a pre-existing level of output molecules. For example, a 10 % increase in the concentration of signal molecules causing a 50 % increase in the concentration of responding species yields a sensitivity amplification factor of 5. Sensitivity amplification is known to arise in three circumstances: Positive cooperativity, multiple inputs of a single effector and zero-order ultrasensitivity. The zero-order effect may occur in regulatory systems utilizing cyclic activation-deactivation through covalent modification of an enzyme protein. The regulated enzyme must be present at a concentration higher than the Michaelis constants of either or both of the two converter enzymes.

The cyclic phosphorylation-dephosphorylation process, responsible for regulating the activity of glycogen phosphorylase in skeletal muscle, interconverts the active phosphorylated form, phosphorylase a, and the physiologically inactive dephospho-form, phosphorylase b (Fig 1.) This cyclic system functions under conditions in which it is possible for significant sensitivity amplification of metabolic signals to occur by the zero-order ultrasensitivity effect. The concentration of phosphorylase in muscle has been estimated to be 100 μM (2). The Km values of phosphorylase kinase (30 μM) and phosphorylase phosphatase (16 μM) are well below that concentration.

A graphical presentation of the basis of zero-order ultrasensitivity is given in Fig. 2. These drawings based on those of LaPorte and Koshland are used with permission (3). In panel A is shown the first-order response of the rates of phosphorylase kinase and phosphorylase phosphatase to changes in the mole fraction of phosphorylase in the a form. The intersections of the phosphatase activity lines (ascending from the left) and the kinase line

31

(ascending from the right) represent the steady-state levels of mole fraction phosphorylase a. One can see that a 20 % change in phosphatase activity results in about a 20 % change in the mole fraction phosphorylase a. In
panel B, where both enzymes are nearly saturated with their respective form
of phosphorylase, a 20 % decrease in the level of phosphatase results in
more than a four fold increase in the mole fraction of phosphorylase a.

Equation 1 describes the steady-state mole fraction of phosphorylase a.

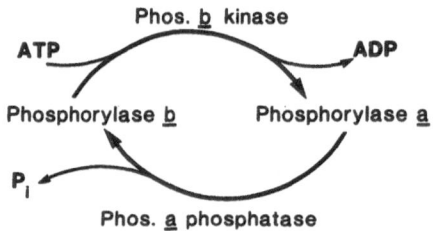

$$W^* = \frac{\left(\frac{V_1}{V_2} - 1\right) - K_2\left(\frac{K_1}{K_2} + \frac{V_1}{V_2}\right) + \left\{\left[\frac{V_1}{V_2} - 1 - K_2\left(\frac{K_1}{K_2} + \frac{V_1}{V_2}\right)\right]^2 + 4K_2\left(\frac{V_1}{V_2} - 1\right)\left(\frac{V_1}{V_2}\right)\right\}^{\frac{1}{2}}}{2\left(\frac{V_1}{V_2} - 1\right)}, \qquad [1]$$

W^* is the mole fraction phosphorylase a, K_1 = Km(kinase)/[total
phosphorylase], K_2 = Km(phosphatase)/[total phosphorylase], V_1 = kinase
rate (v_k) and V_2 = phosphatase rate (V_p). Experimental data were treated
with curve-fitting computer routines to obtain the constants K_1 and K_2.

Experimental evaluation of the phosphorylase system for zero-order ultrasensitivity was carried out by establishing the steady-state systems
shown in Fig 3. We found a steady-state level of phosphorylase a was
reached in 30 min and remained stable for up to 2 hours. HPLC analysis of
the nucleotide concentrations in the reaction mixtures showed the ATP
regenerating system was functional, with less than 10 % of the nucleotide
present as ADP and no detectable AMP for the duration of the experiment.

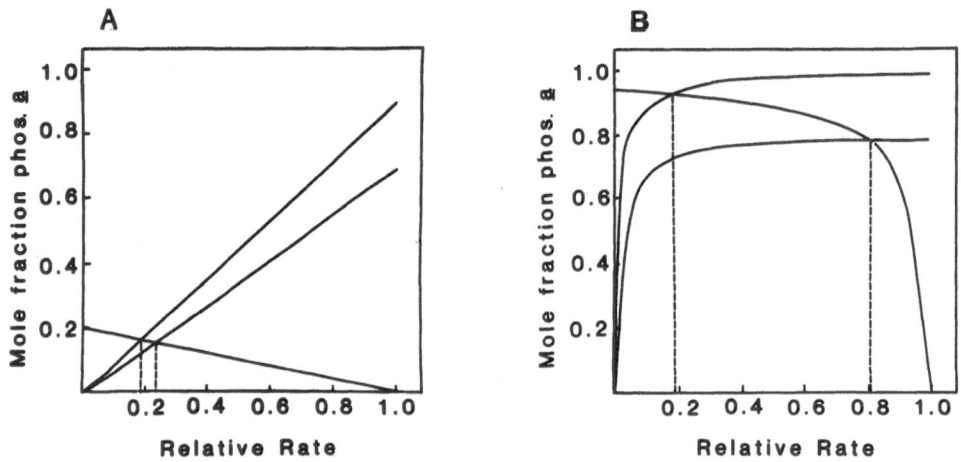

Fig. 1. The cyclic interconversion
of phosphorylases a and b.

Fig. 2. Representation of first-order and zero-order sensitivity.

The effect of changing the kinase/phosphatase ratio on the mole fraction phosphorylase a was determined at different concentrations of phosphorylase as described previously (4). The results are shown in Fig 4. When the kinase/phosphatase activity ratio was varied over the range of 0.1 to 10, the sensitivity of the steady state level of phosphorylase a to those changes can be seen to be dependent on the concentration of phosphorylase. At 70 μM phosphorylase, well above the Km values for the kinase and phosphatase, the sensitivity was much greater than at 20 μM. This enhanced response is a clear demonstration of zero-order ultrasensitivity.

The sensitivity of the system can be quantitatively described by a response coefficient, R_v (Eq 2.) which is the ratio of the $[V_k/V_p]$ giving 90 % of phosphorylase in the active form to the $[V_k/V_p]$ value required for 10 % active phosphorylase. Although zero-order ultrasensitivity and cooperativity are both biochemical phenomena that result in sensitivity amplification, the Hill equation describing cooperative enzymes does not strictly apply to the case of zero-order ultrasensitivity. One can define a constant, n', which allows a comparison of the sensitivity enhancement of the

Steady–State Experimental Design

A. Steady states, at various kinase/phosphatase levels, are established:

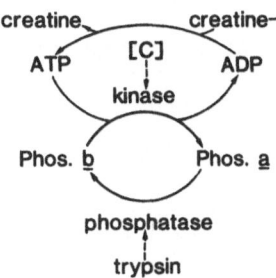

B. The mole fraction phos. a (W*) at steady state is determined:

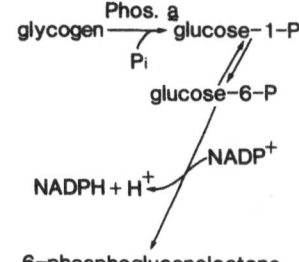

Fig 3.   Steady-state reaction mixture and assay procedure

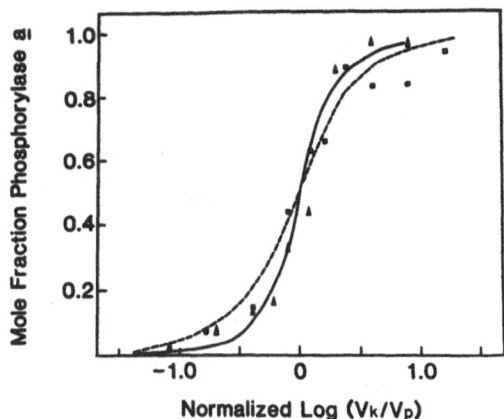

Fig. 4. Effect of kinase/phosphatase ratio on the mole fraction of phosphorylase a at 20 μM (■) and 70 μM (▲) phosphorylase.

zero-order effect with that of a cooperative enzyme. Both the response coefficients and the pseudo-Hill coefficients are presented in the table. Also in that table are the Michaelis constants derived from fitting the data to Eq. 1. as well as those constants determined by the usual studies of the effect of substrate concentration on the initial rate of the reactions.

$$R_v = \frac{[V_k/V_p]_{90\% \text{phos}.\underline{a}}}{[V_k/V_p]_{10\% \text{phos}.\underline{a}}} = \frac{81(K_k+0.1)(K_p+0.1)}{(K_k+0.9)(K_p+0.9)} = 81^{1/n'}, \quad [2]$$

An alternative method of describing the sensitivity of such a system is through the use of finite interval amplification factors. In this case any interval may be chosen for the change in signal (V_k/V_p ratio) and the corresponding change in phosphorylase activation related to it through the amplification factor A_s described in Eq. 3.

$$A_s = \frac{[(W_f^*/W_i^*)-1]}{[(\alpha_f/\alpha_i)-1]}, \quad [3]$$

W_f^* and W_i^* are the final and initial values of the mole fraction of phosphorylase $\underline{a}$; α_f and α_i are the final and initial values of V_k/V_p ratios required to achieve W_f^* and W_i^*. Amplification factors were calculated as a function of the values of (α_f/α_i) for the two phosphorylase concentrations where α_i was arbitrarily chosen to be 0.05 and α_f ranged from 0.06 to 10.0. The maximum observed values of A_s are in the table.

An additional consideration in evaluating the phosphorylase system for zero-order ultrasensitivity is that in vivo the enzymes are functioning in association with the phosphorylase substrate, glycogen, which may be present in the tissue largely in particulate form. In solutions containing glycogen, the Km of phosphorylase kinase has been reported to be reduced about ten-fold (5), giving a potential for further enhancement to sensitivity amplification brought about by the zero-order effect. Since our assay for phosphorylase depends on the production of glucose 1-phosphate from glycogen, we could not add glycogen to the steady-state reaction. As a preliminary attempt at evaluating the effect of glycogen, we used phosphorylase limit dextrin. From the data in the table, one sees that added dextrin enhanced the sensitivity at 20 μM but not at 70 μM. The 2 to 4 fold reduction in the phosphatase Km resulted in that enzyme approaching the zero-order region even at 20 μM. The unexpected increase in the kinase Km may be due the differences between phosphorylase limit dextrin and intact glycogen.

Table 1

[Phosphorylase]	Km(μM)		Rv	n'	As
	Kinase	Phosphatase			
20 μM	16	22	26	1.4	1.2
+ 1% dextrin	30	4.8	16	1.6	-
70 μM	22	10	6.5	2.4	3.6
+ 1% dextrin	39	4.8	6.4	2.4	-
Initial rate studies	30	16	-	-	-

While we showed that zero-order ultrasensitivity could be observed in the phosphorylase system by varying the amounts of phosphatase or kinase added to the steady-state incubations in order to change the V_k/V_p ratio, it was important to demonstrate an effect using a more physiologically relevant method of changing that ratio. For this purpose we varied the V_k/V_p ratio by adding a modifier of the phosphatase. Phosphorylase phosphatase may be regulated by interactions with the inhibitory proteins generally referred to as I-1 and I-2 (6). I-1 is inhibitory after it has been phosphorylated by cAMP-dependent protein kinase while I-2 is inhibitory unless it has been acted on by ATP-Mg^{2+} and a specific kinase known as Fa kinase (6). Two steady state reaction mixtures were prepared at 70 μM phosphorylase and a V_k/V_p ratio of 0.167 (α_i). One of the mixtures was prepared using sufficient I-2 to cause a 40 % inhibition of the phosphatase thus raising the V_k/V_p ratio to 0.278 (α_f). The mole fractions of phosphorylase in the a form after steady state was reached were 0.210 (W_i^*) and 0.537 (W_f^*) respectively. From these results an amplification factor of 2.4 was obtained by using Eq 3. At 1 μM phosphorylase and the same change in V_k/V_p ratio an amplification factor of 1.0 can be calculated. At the low phosphorylase concentration, no amplification occurs and the fractional change in phosphorylase a is the same as the fractional change in phosphatase activity caused by I-2. Thus there is an enhancement of response to regulation by I-2 due to the zero-order effect.

Summary and Conclusions

In vitro experiments have been performed with a reconstituted multi-enzyme system, consisting of the purified muscle enzymes phosphorylase a, phosphorylase b, phosphorylase kinase, phosphorylase phosphatase, cAMP dependent protein kinase (catalytic subunit) and creatine phosphokinase together with ATP and creatine phosphate. In that system we have demonstrated that the steady-state fraction of phosphorylase present in the activated a form responds to changes in the activities of the interconverting enzymes with an enhanced sensitivity due to the the zero-order ultrasensitivity effect. We have also shown that the presence of phosphorylase limit dextrin enhances the zero-order effect at otherwise sub-saturating phosphorylase concentrations. Additionally the phosphatase inhibitor, I-2, was shown to regulate the steady-state level of phosphorylase with sensitivity enhancement due to the zero-order effect.

Supported by the American Diabetes Association - Minnesota Affiliate and the National Institutes of Health (GM-32895).

References

1. Koshland, D. E., Jr., Goldbeter A. and Stock, J. B., Amplification and adaptation in regulatory and sensory systems, Science, 217:220 (1982).
2. Fischer, E. H., Heilmeyer, L. M. G., Jr., and Haschke, R. H., Phosphorylase and the control of glycogen degradation, Curr. Top. Cell. Regul., 4:211 (1971).
3. LaPorte D. C. and Koshland, D. E. Jr., Phosphorylation of isocitrate dehydrogenase as a demonstration of enhanced sensitivity in covalent regulation, Nature, 305:286 (1983).
4. Meinke, M. H., Bishop, J. S. and Edstrom, R. D., Zero-order ultrasensitivity in the regulation of glycogen phosphorylase, Proc. Nat. Acad. Sci. USA, 83:2865 (1986).
5. Krebs, E. G., Love, D. S., Bratvold, G. E., Trayser, K. A., Meyer, W. L. and Fischer, E. H., Purification and properties of rabbit skeletal muscle phosphorylase b kinase, Biochemistry, 3:1022 (1964).
6. Ingebritsen, T. S. and Cohen, P., Protein phosphatases: Properties and role in cellular regulation, Science, 221:331 (1983).

PROTEIN SEQUENCING AND COVALENT PROCESSING

Kenneth A. Walsh

Department of Biochemistry, SJ-70
University of Washington
Seattle, Washington

To understand the control of cellular processes at the molecular level, it is important to ascertain the molecular anatomy of constituent receptors, transducers, messengers and enzymes. Although an ultimate goal of such studies is to visualize the three-dimensional interactions of these complex proteins with their regulators and their targets, this level of understanding is only available among a small number of proteins that have been crystallized and analyzed by X-ray diffraction techniques. Amino acid sequence information is available in a much larger number of cases, as determined either by protein chemistry or by analysis of the corresponding DNA. At this level these findings may provide knowledge of the specific location of binding sites within the linear sequences, information concerning modes of regulation (e.g. by phosphorylation or by limited proteolysis), understanding of the domain or subunit sub-structural arrangements, and an inkling of the evolutionary history in their background.

Our growth of knowledge of amino acid sequences has been expo-nential during the last thirty years with two key improvements in the methodology (cf. review by Walsh et al., 1981). First, in about 1970, sequencers that automated the Edman degradation became generally avail-able. This greatly improved the efficiency of this procedure in struc-tural analyses and made it possible to determine protein structures by analyzing a small number of rather large fragments instead of a much larger number of small fragments (Hermodson et al., 1972). Carriers were found, particularly polybrene, that helped to retain peptides in the instumentation during the degradation (Tarr et al., 1978). Still, a prin-cipal difficulty was to obtain enough pure protein to generate peptide fragments in adequate quantities. A little later, with the introduction of microscale, high-resolution methods of purification by HPLC and of highly specific cleaving enzymes, it became possible to solve large protein structures (e.g. 500 to 2,000 residues) using only 5-20 nmol per enzy-matic digest. These increases in efficiency and sensitivity (Table I) have facilitated the solution of protein structures in months rather than years, often with a single milligram of protein or less. Recent reviews of microsequencing capability (Hunkapiller & Hood, 1983; Hunkapiller et al., 1984) emphasize the role of the "gas phase" sequencer and its use-fulness in generating data that can lead to cloning of the corresponding cDNA and rapid determination of the cDNA sequence.

Table I

**Recent Improvements in Sensitivity and Efficiency
of Sequence Analysis**

	1980	1986[a]
nmol per digest	50-100	1-10
Residue-specific cleavage at	Met, Arg	Met, Lys[b], Arg, Glu
Separation media	Sephadex (formic acid)	TSK-SW columns (guanidine)
	Dowex ion exchange	Reversed phase HPLC
Amino acid composition	Dowex and ninhydrin (1 nmol, 90 min)	PTC-derivatives (10 pmol, 15 min)
Edman degradation	Spinning cup (5-10 nmol)	Spinning cup (1 nmol); Gas phase (10-500 pmol)
Detection of phenyl-thiohydantoins	Various gradient reversed phase HPLC systems (10 pmol)	Brownlee on-line system (Applied Biosystems) 1 pmol[d]
		Isocratic separation on Zorbax PTH[e] (1 pmol, 20 min)

[a]For example, see Titani et al. (1986).

[b]Masaki et al. (1981).

[c]Bidlingmeyer et al. (1984).

[d]Hunkapiller et al. (1984)

[e]Glajch et al. (1985).

Although it is clear that alternative methods of DNA sequencing are simple, rapid and less expensive than automated protein sequencing, there may well be problems finding the cDNA corresponding to a trace protein in even the best cDNA library. Furthermore, as will be discussed below, the vast majority of proteins are covalently modified during their lifetime either by proteolytic editing or by side chain modification for regulatory or catalytic purposes. Although the sites and nature of these modifications are unquestionably intrinsic in the DNA sequence, they are of course invisible at that level. For this reason, especially when considering proteins that are regulated by covalent control (e.g. phosphorylation), it is important to maintain the ability to examine the details of the protein itself rather than of its encoding DNA. Under ideal circumstances one would prefer to have not only the mature amino acid sequence, but also the nucleotide sequence of the encoding DNA, to insure the validity and completeness of the analysis and to examine the various events by which both the RNA and the protein were processed during maturation or regulation.

Another general procedure is of increasing importance in analyses of protein structures. Mass spectrometric techniques now permit examination of fragments of 20 residues or more and of complex mixtures of peptides (Carr & Biemann, 1984). As the sensitivity of these procedures has increased, and the speed and precision of the newer techniques has been recognized, mass spectrometric analysis has found wider application among proteins. For example, it has been proposed as a rapid method to screen mixtures of peptides from HPLC columns for consistency with sequences determined by other techniques (Gibson & Biemann, 1984). In addition, mass spectrometry appears to be the most universally applicable technique to detect posttranslational modifications, many of which can be lost during acidic steps in more conventional analytical procedures.

In the specific case of protein phosphorylation, virtually all known phosphorylation sites have been identified by protein chemistry. Even here, it is not as simple as it sounds. Although ^{32}P can be introduced from radiolabeled ATP by the appropriate protein kinase so that a labeled peptide can be isolated from an enzymatic digest, Edman degradation of that peptide does not give a quantitative estimation of the percentage of phosphorylation of a particular residue. This is because the products of Edman degradation of an O-phospho-serine residue are inorganic phosphate and the phenylthiohydantoin of the beta-elimination product dehydroalanine (Proud et al., 1977). The latter is unstable and is also generated in variable yield from every serine residue, whether phosphorylated or not. The radioactive inorganic phosphate is very poorly extracted by the organic solvents used in the Edman degradation and it is not unusual to recover only 1-5% of the label from the residue in question. Matters become complicated if there are multiple phosphorylation sites within one peptide, as for example in the case of glycogen synthase (Kuret et al., 1985). Moreover, if endogenous, unlabeled phosphate is present on serine or threonine residues, it can be missed entirely as the usual dehydroalanine residues are observed. In these cases mass spectrometric methods offer advantages in revealing phosphorylation sites.

Consider the problems facing a sequence analyst at the initial stages of analysis, as for example in our experience with von Willebrand Factor, a 2,050-residue multifunctional plasma protein that forms a platelet plug upon rupture of the endothelial layer in blood vessels (Titani et al., 1986). With such a large protein, it was clear from the amino acid composition (Table II) that tryptic or chymotryptic digests

Table II

Choice of Cleavage Techniques in Sequence Analysis of von Willebrand Factor (Titani et al., 1986)

Enzyme	Susceptible Bonds	No. of Peptides Expected[a]	Mean Size of Peptides (residues)	No. of Peptides Isolated
Trypsin	R + K	192	11	
Chymotrypsin	L, Y, F, W	280	7	
SP V8 Protease	E	138	15	
Weak Acid (pH 2, 112°C, 4 h)	D	114	18	
Weak Acid (pH 2.4, 37°C, 3 days)	D-P	5	410	
NH₂OH	N-G	8	256	
BNPS-Skatole	W	19	108	
Typsin/Blocked Lys	R	102	20	6[b]
Achromobacter Protease	K	89	23	24[b]
CNBr	M	42	50	42
Limited Proteolysis SP V8 Protease	E	?	?	3

[a]Calculated on the basis of the composition of the final sequence.

[b]Met-containing peptides only.

would give too complex a mixture of small peptides to be resolvable, even in the best separation systems. Moreover, their sequences, even if determined, would be short and it would be very difficult to obtain complete sets of unambiguous overlaps. Other potential cleavage techniques (e.g. at tryptophan, at aspartic acid or at Asn-Gly) were unsatisfactory because those cleavages are not stoichiometric and the complex mixtures of large overlapping fragments would be refractory to resolution. In the specific case of von Willebrand Factor, the figures in Table II indicate that successful separation of overlapping peptides was obtained by virtually quantitative cleavage at methionine and either lysine or arginine. Each of these amino acids occurred infrequently enough to provide separable mixtures of fragments of intermediate size. Perhaps the most useful enzymatic cleavage technique that has been introduced in recent years is on the carboxyl side of lysine residues (Masaki et al., 1981). The enzyme appears to be completely specific toward lysine and it can be used in 3 M urea if the substrate is not soluble in water.

COVALENT MODIFICATION OF PROTEINS

It is not generally realized that virtually every protein in living cells is not the simple translation product of its encoding gene. Most proteins, after all, do not have an amino-terminal methionine; in fact, most proteins are blocked at their amino terminus, usually by an acetyl group. Some other proteins are blocked at their carboxyl termini by an amide derived from a glycyl residue (Sakata et al., 1986). In between one finds intra- or intermolecular crosslinks, as well as chemical modification by methyl, hydroxyl, glycosyl, carboxyl, heme, biotinyl, lipoyl, ADP-ribosyl, and phosphoryl groups, among others (Wold, 1981). In addition, signal peptides are removed during vectorial transport processes (Blobel, 1980), activation peptides are removed during zymogen activation events, and internal peptides are excised as in the processing of polypeptide hormones (Neurath & Walsh, 1976). These processes represent important milestones in the lifetime of a protein as it proceeds from an immature nascent chain through translocation and activation events, with maturation of its active site and perhaps regulation by covalent modification (Table III).

Table III

Stages in the Lifetime of a Protein
Involving Chemical Modification

Immature:	• Seeking cellular destination
	• Gaining prosthetic groups and/or crosslinks
Mature:	• Function may involve covalent intermediates
	• Regulation may be reversible (e.g. phosphorylation) or irreversible (zymogen activation)
Aged:	• Non-enzymatic change (e.g. acetylation by aspirin, glucosylation, deamidation)
	• Enzymatic targeting for disposal (ubiquitin attachment)
	• Proteolysis

A major topic in this course has been the role of phosphorylation/dephosphorylation events in the regulation and control of complex physiological systems. Most of these events are reversible and in turn controlled by other enzymes that respond to yet other signals. As such these covalent changes provide a means of amplifying a signal, often crossing a membrane in the process. These regulatory events are in marked contrast to regulation by proteolysis where non-reversible changes are induced and poised systems are irreversibly turned on (or off) (Neurath & Walsh, 1976).

Clearly it is important to recognize posttranslational modification wherever they occur. Not only do such modifications regulate a wide variety of systems by modulating the activity of protein components, but in principle the introduction of prosthetic groups is unlimited in its capacity to expand the intrinsic chemical potential of a protein. The amino acids that compose the polypeptide backbones contain few chemically reactive side chains. Only cysteine and histidine residues have much potential for interesting chemical interactions at neutral pH. Thus, the ability of a protein to catalyze complex reactions often relies on the chemical novelty of an associated cofactor or of a covalently attached prosthetic group. Hence, a protein can be thought of as a platform with a very precisely shaped binding site in a crevice that provides chemical opportunities in an environment shielded from the surrounding water.

Other consequences of posttranslational modification (Table IV) include the translocation of proteins to specific organelles or extracellular sites (e.g. by attachment of lipid or by removal of a signal peptide), recognition events (e.g. by methylation during chemotaxis or by glycosylation processes), stabilization processes in which crosslinks are introduced or chain ends are protected, and degradation events accompanying the aging and turnover of cellular proteins. Thus, polypeptide chains should not be thought of as static products of their encoding genes, but as flexible molecules that can exist in more than one form in various regulated, chemically modified, assembled and functional states. It is not enough to determine the amino acid sequence of a protein to define its covalent structure. One must seek its various covalently modified states and relate them to the ways in which it is controlled and by which it regulates the physiological function of other molecules in the living cell.

Table IV

Some Consequences of Posttranslational Covalent Modification

Dictates Destination:	**Stabilizes Protein:**
• via signal peptide • in Golgi traffic • via lipid anchors	• via crosslinks • by protecting the ends
Controls Function:	**Destructive Processes:**
• by expanding chemistry with a prosthetic group • by reversible covalent modification • by zymogen activation	• aging (e.g. deamidation) • ubiquitin attachment and proteolysis

REFERENCES

Bidlingmeyer, B.A., Cohen, S.A. & Tarvin, T.L., (1984) J. Chromatogr. Biomed. Appl. 336, 93

Blobel, G. (1980) Proc. Natl. Acad. Sci. USA 77, 1496

Carr, S.A., & Biemann, K. (1984) Methods Enzymol. 106, 29

Gibson, B.W. & Biemann, K. (1984) Proc. Natl. Acad. Sci. USA 81, 1956

Glajch, J.L., Gluckman, J.C., Charikofsky, J.G., Minor, J.M. & Kirkland, J.J. (1985) J. Chromatogr. 318, 23

Hermodson, M.A., Ericsson, L.H., Titani, K., Neurath, H. & Walsh, K.A. (1972) Biochemistry 11, 4493

Hunkapiller, M., Kent, S., Caruthers, M., Dreyer, W., Firca, J., Giffin, C., Horvath, S., Hunkapiller, T., Tempst, P. & Hood, L. (1984) Nature 310, 105

Hunkapiller, M.W. & Hood, L.E. (1983) Science 219, 650

Kuret, J., Woodgett, J.R. & Cohen, P. (1985) Europ. J. Biochem. 151, 39

Masaki, T., Tanabe, M., Nakamura, K., Soejima, M. (1981) Biochem. Biophys. Acta. 660, 44

Neurath, H. & Walsh, K.A. (1976) Proc. Natl. Acad. Sci. USA 73, 3825

Proud, C.G., Rylatt, D.B., Yeaman, S.J., & Cohen, P. (1977) FEBS Lett. 80, 435

Sakata, J., Mizuno, K. & Matsuo, H. (1986) Biochem. Biophys. Res. Commmun. 140, 230

Tarr, G.E., Beecher, J.F., Bell, M. & McKean, D.J. (1978) Anal. Biochem. 84, 622

Titani, K.,Kumar, S., Takio, K., Ericsson, L.H., Wade, R.D., Ashida, K., Walsh, K.A., Chopek, M.W., Sadler, J.E. & Fujikawa, K. (1986) Biochemestry 25, 3171

Walsh, K.A., Ericsson, L.H., Parmelee, D.C. & Titani, K. (1981) Annu. Rev. Biochem. 50, 261

Wold, F. (1981) Annu. Rev. Biochem. 50, 783

PROTEIN CRYSTALLOGRAPHY

David I. Stuart

Laboratory of Molecular Biophysics
University of Oxford
Oxford OX1 3QU, UK

INTRODUCTION

It is over 50 years since the first X-ray photographs of a protein crystal were taken (Bernal & Crowfoot, 1934). Many years were to pass before it was possible to utilize the information inherent in these beautiful diffraction patterns to obtain detailed images of the crystalline proteins.

The physics of the diffraction process may be adequately described in a simple way using principles laid down early in the century. It is continued methodological developments that have ensured a gradual acceleration in the growth of the power of the technique and indeed lead some practitioners of the arcane art to speak of a 'new revolution' in protein crystallography (Dodson, 1986).

This brief discussion will not examine the past achievements of protein crystallography, neither can we expect to cover the fundamentals of the subject in any thorough way. All we shall do is to consider where the subject stands now and how it might develop in the near future. Ultimately the object of interest to the crystallographer, the biochemist and the pharmacologist is the same. It is an article of faith that biological systems are intelligable in terms of the properties of macro-molecules. Perhaps the crystallographer, in contemplating his rigid, static crystal may seem to come closer to the taxonomist perusing stuffed carcases or dry bones than to the zoologist catching glimpses of his elusive subject or the ecologist attempting to unravel a complex and dynamic system. In truth it is clear that in many cases the crystal-lographer is dealing more with a caged creature - still 'alive' and often able to carry out its usual functions. In spite of this the correlation of crystallographic results with biochemical and biological properties usually poses major difficulties. We hope that the developments occurring now, imaginatively applied, will make it possible to address these problems more successfully.

2. Fundamentals

Blundell & Johnson (1976), Lipson & Cochran (1968) & Johnson (1985) cover the fundamentals in a thorough fashion. We will simply present some pertinent points.

X-ray crystallography involves the illumination of a crystal with electromagnetic radiation of a wavelength comparable to the detail to be resolved (about 10^{-10} m). We are interested in the elastic scattering of the light by the electrons of the individual atoms of the elementary object (say a protein molecule). Because the scattering by a single protein molecule is undectably weak we struggle to grow crystals of the protein – beautifully ordered arrays of many billions of molecules. The scattered radiation must be recombined to form the image of the object. To recombine the scattered waves we must know the 'phase' of each wave with resepct to the others. To form an image with visible light we can use a lens which will preserve this information. We cannot at present construct a lens for X-rays and a computer is used as a lens analogue instead. We can only measure the time averaged scattering so that the phase information is utterly lost in the process of making our experimental measurements. The phase and amplitude of a wave (the latter is the measurable quantity) are unfortunately completly independent physical quantities. More unfortunate still is the fact that the phases are more important in forming the image than the amplitudes. The method of 'isomorphous replacement' has frequently proved successful in providing an experimental estimate for the phase (see Blundell & Johnson, 1976). However it is not uncommon for severe difficulties to occur. We shall return to this below.

If we have estimates for the phases of all the scattered X-ray beams then it is straightforward to calculate the electron density at any position x,y,z within the crystal by the relationship:

$$\rho(x,y,z) = 1/V \sum\sum\sum^{hkl} |F_{(hkl)}| \exp{-i(2\pi hx/a + 2\pi ky/b + 2\pi lz/c - \alpha(hkl))}$$

where hkl identifies the index of the scattered beam, $|F_{(hkl)}|$ is the amplitude and $\alpha(hkl)$ the phase of the diffracted beam of index hkl (the structure factor), a,b and c describe the size of the repeating block of the crystal, and V is the volume of this block.

Note that the electron density at any point x,y,z in the crystal is determined by <u>all</u> the observations. Crystallography is thus an all or nothing technique – we cannot choose to monitor just one part of the structure. Since the scattering occurs simultaneously from all molecules in the lattice we see an average structure for the whole volume of the crystal lattice that was illuminated by the X-rays.

In addition to this spatial averaging crystallography provides a composite of a myriad of snapshots of the unit cell since the frequency of X-rays is far higher than even the fastest chemical vibrations within a protein molecule.

We have seen that information about a given point in the protein molecule is 'synthesized' from the complete set of diffraction measurements. In spite of this there is still some segregation of information in the diffraction pattern. The beams that are deflected by only a small angle contribute to the broad features of the image while the quality and extent of the fine detail present in the electron density map depends directly on the quality and completeness of the higher angle diffraction data. This linear inverse relationship between object and diffraction space is described by the well known law of Bragg:

$$\lambda = 2d\sin\theta$$

where λ is the wavelength of the radiation, d is the spacing (commonly called resolution) within the object being examined and θ is the half angle of scattering.

The resolution (d) of a study crucially affects what we can learn of the protein structure. We must remember however that the linear inverse relationship implies that to improve the resolution by a factor of 2 we must measure eight times as many data.

The table below shows the sort of information we can expect to learn from studies at different resolutions.

<u>Resolution</u>

6Å   – Outline of molecule visible.
                  α–helices appear as rods.
                  The polypeptide chain can only be traced for fully α–helical proteins.

3Å   – With luck the polypeptide chain can be traced completely. Amino acid side chains will be visible but not identifiable.
                  It will be difficult to establish the plane of the peptide.

2Å   – Main chain structure certain.
                  Side chains – could probably identify about 50% correctly with no knowledge of the amino acid sequence.
                  Tightly bound waters can be seen.

1.5Å   – Atoms almost resolved.
                  Water structure clear.
                  Isotropic B factors fairly reliable.

1.2Å   – May be see hydrogen atoms?
                  Refine anisotropic parameters for thermal motion.
                  (Few proteins diffract to such high resolution!)

It is important to realize that the 'resolution' quoted for a structure determination is, in the case of a well refined structure, almost an order of magnitude larger than the likely accuracy of the structure. Thus for a structure at 2.0Å resolution the coordinates might be generally accurate to about 0.2Å.

Ideally one would determine all structures to extremely high resolution. In practice this is very rarely possible. This is due to 'smearing' of the structure obscuring the fine details and therefore rendering the high resolution data immeasurably weak. This smearing can arise either from packing faults in the crystal (spatial smearing) or from motion of the atoms (temporal smearing). The effect on the diffraction pattern is indistinguishable in the two cases since X-ray diffraction can give direct information only on spatial dynamics (unlike for example nmr, which probes temporal dynamics).

From this general background to the technique we will now consider the <u>process</u> of solving a structure, illustrating some recent advances as we go.

<u>Current Practice</u>

The figure below presents a very simplfied block diagram of the stages involved:

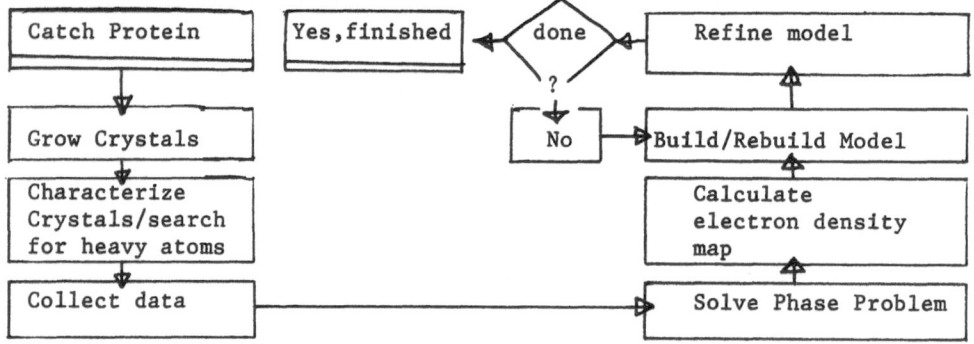

(a) <u>Protein</u>

At least a few milligrams of pure protein are required for a structure
determination and many crystallographic investigations probably eventually
use several grams.   A major cause of the upturn in protein crystallography recently has been the introduction of cloning and expression
techniques to get large amounts of previously rare proteins, possibly
modified to suit crystallization.

We should mention the development of site-directed mutagenesis.
This technique (the basis of so-called protein engineering) has enormous
potential in several directions.   It should widen the scope of crystallographic analyses and encourage collaboration to probe experimentally the
outstanding problems of protein folding and enzyme action.   At present we
are probably far too ignorant to seriously tackle the design of novel
enzymes - as 'stone-age' biological engineers we should recognize our
limitations.

(b) <u>Crystallization</u>

Crystallization is traditionally a black art practised by biochemists
with green fingers.   While we still have little theoretical understanding
that can actually help us with this outstanding problem, over the last few
years new and powerful techniques have emerged and there is now much
accummulated wisdom.   The use of polyethylene glycol as a precipitant has
been of great importance and more recently non-ionic detergents have
widened the field of proteins that can be investigated to include intact
membrane proteins.   The determination of the photosynthetic reaction
centre structure by Michel and Diesenhofer at Munich is a towering
achievement in this area (Diesenhofer et al, 1985).

At present there is a large amount of effort being devoted to putting
these efforts on a thoroughly rational basis.   This is rightly so since
crystallization remains a holdup in many investigations - it is still not
unusual for a project to flounder and die at this stage.   The present
efforts are devoted to comprehensive automation of trials of precipitants
and conditions and to the growth of crystals in space.   There are
plausible reasons for attempting crystallizations in space and it is to be
hoped that at the least there will be considerable spin-off from the
automation and systematization achieved in the ground tests.

Once crystals are obtained it is necessary to thoroughly characterize
them.   They may or may not diffract X-rays.   If they do diffract they
may do so to only very low resoltuion.   If they diffract well it is
necessary to determine the crystallographic and noncrystallographic
symmetry exhibited by them.   Crystallographic symmetry is usually
detected easily and uniquely from the diffraction pattern and is not
generally seen as factor likely to prevent the determination of the
structure.   On the other hand non-crystallographic symmetry can occur in
a completely undisciplined way, may be hard to detect and can render the

structure solution very difficult.   Once the non-crystallographic
symmetry elements are described however they can provide an enormous
amount of phasing power so that the resultant electron density maps are
superior to those obtained from the simpler packing schemes involving only
crystallographic symmetry operations.

It is important to realize that the recent improvements in data
collection methodology described below have extended the power of the
crystallographic method so that it is now possible to collect high
resolution data from very small crystals (e.g. 1.7Å data from crystals of
β-lactamase I measuring 0.5x0.2x0.02mm, R. Todd personal communication,
see Samraoui et al., 1986).

(c)   Data Collection
In general data are collected by rotating a protein crystal within
the X-ray beam while the intensities of the diffracted beams are recorded.
These intensities yield the amplitudes of the structure factors directly.

There are two principle difficulties, (a) the crystals are usually
sensitive to irradiation so that the diffraction pattern fades as the
crystals is exposed to the X-ray beam and (b) the diffraction pattern is
extremely weak anyway since the total diffracted energy is probably not
more than 2% of that of the incident beam and is spread over a large
number of reflections which are stimulated simultaneously.

Synchrotron radiation has made a major impact by providing an
extremely intense and nearly parallel beam of X-rays.   Alongside this has
been fitful progress towards the routine use of electronic area detectors
in the laboratory in place of photographic films.

Synchrotron radiation is produced by the deflection of rapidly moving
electrons or positrons into a (roughly) circular path by intense magnetic
fields.   The radiation emitted at a typical source (there are only a
handful of such facilities around the world) is more than two orders of
magnitude more intense than that produced by a conventional laboratory
source (Wilson et al., 1983).   This brilliance has led to very rapid data
collection while the very high quality of the X-ray optics leads to a
substantial improvement in the signal to noise ratio for the data and a
concomitant increase in the amount of useful data that can be collected
from a crystal.   An additional and largely unforeseen, gain comes from
the apparent time dependency of radiation damage of some protein crystals
so that a short sharp shock allows more data to be collected before the
diffraction fades unacceptably than the usual regime (Wilson et al.,1983).

Potentially the most important feature of synchrotron radiation has
been considered to be the ability to choose the wavelength of the
radiation to be used from a broad spectrum extending to both higher and
lower energies than the usual characterisatic copper Kα radiation used in
the laboratory (λ=1.54Å), (Phillips & Hodgson, 1980).   In practice this
facility has been used more often to reduce experimental errors (for
example by collecting data at short wavelengths to reduce the effects
of absorption of the X-ray beam by the crystal, its liquor of crystallization and glass mounting tube) than to investigate novel ways of
solving the phase problem.   The use of multiple wavelength techniques for
solution of the phase problem is in danger of becoming a technique with a
good future behind it.   It is to be hoped that as measurement facilities
improve to match the quality of the X-ray sources now available it will
assume a significant place in the battery of techniques available (Kahn et
al. 1984 describe a successful application).

It seems likely that an important step forward in data collection methods will come in the next few years as sophisticated two dimensional imaging systems increasingly replace the older technologies of photographic film and 'zero' dimensional electronic detectors. As was the case with other developments the maturation of this trend will probably be more limited by the production of sophisticated software systems than the hardware.

A final mention must be made of the resurgence of the oldest data collection method of all, the so called 'Laue' method. In this technique the illuminating radiation is not monochromatic as is usual but is more like white light in that it consists of a rather broad spectrum of wavelengths. The effect of this is to stimulate a massive number of diffracted beams simultaneously so that given a suitable crystal system a large fraction of a complete data set may be obtained in one exposure.

The potential increase in the rate of data collection with this method is enormous, of the order of 3 orders of magnitude and the technique offers the hope of being able to collect data sets in much less than one second (see Hajdu et al. 1987).

The following table presents rough estimates of the data collection rates for some currently available methods.

Table 1

Source	Radiation	Detector	Time for a 3Å Phosphorylase Data Set	Data rate
Sealed Tube (laboratory)	Cu Kα (λ=1.54Å)	Single counter diffractometer	1 month	1 reflection /minute
Rotating anode (laboratory)	CuKα	Photographic film (rotation method)	5 days	10 reflections /minute
Rotating anode (laboratory)	Cukα	Electronic area detector	1 day	1 reflection /second
Synchrotron	Mono-chromatic (λ∿1Å)	Photographic film	30 mins	50 reflections /second
Synchrotron	White (λ=0.5-2.5Å)	Photographic film	1 second	40,000 reflections/ second

(d)   The Phase Problem

The method of isomorphous heavy atom substitution remains the mainstay of the protein crystallographer but alternatives are slowly emerging.

Isomorphous replacement is a classic technique well described in the general references given earlier. For our purposes it is sufficient to say that it provides indirect but genuinely experimental estimates of the protein phase angles by observing the interference effects on the intensities of the scattered beams when hevy atoms are added to the protein. If measurements can be made with sufficient precision (the interference often affects the amplitudes by less than 20%) and if the heavy atom substitution does not affect the surrounding protein structure and if more than one pattern of heavy atom binding can be produced then it is possible to obtain estimates of the protein phase angles accurate to perhaps better than 50°. These conditions are often difficult to meet and there are now fairly standard methods for assisting in the process (see for instance Wang, 1985). As an alert consumer of crystallographic results you should be aware that, honestly calculated, the mean figure of merit, is the best guide to the likely reliability of an isomorphous replacement analysis. The figure of merit, m, corresponds to the cosine of the likely mean error in the phase angle. Thus m = 0; no phase information, m = 0.4; 66° error, m = 0.6; 53°, m = 0.8; 37°. As a rule of thumb where m is less than 0.6 there will often be difficulties of map interpretation and the consumer should beware.

When the protein is rather small it can be particularly difficult to prepare isomorphous heavy atom derivatives. In this situation it may be possible to determine the protein phase angles experimentally from X-ray data for the native protein crystals. This may be achieved because the scattering properties of atoms change when the incident radiation is of an energy close to an absorption edge of the atom. For the usual wavelengths used this means that quite significant 'anomalous scattering' effects occur for atoms such as Fe and Cd while the effect is much smaller for S and virutally non-existent for C,N, O and H. These effects have been used successfully in rather few cases, the outstanding example is that of Hendrickson and Teeter (1981) who used the scattering from the sulphur atoms in Crambin to solve the structure of this small protein. Another example is the use of a single cadmium ion bound to des-pentapeptide insulin to determine the atomic structure of the protein (Stuart et al. 1986b).

The method of molecular replacement has developed over the last twenty five years into a technique of great power that now makes it possible (a) To solve structures in the absence of isomorphous derivatives where information is available about the structure of an homologous protein (see for example Stuart et al., 1986a where the homologous protein was an enzyme of different function) (b) To solve structures of enormous size where the traditional methods would fail (see for example Rossmann et al., 1985 for the first determination of the structure of an intact animal virus, a particle of weight 8×10^{6} Daltons). Many structural studies now use these techniques, either alone or together with the method of isomorphous replacement and it is the case that the presence of more than one protein molecule within the asymmetric unit of the crystal, once considered a major problem, can lead to a substantial improvement in the eventual quality of the phase information, to the extent that the clarity of the maps obtained for viruses is usually at least as good as for proteins smaller by more than two orders of magnitude.

The generation of phase information from non-crystallographic symmetry is conceptually straightforward. Imagine that we have (for instance) a tetramer where the 222 symmetry of the molecule is expressed solely as non-crystallographic symmetry. If we know the location of the non-crystallgoraphic symmetry axes and the boundary of the molecule we may improve our electron density map by averaging the images of the four subunits.

The averaged map, upon Fourier transformation, will give a set of structure factors. These calculated structure factors will deviate from the starting structure factors used to create the original map - the phases should be closer to the truth but the amplitudes will be distorted. An improved map can now be calculated by using the measured amplitudes with the 'averaged' phases. In practice the process is iterative but converges quickly. It has been shown (Arnold & Rossmann, 1986) that the phasing power available with this method is related to the square root of the number of structural copies in the asymmetric unit of the crystal.

It is now standard to improve molecular models by direct refinement of the atomic positions against the X-ray observations (see below). This refinement leads in turn to greatly improved phase information so that an isomorphous replacement map, once the end of the road, is now simply the starting point for a process of gradual phase improvement.

Finally direct mathematical methods are being developed to solve the phase problem. These have yet to prove themselves in practice but promise to replace experimental observations with crystallographer's prejudice. The mathematically inclined reader is referred to Bricogne, 1984. We have already seen that the phase problem is intractable in so far as the experimentally observed scatter amplitudes are physically completely separate from the phases of the diffracted beams. If we are to feed information from the amplitudes into the phases it can only be indirectly, via our knowledge (prejudices) about the structure of the protein crystal. The so-called 'maximum entropy' method offers the possibility of feeding in such information (such as the knowledge that the electron density should never become negative) and producing a set of estimates for the phase angles that are in some sense as unbiased as possible. We look forward to rapid developments in this area but it is still unclear whether mathematical tricks can ever provide a general solution to the phase problem.

(e) Model Building and Map Interpretation

A computer graphics workstation is now a virtual necessity in a protein crystallography laboratory and whereas 10 years ago molecular models were constructed of metal or plastic parts, they now exist implicitly in the list of coordinates used by the computer to draw a two dimensional image. The graphics systems available are now sufficiently powerful that we may display one or more structures and/or one or more electron density maps simultaneously in several colours. Furthermore the structures may be rotated and manipulated in real time with at least the facility of mechanical models. The maintenance of the model as a data structure on the computer allows constant monitoring of the structure, rapid analysis, comparison and real-time application of stereochemical restraints.

The next generation of programs (already available in parts) will allow semi-automatic model construction on the basis of simple rules by reference to the data base of known protein structures. This should make it possible to construct a complete model for a medium sized protein within a few hours of obtaining a reasonable electron density map (Jones & Thirup, 1986). Just as the first generation of molecular graphics programs enabled the routine use of refinment methods by allowing the easy and rapid display of updated models so this new generation of programs will probably open up new approaches.

(f) Refinement

We have mentioned already that refinement of a molecular model against the X-ray observations is now almost mandatory. This reflects the use of molecular graphics devices, the improvement of computer algorithms and computational power. In the early days it was thought that the paucity of X-ray observations precluded the direct refinement of molecular models, now there exist powerful programmes which solve this problem either (a) by simplifying the model to be refined (reducing the number of parameters via constraints) (Sussman, 1985) or (b) by augmenting the X-ray data by chemical data, knowledge of bond lengths, angles etc. (increasing the number of observations via restraints) (Hendrickson, 1985).

Since these procedures are so widespread it may be helpful to the non-crystallographic consumer to be aware of the measures used to judge the process.

The major index of success is the R-factor

$$R = \frac{\sum ||Fo| - |Fc||}{\sum Fo}$$

where the sum is over all of the X-ray data and $|Fo|$ is the observed amplitude and $|Fc|$ is that calculated from the current model.

For a model derived from an isomorphous replacement map R will often be >0.4 and sometimes >0.5. The value for a compeletely random structure would usually be about 0.59. A refinement is extremely good if R<0.15 and many analyses end with R~0.20. Note that even when R < 0.25 there may be substantial regions of the model containing large errors - the R value can only give a general feel for the average errors in the model.

At a resolution of better than about 2.5Å it is customary to attempt to describe the structure of the tightly bound waters. With care those waters hydrogen bonded to the protein can be positioned fairly reliably, however it is wise to be sceptical of attempts to model a 'second shell' of waters unless the study has been done in the most painstaking way.

In addition to the R-factor an essential check on the reliability of a refinement is the quality of the chemistry of the final model. Perhaps the most useful overall number is the rms deviation from ideal covalent bond lengths. It is to be expected that this will be no more than about 0.035Å (poor stereochemistry will be reflected in an artificially low R value).

Conclusion

We have omitted some rather important areas from the above discussion and we should at least note in passing that there is now a vast body of information available confirming the relevance of crystallographic results to the properties of the molecule in solution (see Johnson, 1985).
Indeed this should not surprise us since a typical protein crystal contains as much liquid of crystal- lization as crystallized protein. We have neglected the ares of low temperature crystallography and, related to this, the extraction of dynamic information from X-ray diffraction results, both rather important, but specialized, areas. The interested reader is referred to Johnson (1985) and Stuart and Phillips (1985) for these two areas. A list of specific omissions would of course be long and tedious. Instead we will end by restating the continued vigour of this subject, one that has a foot in many camps and is susteained by not only its own methodological improvements but by the efforts of other scientists who are continually attempting to open up new veins for the crystallographer to mine.

References

Arnold, E. and Rossman, M.G., 1986, Proc. Natl. Acad. Sci. USA, 83:5489.
Bernal, J.D. and Crowfoot, D.C., 1934, Nature, 133:794.
Blundell, T.L. and Johnson, L.N., 1976, "Protein Crystallography",
    Academic Press, London.
Bricogne, G., 1984, Acta Cryst., A40:410.
Diesenhofer, J., Michel, H. and Huber, R., 1985, Trends Biochem. Sci.,
    10:243.
Dodson, G., 1986, Trends Biochem. Sci., 11:309.

Hajdu, J., Acharya, K.R., Stuart, D.I., McLaughlin, P.J., Barford, D.,
    Oikonomakos, N.G., Klein, H. and Johnson, L.N., 1987, EMBO
    Journal, 6:539.
Hendrickson, W.A., 1985, in 'Methods in Enzymology', C.H.W. Hirs and S.N.
    Timasheff eds., Academic Press, New York, 115:252. Hendrickson,
W.A. and Teeter, M.M., 1981, Nature, 257:107.
Johnson, L.N., 1985, in 'Modern Physical Methods in Biochemistry,
    Part A', Neuberger and Van Deenen Eds.., Elsevier, Amsterdam.
Jones, T.A. and Thirup, S., 1986, EMBO Journal, 5:819.
Kahn, R., Fourme, R., Bosshard, R., Weng, J.P., Dideberg, O.,
    Risler, J.L., Brunie, S. and Janin, J., 1984, in 'Structural
    Biology', Bartunik and Chance Eds., Academic Press, London.
Lipson, H., and Cochran, W., 'The Determination of Crystal Structures',
    Bell, London.
Phillips, J.C. and Hodgson, K.O., 1980, Acta Cryst., A36:856.
Rossmann, M.G., Arnold, E., Erickson, J.W., Frankenberger, E.A.,
    Griffiths, Hecht, H-J., Johnson, J.E., Kamer, G., Luo, M., Mosser,
    A.G., Rueckert, R.R., Sherry, B. and Vriend, G. 1985, Nature,
    317:145.
Samraoui, B., Sutton, B.J., Todd, R.J., Artymiuk, P.J., Waley, S.G. and
    Phillips, D.C., 1986, Nature 320:378.
Stuart, D.I., Acharya, K.R., Walker, N.P.C., Smith, S.G., Lewis, M. and
    Phillips, D.C., 1986a, Nature, 324:84.
Stuart, D., Liang, D-C., Dai, J-B., Todd, R., Lou, M-Z., You, J-M.,
    Li, J-Y., and Wang, J-H., 1986b, in 'Structural Biological
    Applications of X-ray Absorption, Scattering and Diffraction'
    Bartunik and Chance Eds., Academic Press, London.
Stuart, D.I. and Phillips, D.C., 1985, in 'Methods in Enzymology' C.H.W.
    Hirs and S.N. Timasheff eds., Academic Press, New York, 115:117.
Sussman, J.L., 1985, in 'Methods in Enzymology', C.H.W. Hirs and S.N.
    Timasheff eds., Academic Press, New York, 271:303.
Wang, B-C., 1985, in 'Methods in Enzymology', C.H.W. Hirs and S.N.
    Timasheff eds., Academic Press, New York, 90:112.
Wilson, K.S., Stura, E.A., Wild, D.L., Todd, R.J., Stuart, D.I., Babu,
    Y.S., Jenkins, J.A., Standing, T.S., Johnson, L.N., Forume, R.,
    Kahn, R., Gadet, A., Bartels, K.S., and Bartunik, H.D., 1983,
    J. Appl. Cryst., 16:28.

SIGNAL TRANSDUCTION BY THE ADENYLATE CYCLASE SYSTEM

Karl H. Jakobs*, Peter Gierschik*, Rüdiger Grandt*,
Rainer Marquetant*, and Ruth H. Strasser[#]

*Pharmakologisches Institut and [#]Medizinische Klinik
der Universität Heidelberg, D-6900 Heidelberg
Federal Republic of Germany

INTRODUCTION

Regulation of cellular functions by extracellular hydrophilic signal molecules such as neurotransmitters, peptide hormones and locally acting hormonal factors requires efficient mechanisms for transmembrane signalling. By these mechanisms, the primary message, i.e., presence of a hormone or neurotransmitter at the outer surface of the plasma membrane, is translated into one or more second messages inside the cell. Formation of the intracellular messages may be induced by the neurotransmitter-occupied receptors themselves, e.g., the nicotinic acetylcholine and $GABA_A$ receptors, being channels for small cations and anions, respectively. In addition, some polypeptide hormone receptors, e.g., those of insulin and epidermal growth factor, are protein kinases themselves, causing autophosphorylation of the receptors and phosphorylation of other cellular substrates. This type of agonist-activated receptors is thought to regulate cellular functions, at least in part, by these phosphorylation reactions. The overwhelming majority, however, of plasma membrane-located receptors for hormones and neurotransmitters induces formation of intracellular messages by activating multi-component signal transduction systems, apparently located within the lipid bilayer of the plasma membrane. Out of the signalling systems studied so far, the hormone-sensitive adenylate cyclase system, which is responsible for the control of intracellular levels of cyclic AMP, is one of the best characterized examples of transmembrane signal transduction systems.

COMPONENTS AND MECHANISMS INVOLVED IN ADENYLATE CYCLASE STIMULATION

At least three different proteins, all of them intrinsic plasma membrane proteins, are necessary for stimulation of intracellular cyclic AMP formation by extracellular messengers: receptor proteins (R_s) facing the extracellular space and acting as discriminators for the extracellular messengers, the stimulatory guanine nucleotide-binding regulatory protein (G_s) at the inner surface of the plasma membrane and the adenylate cyclase catalytic moiety (C), which possibly spans the plasma membrane and catalyzes the formation of cyclic AMP from MeATP at its cytoplasmic pole (Fig. 1). All of these individual components, including a stimulatory hormone receptor, i.e., the β-adrenoceptor, have been purified to homogeneity and characterized (1-4). Furthermore, the reconstitution of these three purified proteins into phospholipid vesicles has recently been accomplished (5,6). Whereas the receptor (β-adrenoceptor) and the adenylate cyclase catalyst are apparently monomeric proteins with apparent molecular weights of about 60 and 120-150 kDal, respectively, the G_s protein is a heterotrimer, consisting of a larger α-subunit (45 or 52 kDal), which binds guanine nucleotides and possesses GTP-hydrolyzing activity, and β- and γ-subunits with apparent molecular weights of 35/36 kDal and about 8 kDal, respectively.

Hormonal stimulation of cyclic AMP formation by the adenylate cyclase is thought to result from a series of sequential interactions of these protein components and their respective regulatory ligands. First, binding of a hormone agonist to its specific receptor leads to a not yet defined conformational change of the receptor molecule and a subsequent interaction of the agonist-activated receptor with the heterotrimeric G_s holoprotein. In its inactive state, GDP is bound to the α-subunit of the holoprotein. The hormone-receptor complex apparently induces a decrease in the GDP affinity of $G_{s\alpha}$. In the intact cell, where the GTP concentration is much higher than that of GDP, or in the presence of GTP in the *in vitro* assay system, GTP will now bind to $G_{s\alpha}$. In addition to binding of GTP to $G_{s\alpha}$, activation of G_s hypothetically involves its dissociation from the hormone receptor, which then apparently exhibits low affinity for the agonist, as well as its own dissociation into a free GTP-liganded α-subunit and the βγ-dimer. The activation reaction of G_s is critically dependent on the presence of magnesium ions. The activated α-subunit of G_s, then, interacts with the adenylate cyclase catalyst to stimulate its cyclic AMP-forming activity. Studies in intact membranes suggested that activation of the adenylate cyclase by the activated G_s protein is accompanied by a large increase in the affinity of the adenylate cyclase for its activating cation magnesium (7). It is not known yet whether this increase in metal ion affinity is also observed with purified components. Deactivation of the system is thought to result mainly of two mechanisms: hydrolysis of bound GTP by $G_{s\alpha}$ to GDP and reassociation of the α-subunit of G_s with the βγ-dimer, thus forming the inactive heterotrimeric G_s complex.

Several steps of the above described activation-deactivation cycle are not fully understood yet. For example, it is not known whether the association-dissociation reactions of the subunits of G_s, which have been observed with purified components in detergent solutions in the presence of stable GTP analogs, occur at all in native membranes and with naturally occurring guanine nucleotides. Furthermore, there is no strong evidence available, as suggested by the above mentioned hypothesis, that the G_s protein or one of its subunits (α) really "shuttles" between the hormone receptor and the adenylate cyclase. In addition, it is not known whether one hormone-activated receptor molecule activates only one or several G_s molecules and whether one activated G_s molecule, subsequently, activates only one or several adenylate cyclase molecules. In the visual transduction system in vertebrates, which uses a similar receptor (rhodopsin) and a similar guanine nucleotide-binding protein (transducin) as the hormone-sensitive adenylate cyclase system, but a distinct effector enzyme (cyclic GMP phosphodiesterase instead of adenylate cyclase), it has been shown that one light-activated rhodopsin molecule can activate several transducin molecules, which in turn activates several phosphodiesterase molecules (see ref. 8 for a review).

In vitro studies of adenylate cyclase stimulation have utilized four classes of agents in addition to hormones and naturally occurring guanine nucleotides. Fluoride ions (complexed by Al^{3+} to AlF_4^-) and hydrolysis-resistant GTP analogs, e.g., guanosine 5'-[γ-thio]triphosphate (GTP[S]) and guanosine 5'-[β,γ-imido]triphosphate (Guo*PP*[NH]*P*), both can persis-

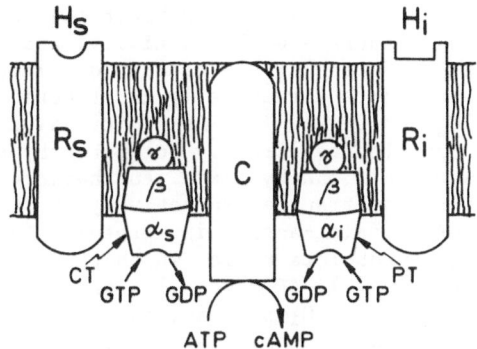

Fig. 1. Components of the hormone-sensitive adenylate
cyclase system. H_s and H_i, stimulatory and inhibitory hormones, respectively; R_s and R_i,
stimulatory and inhibitory receptors, respectively; C, adenylate cyclase catalyst; CT and
PT, cholera and pertussis toxin, respectively.

tently activate the G_s protein and lead to persistent activation of adenylate cyclase. The third agent which interferes with the stimulatory part
of the adenylate cyclase system is cholera toxin. This toxin has been
shown to permanently activate G_s by catalyzing the covalent linkage of
$G_{s\alpha}$ with an ADP-ribose moiety from NAD. The covalently modified $G_{s\alpha}$, then,
exhibits decreased GTP-hydrolyzing activity and decreased affinity for
the inhibitory $\beta\gamma$-dimer. Interestingly, for the ADP-ribosylation of $G_{s\alpha}$
by cholera toxin an additional membrane protein is required, which has
recently been purified to homogeneity (9). This protein, termed ADP-ribosylation factor (ARF), exhibits an apparent molecular weight of 21 kDal
and is also a GTP/GDP-binding protein. Immunological data suggest that
this protein is not identical with another class of 21 kDal membrane
proteins which are also guanine nucleotide-binding proteins, i.e., the
ras protooncogene products (p21 proteins) (9). It is feasible that the
ARF protein is not just present in the membrane for facilitating the ADP-
ribosylation of $G_{s\alpha}$ by cholera toxin, but that this GTP-binding protein
has an additional, not yet identified physiological function in the hormone-sensitive adenylate cyclase system. Finally, the diterpene forskolin
has been shown to stimulate cyclic AMP formation in almost all hormone-
sensitive adenylate cyclase systems studied so far (10). As recently reported with purified adenylate cyclase preparations, the stimulatory action
of forskolin is apparently due to a direct interaction of the compound
with the adenylate cyclase catalytic moiety (4,11).

COMPONENTS AND MECHANISMS INVOLVED IN ADENYLATE CYCLASE INHIBITION

Similar to adenylate cyclase stimulation by hormones, inhibition of
the enzyme by hormonal factors requires the interaction of at least three
different membrane proteins: the inhibitory hormone receptor (R_i), the
inhibitory guanine nucleotide-binding regulatory protein (G_i) and the
adenylate cyclase catalyst (C) (Fig. 1). Similar to G_s, G_i is a hetero-

trimeric protein, composed of a larger α-subunit (41 kDal), which binds
guanine nucleotides and possesses GTPase activity, and β- and γ-subunits.
Whereas the α-subunits of G_s and G_i are clearly distinct polypeptides,
the β- and γ-subunits are both structurally and functionally very similar,
if not identical, in G_s and G_i (12-16). An additional similarity between
G_s and G_i is that G_i can also be directly activated by stable GTP analogs
such as GTP[S] and GuoPP[NH]P and by AlF_4^-. Furthermore, $G_{i\alpha}$ is also a
substrate for the ADP-ribosyltransferase activity of a bacterial toxin,
namely of pertussis toxin (islet-activating protein) (17). In contrast to
ADP-ribosylation of $G_{s\alpha}$ by cholera toxin, which requires the additional
presence of the ARF protein (9), ADP-ribosylation of $G_{i\alpha}$ by pertussis
toxin is seen with the purified G_i protein, but only in the concomitant
presence of the α_i-subunit and the βγ-dimer. The covalent modification of
$G_{i\alpha}$ by pertussis toxin leads to an increased association of $G_{i\alpha}$ and βγ-
dimer. Thus, the inactive holoprotein conformation of G_i with GDP bound
to its α-subunit is favoured by the ADP-ribosylation. Following the ADP-
ribosylation, inhibitory hormone receptors are apparently "uncoupled" from
the G_i protein, which means that the signal transduction from the inhibitory hormone receptor to the adenylate cyclase is blocked (17).

Inhibition of adenylate cyclase caused by an agonist-activated inhibitory hormone receptor is thought to proceed in a manner similar to
agonist-induced stimulation of adenylate cyclase, at least with regard to
the interaction of the receptor with the G-protein and the hypothetical
dissociation-association behavior of the G-protein. A major, not yet resolved problem is how the GTP-liganded, activated G_i protein induces inhibition of adenylate cyclase activity. At least four different mechanisms
have been proposed for this inactivation reaction: i) The dissociated GTP-
bound α_i-subunit inhibits the adenylate cyclase directly; ii) the activated
α_i-subunit interacts competitively with the activated α_s on the adenylate
cyclase; iii) the dissociated βγ-dimer inhibits the adenylate cyclase
directly; and iv) the βγ-dimer of G_i released during activation of G_i
associates with free α_s-subunit and, thereby, inactivates α_s and, subsequently, also the adenylate cyclase, by forming the inactive holoprotein
confirmation of G_s. For all of these proposed mechanisms of adenylate cyclase inhibition by activated G_i not only pro but also contra arguments
have been presented (11,18,19,20). Thus, the final answer to the question
whether in the intact membranes one or several of these proposed mechanisms are operating for inducing receptor-mediated inhibition of adenylate
cyclase or an even completely different one awaits further studies. Furthermore, it has to be studied whether the reduction in apparent metal ion
(magnesium) affinity of the adenylate cyclase caused by hormone and guanine
nucleotide-activated G_i, which has been observed in membrane preparations
(7), can also be found with purified components. Finally, the unresolved
questions mentioned above for adenylate cyclase stimulation by hormones
have also not been answered or have even not been addressed with regard
to adenylate cyclase inhibition by hormones.

In addition to pertussis toxin causing inactivation of G_i by ADP-
ribosylating its α-subunit, several agents have been reported to interfere rather selectively with G_i-mediated adenylate cyclase inhibition (see
ref. 21 for a review). These agents include SH-group reagents such as N-
ethylmaleimide and diamide and proteolytic enzymes such as trypsin, chymo-
trypsin and sperm proteases. Whereas the site of action of N-ethylmaleimide
is apparently at the α-subunit of G_i, the target protein(s) of the proteolytic enzymes is not yet known and it may even be different for different
proteases. Furthermore, manganese ions have been reported to block G_i-
mediated adenylate cyclase inhibition at concentrations having no effect
or causing even an increase in G_s-mediated adenylate cyclase stimulation.
It is not yet known whether this divalent cation action, which is also

seen with magnesium but at much higher concentrations than with manganese, is due to an action at the G_i protein or at the adenylate cyclase catalyst, rendering it less susceptible to G_i-induced inhibition.

An additional factor which will be considered here is the regulation of adenylate cyclase activity by monovalent cations (22). These ions, in the potency order, sodium > lithium > potassium, interfere with both stimulation and inhibition of adenylate cyclase by hormones in a very similar manner. First, the binding of agonists to both inhibitory and stimulatory hormone receptors can be reduced by these ions, whereas the binding affinity of antagonists is rather increased. Second, the regulation of agonist-receptor binding by sodium ions is also observed in membranes with defective α-subunits of G_s and G_i and in solubilized receptor preparations, where the regulation of agonist-receptor binding by guanine nucleotides is lost. These data suggest that at least the α-subunits of G_i and G_s are not required for the expression of the sodium regulation of agonist receptor binding. Third, both stimulation and inhibition of adenylate cyclase by hormones and guanine nucleotides can be reduced by sodium ions. This sodium action can be overcome by increasing the guanine nucleotide and/or hormone concentration. Based on all of these data, which have been accumulated in various membrane systems and with various stimulatory and inhibitory hormone receptors, and because of the similarity in the sodium action on the stimulatory and inhibitory adenylate cyclase system, it has been hypothesized that the βγ-subunits of G_s and G_i, which are very similar, if not identical, in these proteins, may be the target proteins for the observed sodium action (22). However, direct evidence for this proposal has not been presented yet. Furthermore, the possible physiological significance of the *in vitro* observed actions of sodium ions on the hormone-sensitive adenylate cyclase system, which are usually obtained at extracellular rather than at intracellular sodium concentrations, needs to be clarified.

ADAPTATION OF SIGNAL TRANSDUCTION BY THE ADENYLATE CYCLASE SYSTEM

Cellular adaptation to extracellular signal molecules is a widespread phenomenon in biological regulation. Although adaptation has been described for both receptor-mediated stimulations and inhibitions of adenylate cyclase, the molecular mechanisms underlying the desensitization of β-adrenoceptor-mediated stimulation of adenylate cyclase are so far best characterized (23). This and any other form of hormone-induced desensitization is usually divided into two general categories, referred to as agonist-specific or homologous desensitization and agonist-nonspecific or heterologous desensitization. Heterologous desensitization of receptor-mediated stimulation of adenylate cyclase apparently involves, at least in part, actions of cyclic AMP-dependent protein kinases at the stimulatory receptors and possibly also at other components of the system, i.e., at the level of the guanine nucleotide-binding proteins and the adenylate cyclase catalytic moiety (Fig. 2). In contrast, homologous desensitization is cyclic AMP-independent and is apparently also primarily independent of the coupling of the stimulatory receptor to the G_s protein. Recent evidence presented by Lefkowitz and coworkers indicates that an initial reaction in this form of desensitization is the phosphorylation of the agonist-occupied receptor by a specific protein kinase, referred to as "β-adrenergic receptor kinase" (24,25). This phosphorylation reaction is reminiscent of the well-established phosphorylation of light-bleached rhodopsin by the specific rhodopsin kinase. This covalent modification of rhodopsin is thought to represent the initial step in the termination reaction of the "light signal" transduction from rhodopsin to the retinal cyclic GMP phosphodiesterase (see ref. 8 for a review). Because of the similarity in the structure of rhodopsin and the β-adrenoceptor, it was, thus, not completely surprising to note that rhodopsin kinase and the

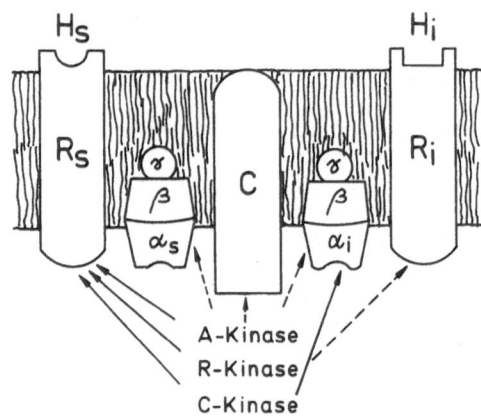

Fig. 2. Phosphorylation of components of the hormone-
sensitive adenylate cyclase system as a basis
for adaptation. Reported phosphorylation reactions are indicated by solid lines, while broken lines indicate hypothetical reactions. A-
Kinase, cyclic AMP-dependent protein kinase;
R-Kinase, "β-adrenergic receptor" kinase; C-
Kinase; protein kinase C.

"β-adrenergic receptor kinase" can substitute each other, at least partially, in phosphorylation of their physiological receptor substrates
(26). Accordingly, it is feasible, as suggested in Fig. 2, that the "β-
adrenergic receptor kinase" is not specific for the β-adrenoceptor, but
that this receptor kinase will phosphorylate other agonist-occupied receptors as well and induce desensitization of these receptors too. However, as phosphorylation of rhodopsin by the rhodopsin kinase is apparently
not sufficient to terminate its light-induced activation (8), phosphorylation of the hormone receptors linked to adenylate cyclase by the receptor
kinase may also not be the sole, but apparently is a very early step in
the homologous desensitization reaction.

Another type of cellular adaptation to signal molecule formation can
apparently be caused by the interaction of different signal transduction
systems. This interaction appears to involve phosphorylation of components
of one signal transduction system by protein kinases activated by intracellular signals formed by the other signalling system. For example, it
has been reported that protein kinase C, which is activated by diacylglycerol formed upon activation of the polyphosphoinositide system, can
phosphorylate and partially inactivate the avian erythrocyte β-adrenoceptor (23,27). Furthermore, activated protein kinase C has recently been
shown to phosphorylate the α-subunit of the inhibitory guanine nucleotide-
binding regulatory protein G_i. This covalent modification apparently suppresses G_i's function in hormonal inhibition of adenylate cyclase (28,29).
It is feasible that other types of mutual interactions between different
signal transduction systems may occur as well, which, generally speaking,

may represent a cellular control system against overexpression of one specific intracellular signal.

CONCLUSIONS AND PERSPECTIVES

The hormone-sensitive adenylate cyclase system is one of the best characterized multi-component transmembrane signalling systems used by a large variety of hormones and neurotransmitters. Several of the components of the system have been purified to homogeneity and characterized, and even the amino acid sequences of some of the components have recently been reported ($G_{i\alpha}$, $G_{s\alpha}$, β-adrenoceptor) (30-33). Nevertheless, it is obvious from the foregoing discussion that presently only some minimum essential components and reactions are established and that many important questions still exist and need to be addressed in the future.

Acknowledgements

The authors' studies reported herein were supported by the Deutsche Forschungsgemeinschaft.

REFERENCES

1. J. K. Northup, P. C. Sternweis, M. D. Smigel, L. S. Schleifer, E. M. Ross, and A. G. Gilman, Purification of the regulatory component of adenylate cyclase, *Proc. Natl. Acad. Sci. USA* 77:6516 (1980).
2. R. G. L. Shorr, R. J. Lefkowitz, and M. G. Caron, Purification of the β-adrenergic receptor, *J. Biol. Chem.* 256:5820 (1981).
3. M. Hekman, D. Feder, A. K. Keenan, A. Gal, H. W. Klein, T. Pfeuffer, A. Levitzki, and E. J. M. Helmreich, Reconstitution of β-adrenergic receptor with components of adenylate cyclase, *EMBO J.* 3:3339 (1984).
4. E. Pfeuffer, S. Mollner, and T. Pfeuffer, Adenylate cyclase from bovine brain cortex: purification and characterization of the catalytic unit, *EMBO J.* 4:3675 (1985).
5. D. C. May, E. M. Ross, A. G. Gilman, and M. D. Smigel, Reconstitution of catecholamine-stimulated adenylate cyclase activity using three purified proteins, *J. Biol. Chem.* 260:15829 (1985).
6. D. Feder, M.-J. Im, H. W. Klein, M. Hekman, A. Holzhöfer, C. Dees, A. Levitzki, E. J. M. Helmreich, and T. Pfeuffer, Reconstitution of β_1-adrenoceptor-dependent adenylate cyclase from purified components, *EMBO J.* 5:1509 (1986).
7. K. H. Jakobs, G. Schultz, B. Gaugler, and T. Pfeuffer, Inhibition of N_s-protein-stimulated human platelet adenylate cyclase by epinephrine and stable GTP analogs, *Eur. J. Biochem.* 134:351 (1983).
8. L. Stryer, Molecular design of an amplification cascade in vision, *Biopolymers* 24:29 (1985).
9. R. A. Kahn and A. G. Gilman, The protein cofactor necessary for ADP-ribosylation of G_s by cholera toxin is itself a GTP binding protein. *J. Biol. Chem.* 261:7906 (1986).
10. K. B. Seamon and J. W. Daly, Forskolin: A unique diterpene activator of cyclic AMP generating systems, *J. Cyclic Nucleotide Res.* 7:201 (1981).
11. M. D. Smigel, Purification of the catalyst adenylate cyclase, *J. Biol. Chem.* 261:1976 (1986).
12. B. M. Bokoch, T. Katada, J. K. Northup, M. Ui, and A. G. Gilman, Purification of the inhibitory guanine nucleotide regulatory component of adenylate cyclase, *J. Biol. Chem.* 259:3560 (1984).

13. J. Codina, J.D. Hildebrandt, R. Iyengar, L. Birnbaumer, R. D. Sekura, and C. R. Manclark, Pertussis toxin substrate, the putative N_i component of adenylyl cyclases, is an $\alpha\beta$ heterodimer regulated by guanine nucleotide and magnesium, *Proc. Natl. Acad. Sci. USA* 80:4276 (1983).

14. D. R. Manning and A. G. Gilman, The regulatory components of adenylate cyclase and transducin. A family of structurally homologous guanine nucleotide-binding proteins, *J. Biol. Chem.* 258:7059 (1983).

15. P. Gierschik, J. Codina, C. Simons, L. Birnbaumer, and A. Spiegel, Antisera against a guanine nucleotide binding protein from retina cross react with the β subunit of the adenylyl cyclase-associated guanine nucleotide binding proteins, N_s and N_i, *Proc. Natl. Acad. Sci. USA* 82:727 (1985).

16. J. D. Hildebrandt, J. Codina, W. Rosenthal, L. Birnbaumer, E. Neer, A. Yamazaki, and M. W. Bitensky, Characterization by two-dimensional peptide mapping of the γ subunits of N_s and N_i, the regulatory proteins of adenylyl cyclase, and of transducin, the guanine nucleotide-binding protein of rod outer segments of the eye, *J. Biol. Chem.* 260:14867 (1985).

17. M. Ui, Islet-activating protein, pertussis toxin: A probe for functions of the inhibitory guanine nucleotide regulatory component of adenylate cyclase, *Trends Pharmacol. Sci.* 5:277 (1984).

18. A. G. Gilman, Guanine nucleotide-binding regulatory proteins and dual control of adenylate cyclase, *J. Clin. Invest.* 73:1 (1984).

19. T. Katada, M. Oinuma, and M. Ui, Mechanisms for inhibition of the catalytic activity of adenylate cyclase by the guanine nucleotide binding proteins serving as the substrate of islet-activating protein, pertussis toxin, *J. Biol. Chem.* 261:5215 (1986).

20. K. H. Jakobs, M. Minuth, S. Bauer, R. Grandt, C. Greiner, and P. Zubin, Dual regulation of adenylate cyclase. A signal transduction mechanism of membrane receptors, *Basic Res. Cardiol.* 81:1 (1986).

21. K. H. Jakobs, K. Aktories, M. Minuth, and G. Schultz, Inhibition of adenylate cyclase, *Adv. Cyclic Nucleotide Protein Phosphoryl. Res.* 19:137 (1985).

22. K. H. Jakobs, M. Minuth, and K. Aktories, Sodium regulation of hormone-sensitive adenylate cyclase, *J. Receptor Res.* 4:443 (1984).

23. D. R. Sibley and R. J. Lefkowitz, Molecular mechanisms of receptor desensitization using the β-adrenergic receptor-coupled adenylate cyclase system as a model, *Nature* 317:124 (1985).

24. R. H. Strasser, D. R. Sibley, and R. J. Lefkowitz, A novel catecholamine-activated adenosine cyclic 3',5'-phosphate independent pathway for β-adrenergic receptor phosphorylation in wild-type and mutant S49 lymphoma cells: Mechanisms of homologous desensitization of adenylate cyclase, *Biochemistry* 25:1371 (1986).

25. J. L. Benovic, R. H. Strasser, M. G. Caron, and R. J. Lefkowitz, β-Adrenergic receptor kinase: Identification of a novel protein kinase that phosphorylates the agonist-occupied form of the receptor, *Proc. Natl. Acad. Sci. USA* 83:2797 (1986).

26. J. L. Benovic, F. Mayor Jr, R. L. Somers, M. G. Caron, and R. J. Lefkowitz, Light-dependent phosphorylation of rhodopsin by β-adrenergic receptor kinase, *Nature* 321:869 (1986).

27. D. J. Kelleher, J. E. Pessin, A. E. Ruoho, and G. L. Johnson, Phorbol ester induces desensitization of adenylate cyclase and phosphorylation of the β-adrenergic receptor in turkey erythrocytes, *Proc. Natl. Acad. Sci. USA* 81:4316 (1984).

28. K. H. Jakobs, S. Bauer, and Y. Watanabe, Modulation of adenylate cyclase of human platelets by phorbol esters. Impairment of the hormone-sensitive inhibitory pathway, *Eur. J. Biochem.* 151:425 (1985).

29. T. Katada , A. G. Gilman, Y. Watanabe, S. Bauer, and K. H. Jakobs, Protein kinase C phosphorylates the inhibitory guanine-nucleotide-binding regulatory component and apparently suppresses its function in hormonal inhibition of adenylate cyclase, *Eur. J. Biochem.* 151: 431 (1985).

30. T. Nukada, T. Tanabe, H. Takahashi, M. Noda, T. Hirose, S. Inayama, and S. Numa, Primary structure of the α-subunit of bovine adenylate cyclase-stimulating G-protein deduced from the cDNA sequence, *FEBS Lett.* 195:220 (1986).

31. T. Nukada, T. Tanabe, H. Takahashi, M. Noda, K. Haga, T. Haga, A. Ichiyama, K. Kangawa, M. Hiranaga, H. Matsuo, and S. Numa, Primary structure of the α-subunit of bovine adenylate cyclase-inhibiting G-protein deduced from the cDNA sequence, *FEBS Lett.* 197:305 (1986).

32. J. D. Robishaw, D. W. Russell, B. A. Harris, M. D. Smigel, and A. G. Gilman, Deduced primary structure of the α subunit of the GTP-binding stimulatory protein of adenylate cyclase, *Proc. Natl. Acad. Sci. USA* 83:1251 (1986).

33. R. A. F. Dixon, B. K. Kobilka, D. J. Strader, J. L. Benovic, H. G. Dohlman, T. Frielle, M. A. Bolanowski, C. D. Bennett, E. Rands, R. E. Diehl, R. A. Mumford, E. E. Slater, E. S. Sigal, M. G. Caron, R. J. Lefkowitz, and C. D. Strader, Cloning of the gene and cDNA for mammalian β-adrenergic receptor and homology with rhodopsin, *Nature* 321:75 (1986).

PROTEIN PHOSPHORYLATION/DEPHOSPHORYLATION AND REVERSIBLE

PHOSPHOPROTEIN BINDING IN RHABDOMERIC PHOTORECEPTORS

Joachim Bentrop and Reinhard Paulsen

Fakultät für Biologie, Tierphysiologie
Ruhr-Universität Bochum
Postfach 102148
D-4630 Bochum, FRG

INTRODUCTION

In the <u>rhabdomeric</u> photoreceptors of invertebrates, the transduction machinery is located in the rhabdomere, a part of the cell which is elaborated into a stack of photosensitive microvilli. The activation of the phototransduction mechanism causes this photoreceptor cell to depolarize, i.e., cation channels open in response to light. Recent progress in the biochemistry of invertebrate photoreceptors suggests that at least the molecular events leading to the activation of the phototransduction process are similar to that verified for the light-triggered enzyme cascade of the <u>ciliary</u> photoreceptors of vertebrates. It is a particular property of many invertebrate visual pigment systems, that the transduction mechanism is triggered by the conversion of rhodopsin (P) into a long-lived (thermostable) metarhodopsin (M)[1,2]. M in turn, is reconverted by light to P. This photoregeneration constitutes one of the main pathways of visual pigment regeneration in the living animal[3]. As photoregeneration is also possible in isolated membranes, invertebrate photoreceptors offer the unique opportunity to study the reversibility of reactions coupling photochemical and enzymatic transduction steps. This for example permits one to investigate the interaction of M with other proteins and to characterize the reactions leading to a reversible inactivation of the photoactivated rhodopsin state.

Photoreceptors R1-6 of flies, e.g. of the blowfly <u>Calliphora erythrocephala</u>, are a favourable system for such studies, particularly because it is possible to create different, but very precise ratios of P and M: maximum conversion of P to M is achieved by irradiation with blue light which establishes a photo-equilibrium of 70% M and 30% P (M-state membranes). By subsequent exposure to red light almost all M can be reconverted into P (P-state membranes). Also, due to the availability of a simple method for the isolation of rhabdoms, it is possible to confine studies to the signal transducing membrane system. The aim of our work is to identify and to characterize enzymes and regulatory components involved in the visual transduction process in this photoreceptor type. The study mainly focusses on light-dependent reversible processes, i.e., reactions which are stimulated by P→M and are inhibited or reversed by M→P.

PHOSPHORYLATION OF OPSIN

The first reversible enzymatic reaction which has been shown to be directly linked to the P→M conversion is the phosphorylation of opsin, the protein moiety of the visual pigment[4,5,6].

$$\text{(I)} \qquad P \underset{}{\overset{h\nu}{\rightleftarrows}} M \xrightarrow[\text{kinase}]{\text{ATP}} M\text{-}Pn \overset{h\nu}{\rightleftarrows} P\text{-}Pn \xrightarrow[\text{phosphatase}]{} P$$

As indicated in (I), the P→M transition apparently converts the visual pigment into a substrate for a cytosolic metarhodopsin kinase, which catalyzes a multiple phosphorylation of at least 4 phosphate binding sites, probably serine or threonine residues located at the C-terminus of opsin. Photoregeneration of phosphorylated M (M-Pn) to P-Pn leads to a rapid dephosphorylation, restoring the initial dark state of the visual pigment. In the presence of 1 mM Ca^{2+} the phosphorylation of M is inhibited and the dephosphorylation of P-Pn is stimulated[5,6]. Since high amounts of calmodulin are associated with blowfly rhabdoms, it is possible that dephosphorylation is catalyzed by a calcineurin-like phosphatase. Both metarhodopsin kinase and rhodopsin phosphatase, however, are not as yet further characterized.

Although the C-terminal regions of fly opsins and vertebrate opsin show little homology in their amino acid sequences[7,8,9], the conservation of phosphate binding site in this domain appears to be an evolutionary tendency. This suggests that multiple phosphorylation of rhodopsin has an important function. It has been proposed that the phosphorylation of invertebrate opsin is a mechanism to turn off activated M (M*), the M-form which is able to trigger the enzymatic steps of the transduction cascade[4,10,11]. M* would then, by phosphorylation, be transformed into an inactive form (M'). Such a function has been recently established for the phosphorylation of vertebrate opsin[12] and has been suggested also for the phosphorylation of the mammalian β-adrenergic receptor, which is structurally related to opsin[13].

As pointed out elsewhere[4,6], the light-induced opsin phosphorylation is most likely a result of structural changes at the cytoplasmic surface of the molecule which occur during the P→M conversion. In addition to opsin - kinase interaction, light-triggered changes in the opsin structure may also enable M to interact with other cytoplasmic or peripheral proteins.

G-PROTEINS IN FLY PHOTORECEPTORS

If transduction in rhabdomeric photoreceptors is mediated by a signal amplifying cascade, the protein interacting with photo-activated M might be a G-protein similar to transducin. The earliest evidence for the activation of a G-protein by light, was found by studying GTPase activity [5, 14, 15, 16]. GTPase activity is stimulated by P→M conversion and is reversible in fly photoreceptors. In homogenates of eyes of the housefly Musca[16] and in isolated blowfly rhabdoms[5] it has been shown that the light-induced GTPase activity can be supressed to the dark level by reconverting M→P.

One candidate for the α-subunit of the proposed G-protein is a 41-kDa protein which is tightly bound to the rhabdomeric membrane[6]. This protein is ADP-ribosylated by cholera toxin in P-state membranes; P→M conversion inhibits the labeling of this protein. Pertussis toxin catalyzed ADP-ribosylation resulted in the labeling of additional proteins (41 kDa and 39 kDa) in homogenates of whole retinae. This indicates the presence of more than one G-proteins in fly photoreceptors.

It is not yet possible to decide whether the light-dependent G-protein of blowfly photoreceptors belongs to one of the known types of G-proteins or if it represents a new kind of G-protein. There are presently two candidates for enzymes activated by this G-protein:
(i) guanylate cyclase: as in rhabdomeric photoreceptor of cephalopods light stimulates the formation of cyclic GMP from GTP[17, 18].
(ii) phospholipase C: as in Limulus and squid photoreceptors light has recently been shown to trigger the release of polyphosphoinositols (IP_3) from polyphosphoinositides[19, 20].

PROTEIN BINDING

It is known from vertebrate photoreceptors, that some of the cytoplasmic or peripherally bound proteins may directly interact with the photoactivated rhodopsin state. These proteins (rhodopsin kinase, transducin and arrestin (S-antigen)) show an increased affinity to bleached vertebrate photoreceptor membranes[21]. Potential sites for interactions of blowfly rhodopsin with cytoplasmic or peripherally bound proteins are the cytoplasmic loops between transmembrane helices. Since these regions, in some amino acid sequences, show a high degree of homology to the corresponding domains of vertebrate opsin[7, 8] we have attempted to identify some proteins interacting with M in fly photoreceptors, by studying the light-dependent binding of proteins to the rhabdomeric membrane. An important point in these studies was to investigate whether photoregeneration of M to P would again decrease the affinity of those proteins to the rhabdomeric membrane. There is now evidence that indeed some proteins of blowfly photoreceptors undergo such a reversible light-modulated binding.

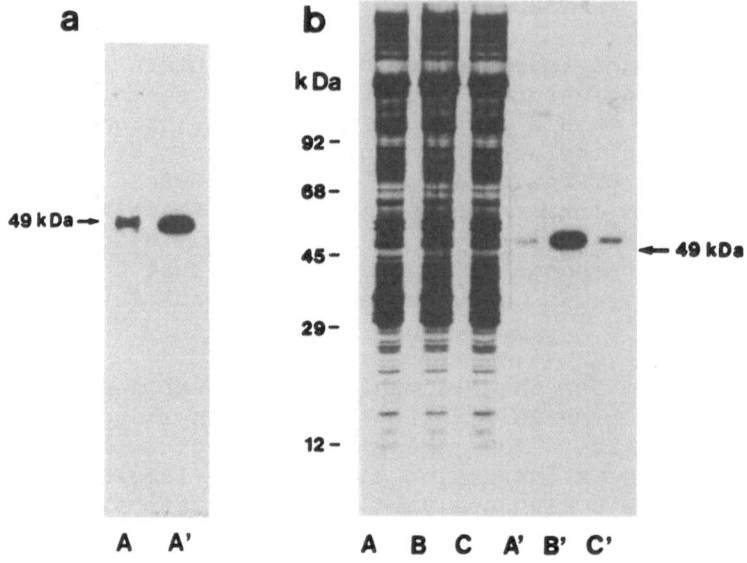

Figure 1. Light-dependent binding of ^{32}P-phospho-
rylated 49-kDa protein to rhabdomeric membranes.
a) Partially purified 49-kDa protein, phosphorylated with ($_\gamma$-^{32}P)ATP. A) Silver stained protein
pattern after SDS-PAGE (9 - 18% gel), A') autoradiograph of A). b) Radiolabeled 49-kDa protein
was incubated with rhabdomeric membranes. The
membranes were A) kept in the P-state, B) converted into the M-state, and C) after P→M conversion as in B) reconverted into the P-state. Unbound proteins were extracted with buffer, the
membranes were solubilized with SDS and subjected
to SDS-PAGE. A-C) protein pattern, A'-C') autoradiographs of A-C, indicating the binding of phosphorylated 49-kDa protein particularly to M-state
membranes.

The predominant protein undergoing light-modulated binding
is a 49-kDa protein [5,6] (M_r between 45 kDa and 50 kDa depending on
the electrophoretic method used for the determination). The 49-
kDa protein is a phosphoprotein which is phosphorylated at serine
and/ or threonine residues. Its phosphorylation in retina homogenates is increased by twofold by light. Fig. 1a) demonstrates the
phosphoprotein nature of this protein. Fig. 1b) shows that the
phosphorylated 49-kDa protein is bound to M-state membranes and
is released after M→P. The amount of 49-kDa protein bound to the
rhabdomeric membrane is related to the amount of M present (not
shown). Thus, it is likely that this protein binding occurs at
molecules converted into the M-state. At the present it seems
that phosphorylation neither of opsin nor of the 49-kDa protein
effects protein binding. The reversibility of the blue-light-in-
duced binding of the 49 kDa-protein by M→P conversion provides a
simple and elegant method for the purification of this protein.

Isolated rhabdomeric membranes are used to bind the 49-kDa protein, other proteins are removed by buffer washes and then the 49-kDa protein is released by irradiation with red light.

This 49-kDa protein may be identical with a 49-kDa protein in Drosophila eyes, which undergoes a rapid light-induced phosphorylation[22]. Like the 48-kDa protein of vertebrate photoreceptors[21] the 49-kDa protein may be a regulatory protein which, for steric reasons, inhibits the interaction of M* with e. g. a G-protein. Another possibility is, that it is an enzyme or a subunit of an enzyme complex activated by M*. Protein binding experiments indicate that blowfly photoreceptors contain in addition to the 49-kDa protein other proteins which undergo light-induced reversible binding. These proteins, e.g. two phosphoproteins of 68 kDa and 200 kDa[6] and two non-phosphorylatable proteins of 36 kDa and 40 kDa might therefore also be involved in the visual process.

CONCLUSION

These results are consistent with the hypothesis that the transduction process in rhabdomeric photoreceptors is based on a signal-amplifying enzyme cascade. This cascade might - in some aspects - be different from that operating in the ciliary photoreceptors of vertebrates. The finding, that protein binding to the rhabdomeric membrane can be induced and reversed simply by irradiation with different lights, makes this photoreceptor type a promising experimental system for studying protein - receptor interactions in visual transduction.

Acknowledgements. This work was supported by the Deutsche Forschungsgemeinschaft (SFB 114, TP B2). A travel grant by the Boehringer Ingelheim Fonds to J. B. is gratefully acknowleged.

REFERENCES

1. K. Hamdorf, The physiology of invertebrate visual pigments, in "Vision in Invertebrates (Handbook of Sensory Physiology, Vol. VII 6 A)" H. Autrum, ed., Springer Verlag, Berlin, New York.
2. P. Hillman, S. Hochstein and B. Minke, Transduction in invertebrate photoreceptors: Role of pigment bistability, Physiol. Rev. 63:668 (1983).
3. J. Schwemer, Pathways of visual pigment regeneration in fly photoreceptors, Biophys. Struct. Mech. 9:287 (1983).
4. R. Paulsen and J. Bentrop, Reversible phosphorylation of opsin induced by irradiation of blowfly retinae, J. Comp. Physiol. A 155:39 (1984).
5. R. Paulsen and J. Bentrop, Light-modulated biochemical events in blowfly photoreceptors. Progr. Zool. 33:299 (1986).
6. J. Bentrop and R. Paulsen, Light-modulated ADP-ribosylation, protein phosphorylation and protein binding in isolated fly photoreceptor membranes, Eur. J. Biochem., 161:61 (1986).

7. J. E. O'Tousa, W. Baehr, R. L. Martin, J. Hirsch, W. L. Pak and M. L. Applebury, The Drosophila nina E gene encodes an opsin, Cell 40:839 (1985).

8. C. S. Zuker, A. F. Cowman and G. H. Rubin, Isolation and structure of a rhodopsin gene from D. melanogaster, Cell 40:851 (1985).

9. A. F. Cowman, C. S. Zuker and G. R. Rubin, An opsin gene expressed in only one cell type of the Drosophila eye, Cell 44:705 (1986).

10. B. Minke, Photopigment dependent adaptation in invertebrates, in "The Molecular Mechanism of Phototransduction" H. Stieve, ed., Springer Verlag, Berlin, Heidelberg, New York (1986).

11. J. Lisman, The role of metarhodopsin in the generation of spontaneous quantum bumps in ultraviolett receptors of Limulus median eye. J. Gen. Physiol. 85:171 (1985).

12. U. Wilden, S. W. Hall and H. Kühn, Phosphodiesterase activation by photoexcited rhodopsin is quenched when rhodopsin is phosphorylated and binds the intrinsic 48-kDa protein, Proc. Natl. Acad. Sci. USA 83:1174 (1986).

13. J. L. Benovic, F. Mayor Jr, R. L. Somers, M. G. Caron and R. J. Lefkovitz, Light-dependent phosphorylation of rhodopsin by β-adrenergic receptor kinase, Nature 321:869 (1986).

14. R. Calhoon, M. Tsuda and T. G. Ebrey, A light-activated GTPase from octopus photoreceptors, Biochem. Biophys. Res. Commun. 94:1452 (1980).

15. C. A. Vandenberg and M. Montal, Light-regulated biochemical events in invertebrate photoreceptors. 1. Light-activated guanosine triphosphatase, guanine nucleotide binding and cholera toxin catalyzed labeling of squid photoreceptor membranes, Biochemistry 23:2339 (1984).

16. A. Blumenfeld, J. Erusalimsky, O. Heichal, Z. Selinger and B. Minke, Light-activated guanosinetriphosphatase in Musca eye membranes resembles the prolonged depolarizing after potential in photoreceptor cells, Proc. Natl. Acad. Sci. USA 82: 7116 (1986).

17. H. R. Saibil, A light-stimulated increase of cyclic GMP in squid photoreceptors, FEBS Lett. 168:213 (1984).

18. P. R. Robinson, E. C. Johnson and J. E. Lisman, A rapid light-induced rise in cGMP levels in squid retinas, Invest. Ophthalmol. Vis. Sci. (Suppl.) 27:218 (1986).

19. J. E. Brown, L. J. Rubin, A. J. Galayini, A. P. Tarver, R. F. Irvine, M. J. Berridge and R. E. Anderson, Myo-inositol polyphosphate may be a messenger for visual excitation in Limulus photoreceptors, Nature 311:160 (1984).

20. E. Szutz, M. Reid, R. Payne, D. W. Corson and A. Fein, Biochemical and physiological evidence for the involvement of inositol 1,4,5-trisphosphate in visual transduction, Biophys. J. 47:202a (1985).

21. H. Kühn, Interactions between rhodopsin and light-activated enzymes in rods, in "Progress in Retinal Research", N. Osborne and J. Chader, eds., Pergamon Press, Oxford, New York (1984).

22. H. Matsumoto and W. L. Pak, Light-induced phosphorylation of retina-specific polypeptides of Drosophila in vivo, Science 223:184 (1984).

UTILIZATION OF A MONOCLONAL ANTIBODY TO STUDY IN VITRO PHOSPHORYLATION OF A LOW Km PHOSPHODIESTERASE

David H. Reifsnyder, Scott A. Harrison,
Colin H. Macphee, and Joseph A. Beavo

Department of Pharmacology, SJ-30
University of Washington
Seattle, WA

INTRODUCTION

The study of cyclic nucleotide phosphodiesterases in tissues is complicated by the presence of isozymes that differ in terms of selective substrate specificity, substrate affinity, and regulatory and physical properties (for review see Beavo et al., 1982 or Wells and Hardman, 1977). Because the phosphodiesterase isozymes are trace cellular proteins which require several thousand-fold purification to apparent homogeneity, the isolation of these isozymes is very difficult. Isozyme specific probes are needed to study the enzymes in crude systems.

Monoclonal antibodies which recognize a low Km, cAMP phosphodiesterase isolated from bovine heart have been produced in this laboratory (Harrison et al., 1986b). This isozyme has been designated as a cGMP-inhibited (CGI) phosphodiesterase since sub-micromolar concentrations of cGMP will inhibit the hydrolysis of cAMP. The monoclonal antibodies directed against the CGI phosphodiesterase (CGI-antibody) are specific for this isozyme and enable the rapid isolation of the CGI phosphodiesterase from crude tissue extracts, thereby minimizing proteolysis during sample preparation. The antibody coupled, CGI-phosphodiesterase displays kinetic parameters similiar to those of the purified enzyme (Harrison et al., 1986a). In addition to bovine tissue, the monoclonal antibodies have been utilized to immunoadsorb low Km phosphodiesterase activity from human platelets (Macphee et al., 1986) and a K30a mutant of mouse lymphoma S49 cells (Reifsnyder et al., 1985). The CGI-monoclonal antibody also has been utilized in immunoblotting analysis, which identified the CGI-phosphodiesterase as a $Mr = 110,000$ peptide following SDS-gel electrophoresis of rapidly prepared samples (Harrison et al., 1986b).

In this study, we utilized the antibody coupled CGI-phosphodiesterase and report that this isozyme can be phosphorylated in vitro by an endogenous kinase activity and by cAMP-dependent protein kinase. Phosphorylation of the CGI-phosphodiesterase increases the cAMP hydrolysis rate 20-50% but does not appear to affect the IC_{50} (half-maximal inhibitory concentration) for milrinone, a selective inhibitor of this isozyme.

EXPERIMENTAL PROCEDURES

Materials

Tritiated cAMP (26 Ci/mmol) was obtained from ICN Pharmaceuticals and ^{32}P-ATP (3000 Ci/mmol) was from New England Nuclear. *Crotalus atrox* snake venom, cAMP, 2-mercaptoethanol, dithiothreitol, and phenylmethylsulfonyl fluoride (PMSF) were purchased from Sigma. Heat inactivated, formalin-fixed cells of the Cowan I strain of *S. aureus* were purchased from Calbiochem-Behring. Rabbit anti-mouse IgG (RAM) antiserum was obtained from Miles. Molecular weight standards for electrophoresis were obtained from Bio-Rad. Milrinone was a gift from Sterling-Winthrop Research Institute. PKI, a heat stable inhibitor of cAMP-dependent protein kinase, and the catalytic subunit of cAMP-dependent protein kinase were generously provided by the laboratory of Dr. E. G. Krebs at the University of Washington.

Preparation of Bovine Tissue Extracts

Fresh bovine heart tissue was obtained from a local slaughter house, trimmed, sliced, and passed through a coarse meat grinder. These and all subsequent operations were performed at 4°. An extract was prepared by homogenizing 1 part tissue with 2 parts buffer containing 50 mM Tris-HCl, pH 7.8, 15 mM 2-mercaptoethanol, and 0.2 mM crystalline PMSF. The homogenate was centrifuged at 4000 x g for 20 min and the resulting supernatant was adjusted to pH 7.8.

Isolation of Monoclonal Antibody-CGI Phosphodiesterase Complex

Solid phase antibody reagents were prepared using heat inactivated, formalin-fixed cells of the Cowan 1 strain of *S. aureus* coupled to rabbit anti-mouse IgG antiserum (Staph A-RAM) as described by Mumby et al.(1982). Ascitic fluid containing monoclonal antibodies directed against the CGI-phosphodiesterase was then incubated with the Staph A-RAM complex previously described (Harrison et al., 1986a). Rapidly prepared tissue extracts described above were centrifuged at 15,000 rpm for 10 min before incubating with the solid phase antibody reagents. After 2 hours, the immune or control pellets (+/- CGI antibody) were centrifuged, washed three times with isotonic Tris-HCl (pH 7.4) containing 0.1% Tween-20, resuspended to the original sample volume in buffer containing 50% glycerol, and stored at -20° for subsequent use. The immobilized CGI-phosphodiesterase can be stored for several months and still retain enzymatic activity.

Phosphodiesterase activity assay

Low Km, cyclic nucleotide phosphodiesterase activity was determined with 1.0 μM cAMP using tritiated substrate as previously described and modified (Harrison et al., 1986a). It is important to note that the phosphodiesterase activity could be assayed directly on the immunoadsorbed pellet and compared to a control pellet which did not contain primary antibody. Non-specific immunoadsorption of low Km phosphodiesterase activity was generally less than 5% of that observed with

CGI-monoclonal antibody. Phosphodiesterase assays in the presence of varying concentrations of milrinone, a selective inhibitor of the CGI-isozyme were performed using 0.25 µM cAMP as previously described (Harrison et al., 1986a). One unit of activity represents the hydrolysis of 1 nmole of cAMP per minute.

Phosphorylation of cGMP-inhibited Phosphodiesterase

An aliquot of the immune or control antibody complex was transferred to a 0.5 ml microfuge vial, centrifuged for 5 min at 15,000 rpm, washed in Tris-HCl buffer devoid of glycerol, and resuspended in this buffer. Phosphorylation reactions were inititated by the addition of $Mg^{2+}[\partial-^{32}P]ATP$ in a final reaction volume of 100 µl and performed at 30°. The reaction also contained 10 mM HEPES, pH 7.0, 0.5 mM dithiothreitol, 50 µM Na_3VO_4, 13 mM $MgCl_2$, 10 µM ATP. In the initial experiments, 10 µM cAMP was also included in the phosphorylation reaction. The purified catalytic subunit of cAMP-dependent protein kinase was added to a concentration of 16 nM. In some studies, a heat stable inhibitor of cAMP-dependent protein kinase (PKI, see Walsh et al., 1971) was used. The specific activity of the $[\partial-^{32}P]ATP$ was generally 500 cpm/pmol.

The phosphorylation reactions were terminated by the addition of 50 µl of ice cold stop buffer (100 mM EDTA, 100 mM $NaPO_4$, and 10 µM Na_3VO_4) and washed with isotonic Tris-HCl buffer. The phosphorylated protein was eluted from the antibody complex by boiling the sample in SDS-containing buffer and loaded on to a 10% acrylamide gel for analysis by SDS-PAGE (Laemmli, 1970). ^{32}P was detected by autoradiography and in some cases, the labeled protein was excised from the dried gel and analyzed by Cerenkov counting. Experiments which examined the effect of phosphorylation on phosphodiesterase activity were performed in the presence of purified catalytic subunit of cAMP-dependent protein kinase and unlabeled ATP. The reaction was terminated after 30 min by placing the samples in ice and an aliquot was assayed for phosphodiesterase activity. Non-phosphorylated samples were obtained by an identical procedure except that no kinase or ATP was added to the reaction mixture. No loss of enzyme activity was observed during these manipulations.

RESULTS

Monoclonal antibodies coupled to a solid phase matrix are useful tools for selective isolation of the CGI-phosphodiesterase from crude tissue preparations. Since the enzyme can be isolated within a few hours, this procedure minimizes the effect proteolysis may have on the enzyme activity or on the intact protein. In this report, the antibodies were utilized to investigate in vitro phosphorylation of the CGI-phosphodiesterase.

Phosphorylation of the immune complex isolated from bovine heart occurs in the presence of an endogenous protein kinase (Fig. 1). The amount of endogenous kinase varied from one preparation to another and may be reduced by more extensive washing of the antibody complex. In addition, phosphorylation of the antibody-adsorbed phosphodiesterase by an endogenous kinase was greatly reduced in the absence of cAMP. Therefore, cAMP was omitted from subsequent phosphorylation experiments.

Phosphorylation of CGI-phosphodiesterase on the control (RAM) pellets is negligible. Addition of the purified catalytic subunit of cAMP-dependent protein kinase greatly enhanced phosphorylation of the CGI-isozyme. Most of the kinase activity can be completely removed by excess PKI, an inhibitor of cAMP-dependent protein kinase.

A linear relationship existed between the amount of phosphate incorporated into the CGI-phosphodiesterase and the amount of immune complex used in the phosphorylation. After identification of the phosphorylated (Mr =110,000) protein band by autoradiography, the band was excised from the gel and analyzed by Cerenkov counting (Fig. 2). The amount of phosphate

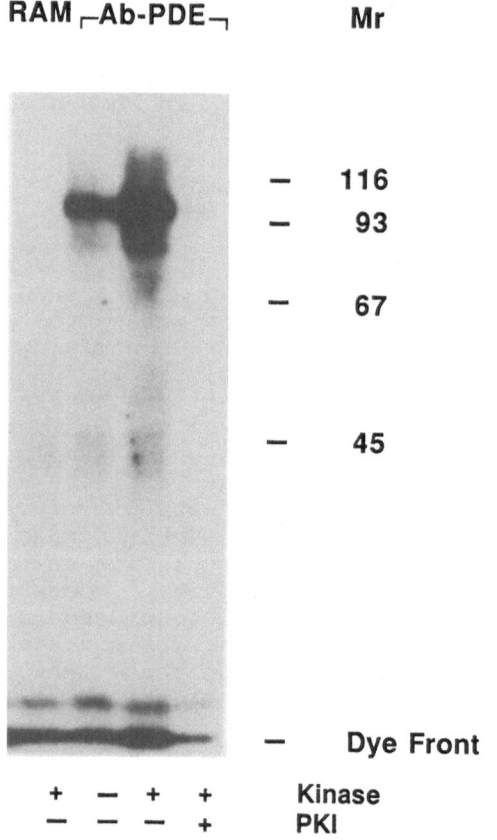

Fig.1. In vitro phosphorylation of CGI-phosphodiesterase. Phosphorylation of immune complex (Ab-PDE) or control (RAM) was performed with purified catalytic subunit of cAMP-dependent protein kinase and a heat stable inhibitor of this kinase (PKI). The antibody-adsorbed enzyme was also phosphorylated by a contaminating kinase activity. After 30 min, the reaction was terminated, boiled in Laemmli sample buffer, and analyzed by SDS-PAGE. An autoradiogram from this gel is shown.

incorporated into the Mr = 110,000 band increased with increasing amounts of resuspended CGI-antibody pellet. Phosphorylation of the CGI-phosphodiesterase by the catalytic subunit of cAMP-dependent protein kinase was increased approximately three-fold relative to that observed with endogenous kinase. The linear relationship was observed with endogenous kinase activity or with cAMP-dependent protein kinase. In the presence of PKI, phosphorylation by the endogenous kinase was reduced from 325 cpm to 100 cpm when 100 μl of the resuspended antibody pellet was used.

The time course of phosphorylation for the immunoadsorbed CGI-phosphodiesterase is shown in Figure 3. Half maximal phosphorylation occurs within 2 min in the presence of exogenous kinase. Unfortunately, treatment of the heart extract with the solid phase CGI-antibody reagent isolates only a trace amount of protein (for example, the phosphodiesterase protein band which can be identified by immunoblotting at Mr = 110,000 is not visible by Coomaisse staining). Therefore, the stoichiometry of phosphorylation can not be determined at the present time.

The effect of phosphorylation on the CGI-phosphodiesterase antibody complex is currently under investigation. The enzyme was isolated from heart extracts by adsorption to the solid phase antibody reagent as previously described. This sample was then incubated in the presence or absence of phosphorylation buffer (ATP + catalytic subunit of cAMP-dependent protein kinase). After 30 min, the reaction was terminated and the phosphodiesterase activity was determined using 1.0 μM cAMP. Phosphorylation of the CGI-phosphodiesterase correlated with an

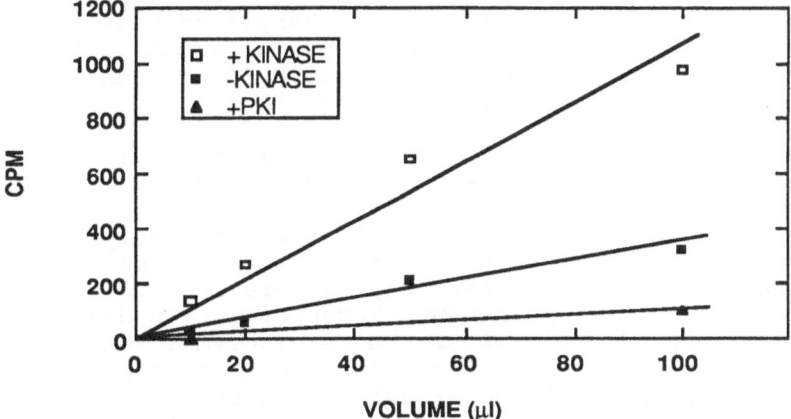

Fig.2. Phosphorylation of the CGI-phosphodiesterase is dependent on the amount of
immune complex used. The indicated amounts
of immune complex was phosphorylated in the
presence of purified cAMP-dependent protein
kinase or by a contaminating endogenous
kinase (+/- kinase) as in Fig.1. Each
reaction was performed at the same final
volume. After locating the labeled
CGI-phosphodiesterase by autoradiography, the
Mr = 110,000 protein band was excised from
the gel and quantitated by Cerenkov counting.

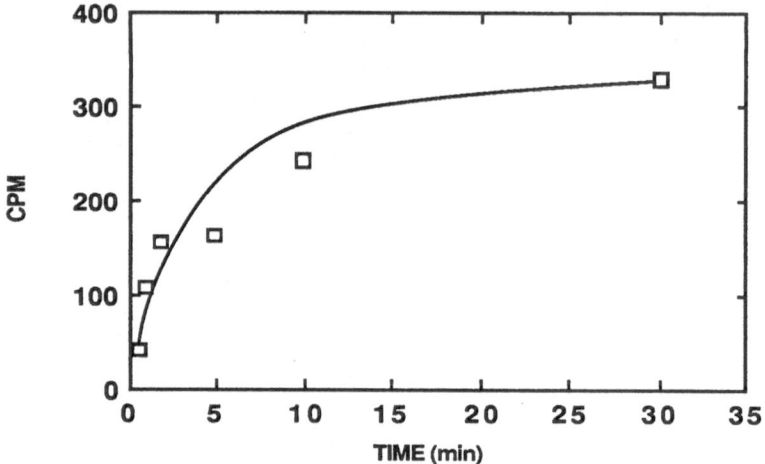

Fig.3. Time course for the in vitro phosphorylation of CGI-phosphodiesterase by cAMP-
dependent protein kinase. Twenty µl of
the resuspended Ab-PDE complex was
phosphorylated for the indicated times.
The amount of phosphate incorporated into
the CGI-phosphodiesterase was determined
as in Fig. 2.

increase in low Km activity, which was 20-50% greater than the
non-phosphorylated samples. The variable increase in phosphodiesterase activity following phosphorylation is currently under
investigation. In two experiments which demonstrated maximal
stimulation, the activity of the phosphorylated enzyme was 1.3 ±
0.2 units per ml of resuspended antibody pellet while identical
samples incubated under non-phosphorylating conditions had an
activity of 0.8 ± 0.2 units per ml (n=3).

The effect of phosphorylation on milrinone inhibition of the
low Km, cAMP phosphodiesterase was investigated. Milrinone was
used since it is a selective inhibitor of the CGI-
phosphodiesterase (Harrison et al., 1986a). Phosphorylated
samples (which exhibited a 20% increase in phosphodiesterase
activity) or non-phosphorylated samples were assayed in the
presence of 0.03-5.0 µM milrinone. Phosphorylation did not appear
to alter milrinone inhibition of the CGI-isozyme. The IC_{50} value
(half-maximal inhibitory concentration) observed for milrinone
was 0.3 µM for phosphorylated or non-phosphorylated isozyme
(Fig. 4).

DISCUSSION

Monoclonal antibodies directed against the CGI-
phosphodiesterase isolated from bovine heart were utilized to
investigate in vitro phosphorylation of this enzyme. There are
several advantages for using this procedure for studying the
CGI-phosphodiesterase. Since the CGI-isozyme is a trace cellular
protein, the use of monoclonal antibodies enables rapid isolation
of this isozyme from crude tissue preparations and thereby
minimizes proteolytic degradation from endogenous proteases
present in the extracts. The monoclonal antibody directed

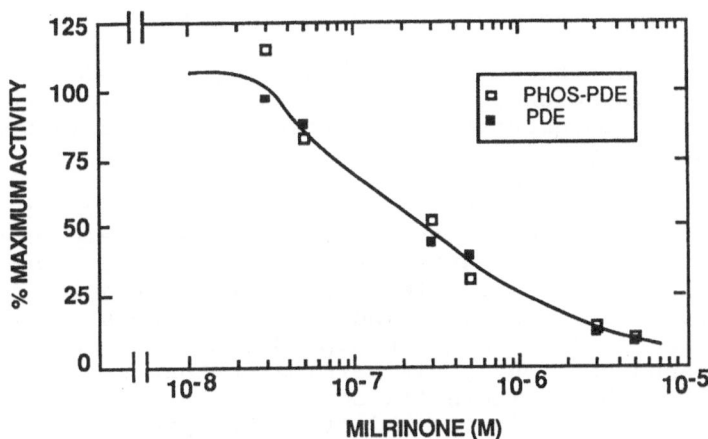

Fig.4. The effect of phosphorylation on milri-
        none inhibition of the CGI-phospho-
        diesterase. The immunoadsorbed isozyme
        was incubated in the presence or absence
        of phosphorylating conditions (+/- ATP
        and cAMP dependent protein kinase). The
        immune complexes were assayed at 0.25 μM
        cAMP in the presence of 0.03-5.0 μM
        milrinone. Data were plotted as a per-
        centage of controls in the absence of
        inhibitor.

against CGI-phosphodiesterase does not inhibit phosphodiesterase
activity. Therefore, activity can be measured directly on the
solid phase antibody pellet. The CGI-monoclonal antibody also can
be used in immunoblotting of this isozyme. Consequently, the
CGI-phosphodiesterase is very amenable to study by the use of
monoclonal antibodies.

     The monoclonal antibody was used to isolate the
CGI-phosphodiesterase from crude tissue extracts and to examine
the possible phosphorylation of this isozyme. The soluble
CGI-isozyme isolated from bovine heart was phosphorylated in
vitro by cAMP-dependent protein kinase with a subsequent increase
in phosphodiesterase activity. No change in the inhibition of
cAMP hydrolysis by milrinone, a selective inhibitor of the
CGI-phosphodiesterase, was observed following phosphorylation.

     Phosphorylation of low Km phosphodiesterase has not been
extensively characterized, though this has been postulated as a
regulatory mechanism. Marchmont and Houslay (1980) reported a
two-fold increase in low Km phosphodiesterase activity obtained
from plasma liver membranes incubated with insulin, ATP, and
cAMP. Furthermore, their data suggested that insulin was
mediating its effect through a cAMP-dependent protein kinase
since addition of the heat stable kinase inhibitor prohibited an
increase in phosphodiesterase activity. Both soluble and
membrane-bound cyclic nucleotide phosphodiesterase have been
found to be stimulated by insulin. Weber and Appleman (1982)
separated soluble insulin dependent and independent, low Km forms
from adipose tissue. The possibility exists that the soluble
forms were derived from the membrane-bound isozyme during

isolation. In fact, antisera directed against the CGI-phosphodiesterase was used to immunoadsorb an insulin-sensitive, low Km phosphodiesterase from rat epididymal fat pads (Reifsnyder et al., 1985). Therefore, it is tempting to speculate that the soluble CGI-phosphodiesterase is related to an insulin-sensitive, membrane-bound isozyme.

Prelimiary results from our laboratory for in vitro phosphorylation of the CGI-phosphodiesterase are very promising. However, there are many question yet to be answered. Of paramount importance is the effect that phosphorylation may have on phosphodiesterase activity. Though we have consistantly observed an increase in low Km phosphodiesterase activity following phosphorylation, it is perplexing that this response varies from 20 to 50%. There are several possible explanations which may account for this variability. The presence of endogenous phosphate on the antibody-isolated, CGI-phosphodiesterase would minimize the effect of in vitro phosphorylation. Therefore, removal of any endogenous phosphate prior to phosphorylation should demonstrate a greater effect on enzymatic activity or other kinetic parameters. Alternatively, the monoclonal antibody itself may be blocking a phosphorylation site or locking the phosphodiesterase into a conformation such that a true change in enzyme activity can not be observed following phosphorylation. Therefore, the isolation scheme may require dissociating the monoclonal antibody from the CGI-phosphodiesterase to measure the full effect of phosphorylation. Other questions yet to be investigated include the number of phosphorylation sites and the stoichiometry of phosphorylation, which will require isolating sufficient amounts of the CGI-phosphodiesterase. Also, future studies should consider the possible regulation by other protein kinases.

The use of the monoclonal antibodies also has facilitated in vivo phosphorylation studies. Using intact human platelets, it has been shown that hormone-induced phosphorylation of the CGI-phosphodiesterase correlated with increased low Km phosphodiesterase activity (C.H. Macphee and J.A. Beavo, manuscript in preparation). Therefore, utilization of the monoclonal antibodies is proving to be very beneficial in determining the physiological significance that phosphorylation may have on low Km phosphodiesterase activity and the subsequent maintenance of cAMP levels within a cell system.

REFERENCES

Beavo, J.A., Hansen, R.S., Harrison, S.A., Hurwitz, R.L., Martins, T.J., and Mumby, M.C., 1982, Identification and properties of cyclic nucleotide phosphodiesterases, Mol. Cell. Endocrinology, 28:387.

Harrison, S.A., Chang, M.L. and Beavo, J.A., 1986a, Differential inhibition of cardiac cyclic nucleotide phosphodiesterase isozymes by cardiotonic drugs, Circulation, 73 (part 2): 109.

Harrison, S.A., Reifsnyder, D.H., Gallis, B., Cadd, G.A., and Beavo, J.A., 1986b, Isolation and characterization of bovine cardiac muscle cGMP-inhibited phosphodiesterase: a receptor for new cardiotonic drugs, Mol. Pharmacol., 29:506.

Laemmli, U.K., 1970, Cleavage of structural proteins during the assembly of the head of bacteriophage T_4, <u>Nature</u> (Lond.), 227:680.

Macphee, C.H., Harrison, S.A. and Beavo, J.A., 1986, Immunological identification of the major platelet low-Km cAMP phosphodiesterase: probable target for anti-thrombotic agents, <u>Proc. Natl. Acad. Sci</u>. (USA), 83:6660.

Marchmont, R.J. and Houslay, M.D.,1980, Insulin triggers cyclic AMP-dependent activation and phosphorylation of a plasma membrane cyclic AMP phosphodiesterase, <u>Nature</u> (Lond.), 286:904.

Mumby, M.C., Martins, T.J., Chang, M.L., and Beavo, J.A., 1982, Identification of cGMP-stimulated cyclic nucleotide phosphodiesterases in lung tissue with monoclonal antibodies, <u>J. Biol. Chem</u>., 257:13283.

Reifsnyder, D.H. Harrison, S.A., Macphee, C.H., McCormack, M.B., and Beavo, J.A., 1985, Antibody and inhibitor analysis of cGMP-inhibited phosphodiesterase in mutant S49 cells and adipocytes, <u>Fed. Proc</u>., 44:1816.

Walsh, D.A., Ashby, C.D., Gonzalez, C., Calkins, D., Fischer, E., and Krebs, E.G., 1971, Purification and characterizatrion of a protein inhibitor of adenosine 3'5'-monophosphate-dependent protein kinases, <u>J. Biol. Chem</u>., 246:1977.

Weber, H.W. and Appleman, M.M., 1982, Insulin-dependent and insulin-independent low Km cyclic AMP phosphodiesterase from rat adipose tissue, <u>J. Biol. Chem</u>., 257:5339.

Wells, J.N. and Hardman, J.G.,1977, Cyclic nucleotide phosphodiesterases, <u>Adv. Cyc. Nuc. Res</u>., 8:119.

SENSITIVITY OF A cAMP PDE TO ROLIPRAM IN DIFFERENT ORGANS

Herbert H. Schneider, Gudrun Pahlke, and
Ralph Schmiechen

Research Laboratories of Schering AG
Berlin (West) and Bergkamen, FRG

Manipulation of cAMP-dependent mechanisms in aminergic transmission by selective phosphodiesterase (PDE) inhibitors may provide a novel therapetic approach for the management of depression (Wachtel, 1983). The PDE inhibitor rolipram has been shown in recent clinical trials to possess antidepressant activity (Horowski and Sastre, 1985). Antidepressant drug therapy needs several days before relief of symptoms is observed. Time-dependent neuronal adaptation processes may be based on an altered function of regulatory proteins, as e.g. resulting from protein phosphorylation. The therapeutic effect following rolipram treatment could therefore be explained by a mechanism involving a cAMP dependent protein kinase.

The cAMP PDE activity in crude organ extracts consists of many different isoenzymes, which depend on the cell type and which have distinct enzyme kinetics and modulator sensitivity (see Beavo, this Vol.). Rolipram inhibits crude PDE preparations with an apparent IC50 in the micromolar range (Schwabe et al., 1976), which contrasts with the nanomolar affinity of rolipram in binding experiments (Schneider et al., 1986). This latter ability of ^{3}H-rolipram to label brain protein constituents with a K_d of 1 - 2 nM was used to extract a PDE isoenzyme from tissue supernatants by affinity chromatography.

In the experiments reported here, we have compared the properties of this rolipram-sensitive PDE isoenzyme in the soluble fraction of some peripheral organs of the rat to those of the isoenzyme extracted from rat and pig central nervous system.

METHODS

Tissue was homogenized in 20 mM Tris HCl, 2 mM $MgCl_2$, 0.1 mM DTT, pH 7.5 (buffer A), with the addition of 0.2 mM PMSF, and centrifuged for 1 h at 100,000 x g. The supernatant was diluted appropriately for affinity chromatography or for the test assay.

Affinity chromatography was performed on 0.5 ml columns of sepharose 4B coupled to a rolipram derivative. Twenty ml aliquots of supernatant from peripheral tissue were made up to 500 mM with NaCl and loaded at 0 °C onto the column at a flow rate of 10 ml/h. After washing with 100 ml of buffer A + 500 mM NaCl the column was eluted batchwise at room temperature in a total of 7 ml buffer A + 1.5 M NaCl + 10 μM (-)-rolipram. Non-bound rolipram and salt was removed by gel chromatography. Further purification was accomplished by exchange chromatography on Polyanion (Fig. 2 curve A, Fig. 3) or MonoQ (both Pharmacia) using a linear NaCl gradient.

Rolipram binding was performed as described by Schneider et al. (1986). ^{3}H-Rolipram (1 nM, 799 GBq/mmol, Dr. Acksteiner, Schering) with or without 1 μM (-)-rolipram (for non-specific binding) or competitor compounds, 200 μl tissue supernatant or eluant fraction, and buffer A up to 1 ml was incubated at 6° C. Protein bound tracer was retained on Whatman GF/B filters treated with 0.3 % polyethylenimine (Bruns et al., 1983).

PDE activity was determined with 0.5 μM ^{3}H-cAMP, and the ^{3}H-AMP formed was separated by PEI cellulose chromatography (Marks and Raab, 1974). Incubation time was usually 30 min, at 30 °C, with 20 - 40 % degradation of ^{3}H-cAMP.

RESULTS

Anion exchange chromatography (MonoQ) of pig brain supernatant revealed two peaks of "low Km" cAMP PDE activity.

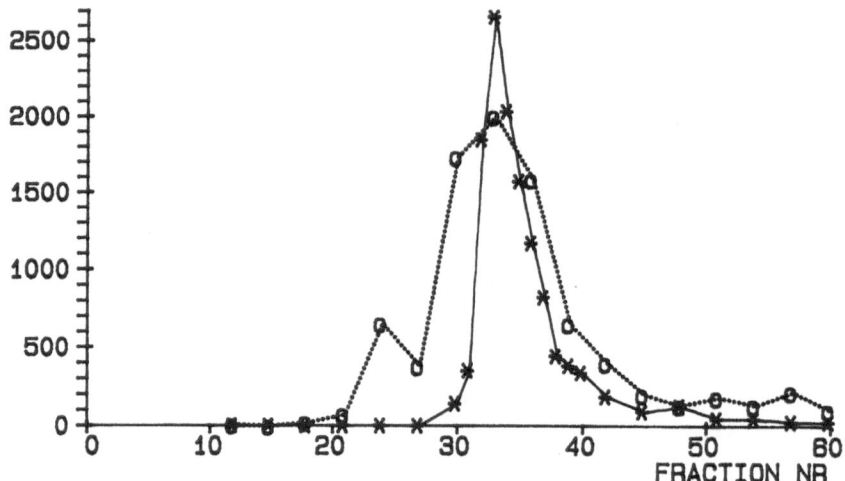

Fig. 1. Anion exchange chromatography on MonoQ. Pig brain supernatant (500 mg tissue) was eluted with a linear salt gradient in buffer A, from 0 M NaCl at fraction 13 to 0.7 M NaCl at fraction 52. Flow rate 1.5 ml/min; 1 ml/fraction. o--o: cAMP PDE activity (nmoles 5'-AMP/ min/ml); *--*: specific ^{3}H-rolipram binding (cpm/ml).

The second peak eluted at ca. 400 mM NaCl. Embedded in this broad peak was a sharper peak of material which bound ^{3}H-rolipram (Fig. 1).

Chromatography of pig brain supernatant (S) on a rolipram affinity column yielded a flow-through fraction free of rolipram-sensitive PDE (Fig. 2,Curve F), which still contained the bulk of PDE activity. Upon elution of the column with (-)-rolipram a fraction was obtained containing rolipram-sensitive PDE binding sites as well as rolipram binding sites (data not shown). Further purification of the eluate on Polyanion enhanced the sensitivity of the PDE activity to inhibition by (-)-rolipram (Curve A) in comparison to the sample (Curve S), the IC50 decreased from 20 µM (S) to 10 nM (A).

This fraction of rolipram-sensitive PDE was used to test 9 PDE inhibitors. The IC50s determined in the PDE test correlated significantly (r= 0.93, p < 0.01) with their ability to compete with the ^{3}H-rolipram binding site (Fig. 3). The PDE isoenzyme of this fraction was neither stimulated nor inhibited by cGMP or Ca^{++}/ calmodulin, and degraded cGMP only marginally.

Isolation of the rolipram-sensitive PDE from tissue extracts of rat peripheral organs by affinity chromatography also increased the potency of rolipram for the "low Km" cAMP PDE. The increase was dependent on the organ examined. IC50s for the supernatant PDE from heart, liver, lung, and uterus were all greater than 1 µM (except muscle: IC50=23 nM), while the IC50s for the extracted PDE were in the range of 10 nM with the exception of liver (66 nM).

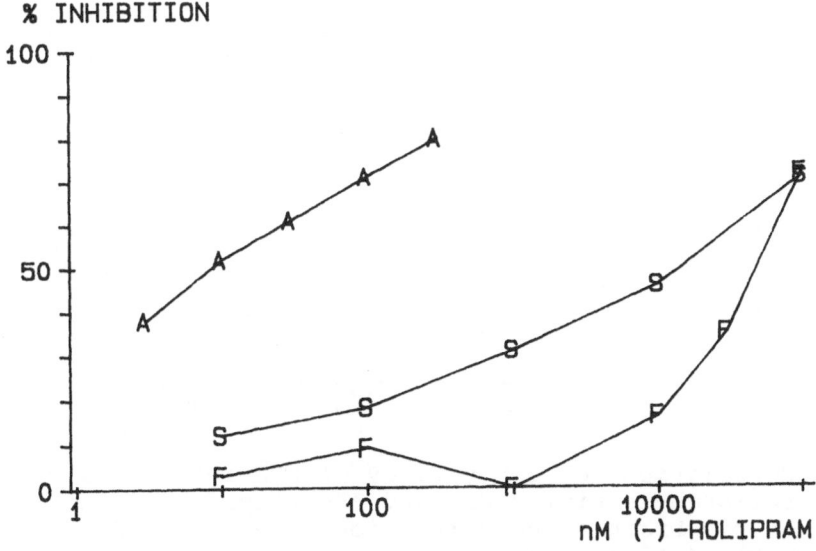

Fig. 2. Inhibition of cAMP PDE activity by (-)-rolipram. Pig brain supernatant was tested before (S) and following affinity chromatography. F: Flow-through; A: Affinity and ion exchange chromatography purified fraction.

DISCUSSION

Much effort has been directed to understanding the role of cAMP in cell function. However, a major problem in doing this has been the focus on the properties of "high Km" PDEs, since the physiological levels of cAMP are in the low micromolar range, when determined avoiding postmortem accumulation (Schneider, 1984). The "low Km" PDEs may therefore have to be considered as representing the more relevant enzyme activity in cell function. In brain, as presumably in most other tissues the fraction of "low Km" to total PDE activity is rather small and depends on different factors. The experiments reported here are based on affinity chromatography using a selective PDE inhibitor as immobilized ligand. The method allows an easy and effective enrichment of the corresponding isoenzyme. The data show that

- the rolipram-sensitive PDE - which may represent the target site of the rolipram antidepressant activity - is accessible only following separation from the bulk of "high Km" PDE isoenzymes.

- PDE with similar sensitivity to rolipram is, besides in the brain, also present in the rat peripheral organs examined.

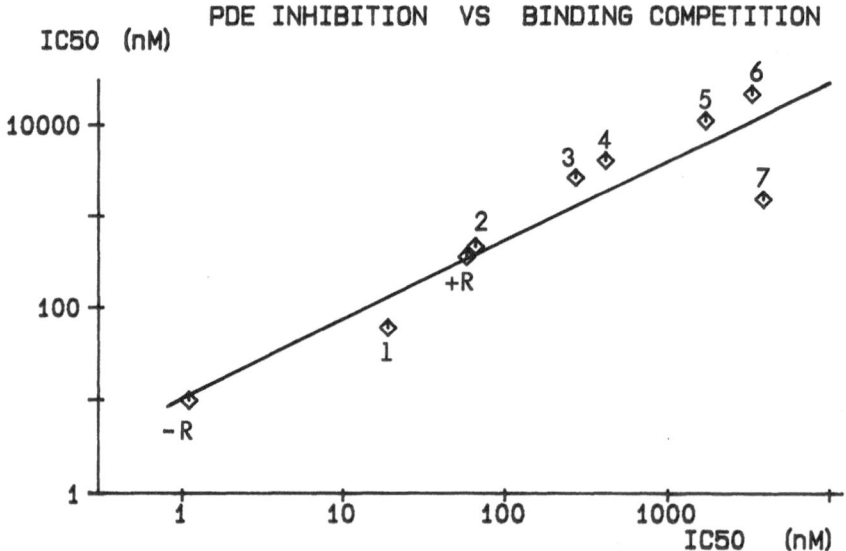

Fig. 3. Correlation between potency of PDE inhibitors for rolipram-sensitive PDE and ^{3}H-rolipram binding site of pig brain soluble fraction. The compounds were: (-R): (-)-rolipram; (+R): (+)-rolipram; (1): TVX 2706; (2): Ro 20-1724; (3): ICI 63.197; (4): medazepam; (5): diazepam; (6): IBMX; (7): SQ 20009; r= 0.93, p < 0.01.

- the excellent correlation between the inhibitory potency of the compounds tested in the ^{3}H-rolipram binding and rolipram- sensitive PDE assay suggests that these two interaction sites are identical.

The findings encourage the search for species - and tissue - specific PDE inhibitors of potential therapeutic value.

ACKNOWLEDGEMENTS

The skilful technical assistence of Simone Fritz, Tina Wegner, Kurt Hamp, Alfred Breitkopf and Jörg Seidler is gratefully acknowledged.

REFERENCES

Bruns, R.F., Lawson-Wendling, K., and Pugsley, T.A., 1983, A rapid filtration assay for soluble receptors using polyethylenimine-treated filters, Anal. Biochem., 132:74.

Horowski, R., and Sastre, M., 1985, Clinical effects of the neurotropic selective cAMP phosphodiesterase inhibitor rolipram in depressed patients: Global evaluation of the preliminary reports, Curr. Therapeutic Res., 38:23.

Marks, F., and Raab, I., 1974, The second messenger system of mouse epidermis, Biochim. Biophs. Acta, 34:368.

Schneider, H.H., 1984, Brain response to phosphodiesterase inhibitors in rats killed by microwave irradiation or decapitation, Biochem. Pharmacol., 33:1690.

Schneider H.H., Schmiechen, R., Brezinski, M., and Seidler, J., 1986, Stereospecific binding of the antidepressant rolipram to brain protein structures, Europ. J. Pharmacol., 127:105.

Schwabe, U., Miyake, M., Ohga, Y., and Daly, J.W., 1976, 4-(3-Cyclopentyloxy-4-methoxyphenyl)-pyrrolidone (ZK 62711): A potent inhibitor of adenosine cyclic 3',5'-monophosphate phosphodiesterase in homogenates and tissue slices from rat brain, Mol. Pharmacol., 12:900.

Wachtel, H., 1983, Potential antidepressant activity of rolipram and other selective cyclic adenosine 3',5'-monophosphate phosphodiesterase inhibitors, Neuropharmacology, 22:267.

IMMUNOLOCALIZATION OF PROTEINS IN SITU BY LIGHT AND ELECTRON MICROSCOPY.

Ute Gröschel-Stewart

Institute of Zoology
Technical University Darmstadt
D-6100 Darmstadt, FRG

It is often desirable to localize proteins in situ in tissues and cells. Not only are we interested in the subcellular and tissue distribution of proteins, at times, the quantity in which a protein is present can be so low that its isolation and characterization is very difficult and tedious. For the histological identifaction of proteins, the most specific reagents are antibodies raised to it or to a closely related crossreacting immunogen - be they mono- or polyclonal. As a prerequiste, the specificity of the antibodies has to be ascertained by standard immunological methods, such as immunoblotting, enzyme immunoassay, or activity inhibition tests. In the following, basic technical recommendations for beginners in the field are given. They include:

I. Choice of Methods, Preparation of Specimen.
II. Antibodies and Antibody-markers.
III. Methods of staining.
IV. Controls.

Ad I. Choice of Methods Preparation of Specimen

A. Light microscopy (LM) is the method of choice for a quick screening of tissues and cells; it also does not require too much previous experience in the field. A large area can be covered fast, colors can be seen, water causes no problems, but the resolution is low. If the antigen in question is a large and rather insoluble protein (e.g. myosin), cryostat sections of unfixed tissue can be used. If the antigen is small and highly soluble (e.g. myoglobin), the tissue has to fixed prior to sectioning. All fixatives have to be such that they cause minimal alteration from the living state. Fixation is a complicated and not easily controlled process, and results depend very much on the nature of the antigen. It is best to begin with the most commonly used fixative, methanol-free formaldehyde, freshly prepared from paraformaldehyde in a suitable buffer, used either alone or in combination with another fixative, e.g. 0.5 % glutardialdehyde or osmiumtetrooxide. One can either remove a small piece of the tissue in question and allow the fixative to permeate, or one can perfuse the animal itself and then remove the fixed tissue. Some antigens and immunostaining methods can even tolerate the classical paraffin embedding method. Cell cultures and cell smears can also be fixed with formalin. After fixation, however, the membranes should be opened for antibody penetration with a low concentration of a mild detergent, or an organic solvent a low temperature (e.g. acetone at -20° C).

B. <u>Electron Microscopy</u> requires a lot of experience, is very tedious and time-consuming, but permits the visualization of minute details. If the antigen is localized on the cell surface, <u>Scanning Electron Microscopy (SEM)</u> can be used for both small and larger objects (10 nm to a few millimeters). <u>Transmission Electron Microscopy (TEM)</u> permits localization of antigen both on the surface and in tissue sections. Samples have to be suitably prepared; almost all require <u>fixation</u> (as in I.A.). For TEM, tissues are then <u>dehydrated</u> (alcohol, polyethylenglykol) and <u>embedded</u> in a resin. Epoxy-resins are often used, although the high temperatures required for their polymerization (60° C) are not tolerated by all antigens. A more benign alternative seem to be the newer acrylate-methacrylate based resins that allow polymerization at low temperatures with indirect U.V. light. At what stage will the immunereaction be performed? There is either the possibility of "<u>Preembedment</u>" staining where the antibodies are allowed to react prior to embedment. One of the drawbacks is usually low antibody penetration, requiring the use of detergents. These, however, may cause relocalization of antigens or other diffusion artefacts. If the antibodies are applied to the final tissue sections (= <u>Post-embedment</u> staining) other problems may arise. The antigenic sites will have to be uncovered in the tissues by removal of the resins with rather drastic methods (e.g. Na-methoxide), this may damage the antigen, if it has not already been lost or damaged by the previous tissue processing itself. The question of pre- or postembedment have to be solved for each individual antigen or groups of antigens. In TEM, sections are usually postfixed with <u>osmiumtetroxide</u>, or counterstained with <u>uranylacetate</u> and/or <u>lead citrate</u>. Such treatments increase contrasts and enhance the tissue-structures. For SEM, fixed or unfixed cells are allowed to react with the antibody, postfixed with osmiumtetroxide, dehydrated, subjected to critical point drying and covered with a 5- 10 nm heavy metal layer by an evaporative method.

Ad II. Antibodies and Antibodymarkers

A. <u>Preparation of antibodies</u>. For immunohistochemistry, it is generally recommended to isolate (if possible) a highly purified immunoglobulin fraction from the antisera and control sera; this will greatly reduce non-specific staining. If the titer of the specific antibodies in the globulin fraction is very low (e.g. 1 - 4 % for polyclonal actin antibodies), the specific antibodies can be enriched by affinity chromatography, using the immobilized antigen as adsorbant. The antigen used has to be of the highest purity in order to remove possible contaminating antibodies from the immunization procedure. Since the antibodies have to be removed from the adsorbant by rather drastic methods (chaotropic salts, buffers of low pH) some antibody reactivity will be lost. In addition, there is the danger that the highest affinity antibodies will remain on the adsorbant.

a) Using unlabelled antibodies.
   Unlabelled antibodies have been used to localize epitopes
   on the myosin molecule. The antigen-antibody complexes
   can be visualized by electromicroscopy after rotaryshadowing
   (e.g. Winkelman, D.A. and Lowey, S., J. Mol. Biol. (1986)
   <u>188</u>, 595-612).

b) Using labelled antibodies.

1. Fluorochromes (for LM only), introduced by Coons, 1941

   The most commonly used fluorescent labels are fluorescein
thiocyanat, FITC (excitation wavelength 495 nm, emission max. 520 -
525 nm; fluorescence apple green), tetramethylrhodamine iso-
thiocyanate, TRITC (excitation 545 - 555 nm, emission 580 nm,
fluorescence orange red). More recently, another red dye has been
recommended: Sulforhodamin 101, Texas red (excitation 570 nm,
emission 620 nm, red fluorescence) which is very suitable for
double labelling procedures with FITC. The fluorochromes are covalently linked to the immunoglobulins. A special fluorescence
attachment to the microscope is needed.

2. Ferritin (TEM) introduced by S. J. Singer, 1959

   This metalloprotein from horse spleen is a spherical molecule
(Ø~12 nm) with a 20 % iron core in a protein shell. It can be
coupled to immunoglobulin by conventional coupling methods.
Ferritin is a "primary electrondense" marker.

3. Enzymes (LM, TM, SEM) introduced by Pierce and Nakane, 1966

   The most commonly used enzymes, coupled to immunoglobulins
while retaining their enzymatic activity, are a) <u>horse-radish
peroxidase</u> and b) <u>alkaline phosphatase</u> (also galactosidase and
glucose oxidase). The reaction is visualized by offering the
enzymes their specific substrates. Since the above enzymes have
high turnover numbers, one achieves a considerable amplification
effects. The substrate has to be chosen such that it will yield an
insoluble reaction product, to be deposited directly at the enzyme
site.

ad a) Horseradish-peroxidase: Inspite of its carcinogenicity,
      1. diaminobenzidine (DAB) is still the most widely used
         substrate. The brownish reaction product, although not
         very photogenic itself, can be deepened by osmication.
         The complex is stable for further processing, thus can be
         used for LM, TEM and SEM.the other two substrates are
         not suitable for further processing due to their
         alcoholsolubility; and are therefore only used for LM.
      11. 4-chloro-1 naphthol, which yields a light-sensitive
         gray-bluish reaction production.
      111. 3-amino-9-ethylcarbazol, which gives a very photogenic
         bright red reaction product.

ad b) Alkaline phosphatase: The substrates used here are the same
      ones as used in conventional histochemistry ("Simultaneous
      azodye method"): 1-Naphthylphosphat in combination with Fast
      Blue BB (blue reaction product) or Fast Red. TR (red reaction
      product). Only usable in LM.

4. Enzyme complexes (LM, TEM, SEM)

An improvement and enhancement of the peroxidase method was introduced by L. Sternberger's "PAP" technique. Instead of co-valently linking the enzyme to the antibody, Sternberger prepared soluble and stable peroxidase-antiperoxidase complexes as indicators.

5. Avidin-biotin complexes (LM, TEM, SEM)

In this method, one makes use of the high affinity between biotin (vitamin B_1) and the egg white protein avidin. Labelling the immunoglobulin with biotin and avidin with a marker (fluorochrom, enzyme etc), one can create small or large complexes for the visualization of the antigen.

6. Large particles (SEM)

Large particles, either rendered electrondense or not, have also been coupled to antibodies, e.g. latex spheres of 30 - 300 nm diameter, plant viruses or invertebrate hemocyanins, and these complexes are used as surface markers in SEM.

7. Colloidal gold (LM, TEM, SEM)

This is at present the most popular label of antibodies. Mono-disperse gold particles, ranging from 5 - 150 nm are usable for multiple markings, since different size particles can be prepared by varying the nature and the amount of the agent to reduce gold chloride. Colloidal gold labelled antibodies are stable and can tolerate further processing.

Ad III. Methods of Staining

1. The direct way

The marker is attached directly to the specific antibody. Although less sensitive, the direct method usually has little non-specific background staining. It is very suitable for double stainings, i.e. simultaneous localization of two antigens with the two antibodies carrying different labels.

2. The indirect way

The specific antibody remains unlabelled, the immune-reaction is enhanced by the application of a second (non-specific) labelled antibody, which is directed against the immunoglobulinfraction of the first (specific) antiserum. The labelled second antibody may be replaced by labelled protein A, a 42 kd protein from the cell walls of certain Staph. aureus strains, which has a high affinity for the F_c region of many immunoglobulins. When double labelling is attempted and the primary antibodies are not derived from two different animal species, it is advisable to use primary anti-body-labelled protein A to localize one antigen and primary antibody-(differently) labelled second antibody for the other antigen. The inclusion of additional antibody layers can further amplify the reaction. The classical example is Sternberger's PAP method, where a sparsely seeded first antibody layer is followed by an excess of secondary antibody (both unlabelled) and, as a third layer, peroxidase-antiperoxidase complexes. The anti-peroxidase is raised in the same animal as the first antibody.

Ad IV. Controls

The specificity of all reactions has to be ascertained by the inclusion of appropriate controls. Causes for "false negative" results have been indicated in section I. B. (pre- versus post-embedding technique). To check for "false positive" results, the following controls are recommended:

1. omit to first specific antibody and check for nonspecific binding of the second antibody;

2. omit both antibodies and check for endgenous enzyme activity when enzyme-labelled antibodies are used;

3. replace first antibody by pre-immuneglobulin or specific antibody twice absorbed with the homologous antigen;

4. replace first specific antibody by another non-related antibody raised by the same immunization schedule (to exclude any non-specific immune responses).

Major sources used for this overview:

1) Bullock, G.R. & Petrusz, P. eds (1982-1985) Techniques in Immunocytochemistry Vol. 1-3, Academic Press.

2) Kuhlman, W.D. (1984) Immuno Enzyme Techniques in Cytochemistry, Verlag Chemie, Weinheim.

3) Marchalonis, J.J. & Warr, G.W. eds (1982) Antibody as a Tool. John Wiley & Sons Ltd., Chicester.

4) Sternberger, L. (1979) Immunocytochemistry, 2nd ed. John Wiley & Sons Ltd, New York.

FLUORESCENT PROBES

Norma Selve, Elke Schröer, Klaus Ruhnau, and
Albrecht Wegner

Institut für Physiologische Chemie I
Ruhr-Universität Bochum
D-4630 Bochum, Federal Republic Germany

INTRODUCTION

Fluorescence is widely used in biochemistry and cell biology
because this method offers many advantages. Fluorescent labels
often bind specifically to one type of molecules or ions. The
concentration of the fluorescently labeled molecules or ions
can be determined without interference from other molecules.
Concentrations in the nanomolar range or below can be accurately
determined. Advanced techniques and instruments have been
developed by which the mobility and turnover of fluorescently
labeled molecules can be observed in a test tube or in living
cells.[1] Rotation and diffusion of fluorescent molecules can be
measured and the locations where fluorescent molecules assemble
or are transported in living cells, can be determined.

1. Fluorescent labels

Most biological molecules do not fluoresce. However, often
synthetic organic compounds can be attached to biological
molecules. These organic compounds consist of a fluorescent
moiety and a reactive group that binds to side chains of
proteins. Reactive groups have been synthetized which are
specific for various side chains of amino acids, particularly
for lysine, cysteine, methionine or tyrosine. By an appropriate

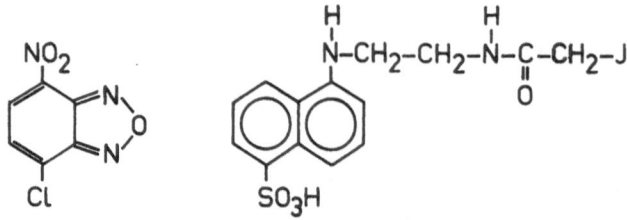

Fig. 1: Fluorescent labels. NBD-chloride[2] (left), 1,5-J-AEDANS[3] (right).

choice of the reactive group specific labeling of one or a few residues on the surface of proteins can be accomplished. Two frequently applied fluorescent labels are depicted in Fig. 1.[2,3]

2. Fluorescence intensity, life time and polarization

Fluorescent labels change their fluorescence intensity when the environment of the labels is altered. Conformational changes, binding of ions and other ligands and assembly of macromolecules can be detected by changes of the fluorescence intensity. The method is highly sensitive. Concentrations in the nanomolar range can easily be determined. By using fluorescently labeled calmodulin binding of Ca^{2+} to calmodulin could be measured.[4]

Following absorption of light the average life time of the excited state of the label used is in the range of 10 ns. This life time of the excited state can be measured by excitation using a nanosecond light pulse. The decay of the fluorescence can be directly determined.[5] The life time of the excited state depends on the environment of the fluorescent label. The life time may change on conformational changes or on binding of ligands and assembly with other macromolecules. Binding of Ca^{2+} to fluorescently labeled calmodulin has been followed by a little change of the life time.[6]

Fluorescent labels absorb preferentially light that is polarized in the direction of the largest dimension of the label. Also the emitted light is polarized in that direction. By excitation using polarized light preferentially those molecules are excited which are oriented in the direction of the polarization. If the excited molecules do not change their orientation during their life time also the fluorescence is polarized because the emitting molecules have a preferential orientation. If however the excited molecules are so mobile that they change their orientation during their life time, the fluorescence becomes depolarized.[7] The fluorescence polarization depends on the rotational mobility of the fluorescent molecules. Small molecules rotate faster than large molecules. Fluorescence polarization is a good method for determination of binding of small fluorescent molecules to large molecules. Association of fluorescently labeled calmodulin with phosphodiesterase has been measured by the change of the fluorescence polarization.[8]

3. Fluorescent indicators

A number of fluorescent molecules have been synthetized which change their fluorescence intensity when they bind Ca^{2+} or H^+. These fluorescent molecules have been used successfully as Ca^{2+} or pH indicators. Indicators have been invented which can be applied for measurements even in living cells. The indicators were rendered hydrophobic by esterification of the charged carboxyl groups of the indicators. These esterified indicators cross the cell membrane. Intracellularly, the ester bonds are hydrolyzed by enzymes. The formed indicator is highly hydrophobic and is not able to penetrate the cell membrane. Thus, the indicator is trapped and accumulated in the cell.[9] Extra-

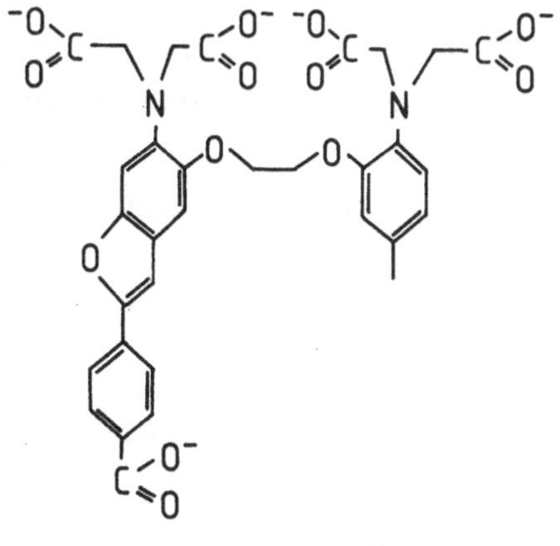

Fig. 2: Fura-2, a Ca^{2+}-indicator [9]

cellular indicator can be removed by centrifugation to minimize the background of extracellular fluorescence. Intracellular Ca^{2+} concentrations and pH values have been determined by using these indicators. Changes of the Ca^{2+} concentration and the pH value following stimulation of cells have been measured. By microscopy the intracellular distribution of Ca^{2+} has been observed. Ca^{2+} gradients between the nucleus and the cytoplasm or between the endoplasmic reticulum and the cytoplasm have been documented.[10] One major problem with these indicators is that they may affect the Ca^{2+} concentration by their buffering power.

4. Fluorescence photobleaching recovery

Fluorescence photobleaching recovery is a relatively new technique for studying the motion of specific components in a complex mixture. The species of interest is labeled by covalent attachment of a fluorescent label. A specific region or pattern is then exposed to a brief pulse of high intensity light that irreversibly photobleaches a portion of the labels in that region. Subsequent monitoring of the same region with a low-intensity light beam detects the re-emergence of fluorescence in the irradiated region as bleached and unbleached species randomize their positions by diffusion. If all labeled species are free to move about in a random fashion, the bleached region will eventually disappear. The time constant for the disappearance of the bleached region will be determined by the translational diffusion coefficient of the labeled species and the square of the characteristic dimension of the bleached region. If some fraction of the bleached species is not free to move, then that fraction of the bleaching will persist, permitting determination of the fraction of immobile species.[11] Individual measurements are generally quite rapid and data interpretation is straightforward. A continued expansion of the list of successful applications can be anticipated.

References

1. Pardue, R. L., Kaetzel, M. A., Hahn, S. H., Brinkley, B. R., & Dedman, J. R. (1981) Cell 23, 533-542.
2. Birkett, D. J., Price, N. C., Radda, G. K., & Salmon, A. G. (1970) FEBS Lett. 6, 346-348.
3. Hudson, E. N., & Weber, G. (1973) Biochemistry 12, 4154-4161.

4. LaPorte, D. C., Keller, C. H., Olwin, B. B., & Storm, D. R. (1981) Biochemistry 20, 3965-3970.

5. Yguerabide, J., Epstein, H. F., & Stryer, L. (1970) J. Mol. Biol. 51, 573-590.

6. Olwin, B. B., & Storm, D. R. (1983) Methods Enzym. 102, 148-157.

7. LaPorte, D. C., Builder, S. E., Storm, D. R. (1980) J. Biol. Chem. 255, 2343-2349.

8. LaPorte, D. C., & Storm, D. R. (1978) J. Biol. Chem. 253, 3374-3377.

9. Grynkiewicz, G., Poenie, M., & Tsien (1985) J. Biol. Chem. 260, 3440-3450.

10. Williams, D. A., Fogarty, K. E., Tsien, R. Y., & Fay, S. F. (1985) Nature 318, 558-561.

11. Lanni, F., & Ware, B. R. (1982) Rev. Sci. Instrum. 53, 905

BINDING OF FLUORESCENT ANALOGS OF CYCLIC GMP

TO cGMP-DEPENDENT PROTEIN KINASE

H. H. Ruf, M. Rack, W. Landgraf and F. Hofmann

Physiologische Chemie, Universität des Saarlandes
D-6650 Homburg-Saar, Fed. Rep. Germany

INTRODUCTION

Cyclic GMP (cGMP) is a second messenger for cellular regulation and activates cGMP-dependent protein kinase (cG-PK). cG-PK, a homo-dimer of 150 kDa, has four partially cooperative binding sites for cGMP with K_D-values in the order of 10 to 200 nM as has been shown by binding studies with ^3H-cGMP[1,2]. Two types of sites have been described, site 1 with high affinity and slow dissociation and site 2 with lower affinity and faster dissociation. The primary structure of the enzyme has been reported and assigned to functional domains[3].

Using spectroscopic probes, we attempted to better characterize the binding of cGMP and its cooperativity. A direct spectroscopic signal for binding will allow direct kinetic measurements of association and dissociation rate constants. These constants should yield information on the binding mechanism. Furthermore, suitable probes can report on the nature of the binding sites. Fluorescence should be a feasible method because of its high sensitivity. In this communication, we report the syntheses of two fluorescent cGMP-derivatives and the changes in fluorescence parameters due to binding to the protein kinase.

METHODS

cGMP-dependent protein kinase was isolated from bovine lung as described[4] and stored with 50 % glycerol at $-18^\circ C$. Protein was determined by the BCA protein assay (Pierce Chemical Co.) with bovine serum albumin as standard. Binding and competition assays with ^3H-cGMP were done by a filter technique at $4^\circ C$[5]. The data were analyzed by the LIGAND program[6] on an IBM PC.

Fluorescence was measured with a SLM 8000S photon-counting fluorometer. Fluorescence anisotropy was obtained in T-format with Glan-Thompson prisms as polarizers. We used 3 x 3 mm cuvettes for samples of 0.1 ml and 10 x 10 mm cuvettes for samples of 1.5 ml. Spectral bandwidths were 2 or 4 nm. Kinetics were started by manual mixing with plunger cuvettes with a mixing time of less than 1 s. Usually, the solvent was 10 mM phosphate buffer, pH 7.0. Fluorescence lifetimes were measured with a SLM 4800 phase fluorometer at 18 and 30 MHz modulation frequency with glycogen as scattering reference.

RESULTS AND DISCUSSION

Syntheses of Fluorescent Analogs

8-(NBD-aminoethylthio)-cGMP (NBD-8-cGMP). 8-Br-cGMP was reacted with 2-aminoethanethiol essentially as described by Dills et al.[7]. The yield of the purified 8-(2-aminoethylthio)-cGMP, however, was only 2 %. This product was then reacted with 7-fluoro-4-nitrobenzo-2-oxa-1,3-diazole (NBD-F). The fluorescent cGMP-derivative was isolated by anion exchange chromatography.

8-(5-Thioacetamidofluorescein)-cGMP (FLU-8-cGMP) was prepared by a modification of the method of Caretta et al.[8].

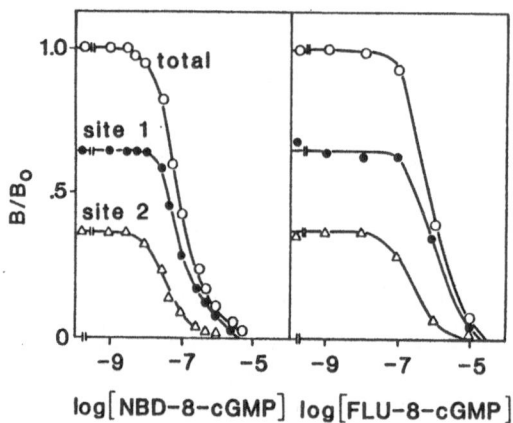

NBD-8-cGMP   Fluorescein-8-cGMP

Equilibrium Binding of the Analogs to cG-PK

Both analogs competed with the binding of ³H-cGMP at cG-PK (Fig. 1). Thus the analogs showed specific binding to cG-PK, NBD-8-cGMP with a preference for the high-affinity site 1 as do other 8-substituted cGMP derivatives[9], whereas FLU-8-cGMP slightly preferred site 2.

Fig. 1. Displacement of ³H-cGMP at cG-PK by fluorescent analogs.

The concentration of cGMP was 0.1 μM. For total binding the assays were incubated for 1 h at 4°C. For site 1 an additional incubation with 1 mM cGMP was carried out for 1 min at 4°C. Site 2 is the difference between total and site 1. Fits with LIGAND yielded:

K_D (nM)	site 1	site 2
NBD-8-cGMP	6.0	210
FLU-8-cGMP	103	52

Binding of the analogs to cG-PK changed their fluorescence properties (Figs. 1 and 2, Table 1). The intensities changed by a factor of about two, thus indicating the transfer of the fluorophors to an environment less polar than water. Compared to 60% ethanol in water, however, the sites at the protein were more polar.

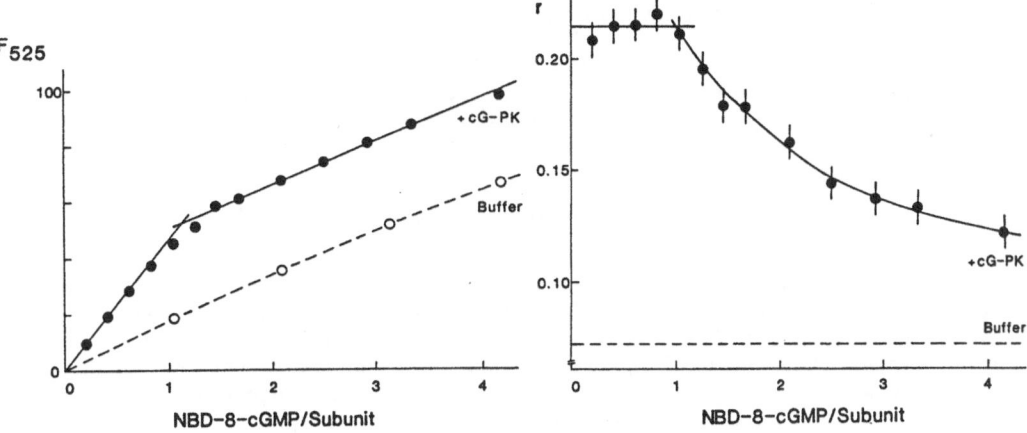

Fig. 2. Fluorescence titrations of cG-PK with NBD-8-cGMP

Fluorescence intensity F (left panel) at 525 nm (excitation at 475 nm) and fluorescence anisotropy r (right panel). The titrations were done at 22°C with a cG-PK subunit concentration of 1.2 μM.

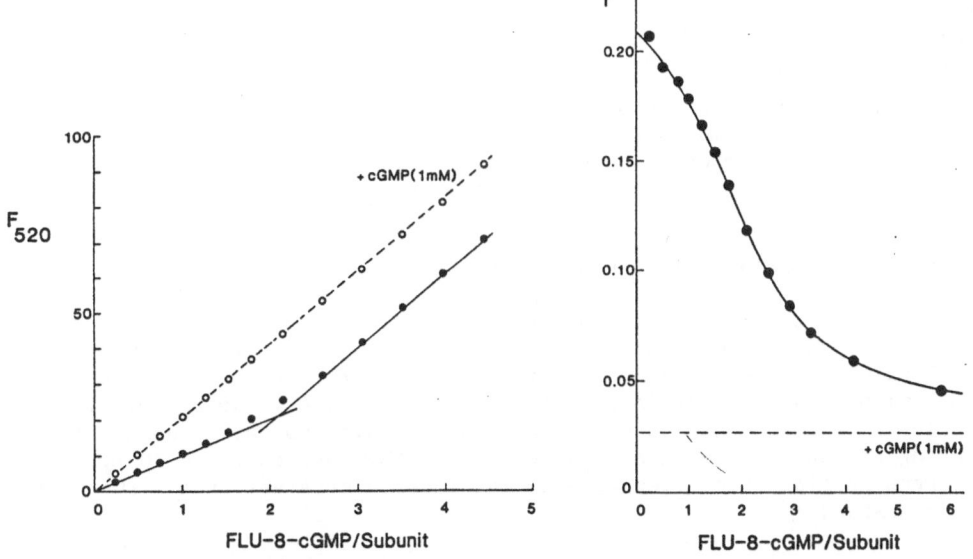

Fig. 3. Fluorescence titrations of cG-PK with FLU-8-cGMP.

Fluorescence intensity (left panel) at 520 nm (excitation at 485 nm) and fluorescence anisotropy (right panel). The titrations were done at 22°C with a cG-PK subunit concentration of 1.2 μM. Fluorescence intensities were corrected for dilution and inner filter effects from added FLU-8-cGMP.

Table 1. Fluorescence Properties of cGMP Analogs

Parameter	NBD-8-cGMP	FLU-8-cGMP	
Maximum of excitation	475	490	nm
Maximum of emission	525	515	nm
Intensity in buffer[a]	0.31	100	
Relative intensity F/F_o			
in buffer	1.00	1.00	
bound to cG-PK	2.05	0.47	
in 60 % ethanol	3.43	0.23	
Fluorescence anisotropy (at 22°)			
in buffer	0.07	0.027	
bound to cG-PK	0.21	0.21	
in glycerol	0.35	0.37	

Binding increased the fluorescence anisotropy which reflected the immobi-
lisation of the fluorophors upon binding. The evaluation of the rotational
correlation times of FLU-8-cGMP (Table 2) showed that the bound fluorophor
moved faster than the whole protein. This could be explained by residual
motion between fluorescein and protein and/or strong segmental mobility
of the protein chains around the binding sites.

Table 2. Rotational Diffusion of FLU-8-cGMP and cG-PK at 22 °C.

		FLU-8-cGMP in buffer	FLU-8-cGMP bound to cG-PK	cG-PK	
Fluorescence anisotropy	r	0.027[a]	0.21	-	
Fluorescence lifetime	τ	4.05	4.02	-[d]	ns
Rotat. correl. time[b]	ϕ	0.32	5.28	56[d]	ns
Rotational volume[c]	V	1.04	17	183[d]	nm³
Radius of sphere	R	0.63	1.6	3.5[d]	nm

[a] r_o = 0.37 in glycerol
[b] Perrin equation $r_o/r = 1 + \tau/\phi$
[c] for rigid sphere $V = \phi RT/\eta$ (viscosity $\eta = 1.25$ cP)
[d] calculated for a spherical protein of 150 kDa

Time-resolved Binding Kinetics of Analogs to cG-PK

The resolution of the association and dissociation kinetics was attempted
in order to characterize the binding sites better than merely by equilibrium binding constants. These kinetics could be measured using FLU-8-cGMP
at 5°C with manual mixing in plunger cuvettes (Fig. 4). The association
kinitics were caried out under second order conditions, at about equal
starting concentrations of enzyme and FLU-8-cGMP between 7 and 30 nM (Fig.
4A). From initial velocities an association rate constant of 4×10^6 $M^{-1}s^{-1}$
was determined.

Dissociation was initiated either by dilution (Fig. 4B) or by displacement
with an excess of cGMP (Fig. 4C). Under both conditions, heterogeneous kinetics were observed consisting of at least two exponentials which were
assigned to the two types of binding sites. The rapid dissociation, assigned to site 2, proceeded with equal rate constants, 0.08 s^{-1}, for both
dilution and displacement. However, the rates of the slow dissociation,
assigned to site 1, were different for both types of experiments. For dilution the rate constant was 0.004 s^{-1} and for displacement by cGMP less
than 0.001 s^{-1}. These data indicated that the occupancy of all binding

sites, accomplished by the high amount of cGMP used for displacement, slowed down the dissociation of site 1. This cooperativity has also been observed for the dissociation of cGMP[5]. The dissociation constants calculated from the rate constants were nearly one order of magnitude lower than the constants determined from equilibrium measurements, between 10 and 100 nM.

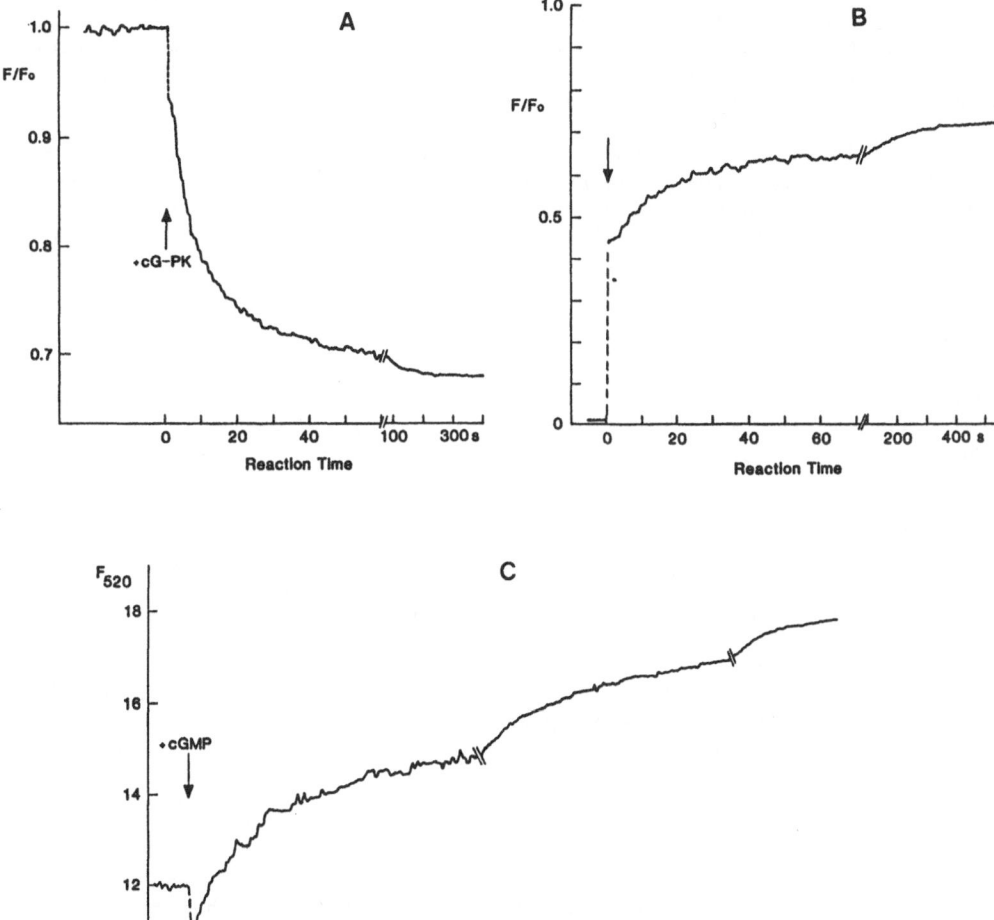

Fig. 4. Association and dissociation kinetics of FLU-8-cGMP and cG-PK at 5°C. Association (panel A); 3.8 µl of cG-PK were mixed to 1.5 ml phosphate buffer with 15 nM FLU-8-cGMP to a final concentration of 15 nM. Dissociation by dilution (panel B); 5 µl of cG-PK with FLU-8-cGMP were diluted 300-fold into 1.5 ml phosphate buffer (final concentrations: cG-PK 122 nM, FLU-8-cGMP 5.3 nM). Dissociation by displacement with cGMP (panel C); 5 µl of cGMP were mixed to 0.1 ml cG-PK with FLU-8-cGMP (final concentrations: cG-PK 122 nM, FLU-8-cGMP 160 nM, cGMP 0.95 mM). Fluorescence was measured at 520 nm (excitation at 485 nm). For relative intensity (F/F_o) F_o corresponded to the intensity of free FLU-8-cGMP. Mixing was done in plunger cuvettes. The traces were not corrected for dilution.

CONCLUSIONS

Two fluorescent analogs of cGMP were synthesized, NBD-8-cGMP and FLU-8-cGMP, which bound to cG-PK with affinities comparable to those of cGMP. The changes in fluorescence properties upon binding can be used as spectroscopic signals for binding as well as for probing the binding sites. Equilibrium binding studies yielded dissociation constants in the range of 10 to 100 nM although binding to the individual sites could not be discriminated. With FLU-8-cGMP time-resolved measurements revealed a heterogeneity of the off-rates due to different binding sites. In addition, from the dependence of the off-rates from site 1 on the occupation of the other sites, cooperative binding was concluded also for the fluorescent analog.

Hence the fluorescent analogs are suitable for characterizing the cyclic nucleotide binding sites of cG-PK and their cooperativity. After having established the fluorescence parameters of the analogs bound to the individual sites, modified cG-PK will be studied as well as the time-resolved kinetics at physiological temperature.

ACKNOWLEDGMENTS

We are greatly indebted to Ms. R. Trautmann and C. Waschow for their expert technical assistance. This work was supported by the Deutsche Forschungsgemeinschaft, Sonderforschungsbereich 246.

REFERENCES

1. T. M. Lincoln, and J. D. Corbin, Characterization and biological role of the cGMP-dependent protein kinase, Adv. Cycl. Nucl. Res. 15:139 (1983).
2. J. D. Corbin, this volume, p.11 (1987).
3. K. Takio, R. D. Wade, S. B. Smith, E. G. Krebs, K. A. Walsh, and K. Titani, Guanosine cyclic 3´,5´-phosphate dependent protein kinase, a chimeric protein homologous with two seperate protein families, Biochemistry 23:4207 (1984).
4. F. Hofmann, and V. Flockerzi, Characterization of phosphorylated and native cGMP-dependent protein kinase, Eur. J. Biochem. 130:599 (1983).
5. F. Hofmann, H.-P. Gensheimer, and C. Göbel, cGMP-dedendent protein kinase - autophosphorylation changes the characteristics of binding site 1, Eur. J. Biochem. 147:361 (1985).
6. P. J. Munson, and D. Rodbard, LIGAND: a versatile computerized approach for characterization of ligand-binding systems, Anal. Biochem. 107:220 (1980).
7. W. L. Dills, J. A. Beavo, P. J. Bechtel, K. R. Myers, L. J. Sakai, and E. G. Krebs, Binding of adenosine 3´,5´-monophosphate dependent protein kinase regulatory subunit to immobilized cyclic nucleotide derivatives, Biochemistry 15:3724 (1976).
8. A. Caretta, A. Cavaggioni, and R. T. Sorbi, Binding stoichiometry of a fluorescent cGMP analogue to membranes of retinal rod outer segments, Eur. J. Biochem. 153:49 (1985).
9. J. D. Corbin, D. Øgreid, J. P. Miller, R. H. Suva, B. Jastorff, and S. O. Døskeland, Studies of cGMP analog specificity and function of the two intrasubunit binding sites of cGMP-dependent protein kinase, J. Biol. Chem. 261:1208 (1986).

II. REGULATION ON MUSCLE CONTRACTION

REGULATION OF ACTOMYOSIN ATPASE

Edwin W. Taylor

Department of Molecular Genetics & Cell Biology
The University of Chicago
Chicago, IL

INTRODUCTION

Muscle cells have an extremely efficient regulation system which reduces actomyosin ATPase activity by more than one thousand fold in relaxed versus active muscle. A surprising fact is that two quite different regulation systems are used in striated and smooth muscles. The actomyosin ATPase of smooth muscles as well as non muscle cells is activated by a calmodulin dependent protein kinase which specifically phosphorylates a myosin light chain (LC-2). Actomyosin is inhibited by dephosphorylation by a specific phosphatase.[1] Activation of enzymes by phosphorylation is a very common control mechanism and its use in muscle regulation is not surprising. The second mechanism apparently evolved to meet the need for faster switching on and off in striated muscles. Although a phosphorylatable light chain has been retained by striated muscle myosin and the level of phosphorylation can be altered by stimulation, phosphorylation no longer activates the actomyosin ATPase. Regulation is obtained by a structural change of the thin filament. The thin filaments in striated muscle contain troponin which is not present in smooth muscle or non-muscle cells.

Until recently much more attention was paid to striated muscle, consequently structural and kinetic mechanisms were developed first for this system. A steric blocking mechanism of regulation was proposed by Huxley[2] based on the observation that calcium binding to thin filaments induces a change in the x-ray diffraction pattern which can be interpreted as a change in position of tropomyosin on the actin filament. It was postulated that in relaxed muscle tropomyosin blocks the myosin binding site on actin. Calcium binding to troponin is presumed to cause the movement of tropomyosin to a position in which it no longer blocks myosin binding to actin.

Since the regulatory site is on myosin in one case and on the thin filament in the other case, the mechanisms appeared to be different. However, recent studies have raised questions about the simple steric blocking mechanism and it is possible that the kinetic mechanisms of regulation are closely related.

The kinetic mechanism of active actomyosin will be reviewed since it provides the basis for a discussion of regulation of the two types of muscles. The actomyosin system is an example of signal transduction that has been studied in detail. What we have learned about this mechanism

may be relevant to understanding other systems in which nucleoside triphosphate hydrolysis is part of the signal transduction mechanism.
Examples of other control mechanisms will be discussed in the last section.

MECHANISM OF ACTOMYOSIN ATPASE

The kinetic mechanism of the active actomyosin is essentially the
same for the striated and smooth muscle enzymes (see Cooke[3] for a recent
review). The kinetic scheme is

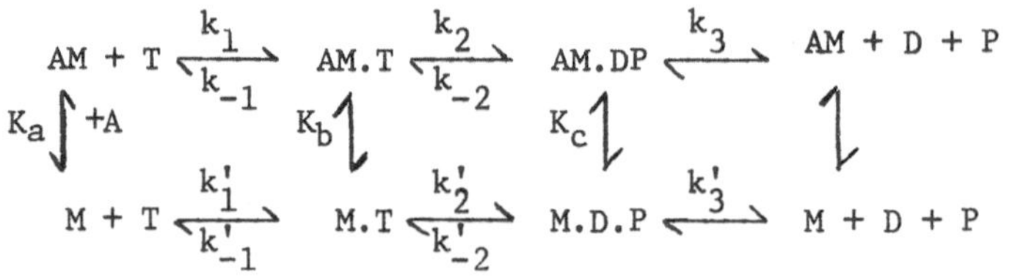

$$AM + T \underset{k_{-1}}{\overset{k_1}{\rightleftharpoons}} AM.T \underset{k_{-2}}{\overset{k_2}{\rightleftharpoons}} AM.DP \overset{k_3}{\rightleftharpoons} AM + D + P$$

$$K_a \uparrow +A \qquad K_b \updownarrow \qquad K_c \updownarrow \qquad \updownarrow$$

$$M + T \underset{k'_{-1}}{\overset{k'_1}{\rightleftharpoons}} M.T \underset{k'_{-2}}{\overset{k'_2}{\rightleftharpoons}} M.D.P \overset{k'_3}{\rightleftharpoons} M + D + P$$

Scheme 1

where A refers to an actin unit of F actin, M refers to a single myosin
head (subfragment 1), T, D and P are ATP, ADP and inorganic phosphate,
and K_a, K_b and K_c are association constants of M, M.T and M.D.P with
actin respectively. The kinetic mechanism is characterized by the values
of the rate and association constants for the various steps. The scheme
is oversimplified since ATP forms a collision complex with M or AM followed
by a very rapid conformation change to form the states M.T or AM.T. Also,
the dissociation of products involves at least two steps which are represented here by a single effective rate constant, k_3 or k'_3.

There are three important features of this mechanism. Myosin has
low ATPase activity under physiological conditions because the rate constant of product dissociation (k'_3) is small (0.1 sec^{-1}) although the
hydrolysis step is relatively fast (k'_2 = 100 sec^{-1} at 20°). Second, the
binding of myosin heads to actin is strong (K_a = 10^8 M^{-1} at low ionic
strength) but ATP binding tends to dissociate actomyosin because M.T and
M.D.P are weakly bound to actin ($K_b \simeq K_c \simeq 2 \times 10^4$ M^{-1} at low ionic
strength). The rate of dissociation is very large (> 1000 sec^{-1}) and the
M.T and M.D.P complexes are in rapid equilibrium with actin. Third, the
hydrolysis step is hardly affected by actin binding (k_2 is reduced two to
three fold compared to k'_2) but the rate constant of product release is
much larger for AM.D.P compared to M.D.P ($k_3 \simeq 30$ to 50 sec^{-1} which is
300 to 500 times larger than k'_3). Thus the activation of myosin ATPase
by actin arises from the large increase in rate constant of product
dissociation. A variety of nucleoside phosphates (ADP, AMPPNP) dissociate
the actomyosin complex and have much larger rate constants of dissociation
from the actomyosin complex compared to the myosin complex.

The regions of myosin which bind actin and ATP are separated by at
least 50 Å.[4] We could think of the mechanism in terms of a conformational
interaction. The binding of ATP induces a change in conformation in which
the substrate is enclosed in a pocket and a change in structure is transmitted to the actin binding site weakening the binding of actin. At equilibrium the system would be largely dissociated into actin plus M.T.
Cleavage of the ATP to ADP and phosphate reduces the stability of the
closed state and actin binding reverses the change in conformation leading
to an open ADP, P site and strongly bound actin. Force development or
relative movement of actin and myosin filaments is associated with this
step.

There are two kinds of answers that could be given to the question, what is the mechanism of regulation? In kinetic terms the answer consists of measuring the rate and equilibrium constants for regulated acto S1 ATPase in the presence and absence of calcium (regulated actin is the complex of actin-tropomyosin-troponin). We could then specify which steps in the mechanism are altered by calcium binding. The second kind of answer consists of determining the change in structure of the proteins induced by calcium binding and inferring the mechanism from the structural change. This is the type of answer given originally by Huxley. Some confusion has arisen from failing to distinguish between the two kinds of answers.

The steric blocking mechanism seems to imply that myosin should not bind to regulated actin in the absence of calcium since the binding site is blocked by tropomyosin. It was shown[5] that the degree of association of acto S1 in the presence of ATP at low ionic strength is hardly affected by the addition of calcium yet the ATPase activity is increased up to 20 fold. It was proposed by Chalovich et al.[6] that product dissociation (k_3) is the step which is regulated rather than binding (K_b and K_c).

A more detailed study[7] of the transient kinetic behavior in which we attempted to measure the rate and binding constants confirmed the proposal of Chalovich et al. The rate constants of product dissociation (k_3) and also substrate dissociation (k_{-1}) were increased 20 fold by addition of calcium while the hydrolysis step was almost unaffected. The binding of M.T and M.D.P to actin was increased by only 25-30% in the presence of calcium. In addition the rate constant of dissociation of ADP from the AM.ADP complex was also increased 20 to 30 fold by the addition of calcium.

The simplest interpretation of the kinetic evidence is that calcium binding increases the rate constant of the conformational change in which the interaction with actin opens the active site pocket. Since whatever is bound in the active site would be released at a higher rate in the presence of calcium this model accounts for the finding that ATP, ADP.P and ADP are all released at a higher rate.

The answer in kinetic terms does not specify the structural mechanism. A partial steric blocking mechanism is compatible with the result if we assume that in the relaxed state tropomyosin blocks part of the myosin binding site. The weakly bound M.T and M.D.P states do not occupy this region of the site while the strongly bound AM state is formed only if tropomyosin is displaced. In order to complete the cycle with release of products the AM state is generated from AM.D.P (step 3). Thus, this step is sterically blocked by tropomyosin. The kinetic evidence could also be explained by a conformational model in which the interaction of tropomyosin with actin affects the rate of the conformational change in which the active site pocket is opened. Although tropomyosin moves to a different position in the active state the two tropomyosin positions on actin need not overlap with the myosin binding site.

The studies have so far been carried out on subfragment 1 at low ionic strength and additional effects on binding may occur for the two-headed molecule (myosin or heavy meromyosin) at physiological ionic strength. Nevertheless an important step in regulation is the change in rate of a conformational change of an actomyosin complex which determines the rate of opening of the substrate pocket.

The same considerations apply to the determination of the kinetic mechanism of regulation of smooth muscle actomyosin. In this case it is also necessary to investigate the effects of phosphorylation on the mechanism of ATP hydrolysis by myosin alone. A difficulty is that subfragment 1 is not regulated by phosphorylation and acto subfragment 1 is in the active state. Kinetic studies have to be carried out using heavy meromyosin (HMM) or myosin.

Smooth muscle and non muscle myosins undergo a structural change which is not observed with striated muscle myosin. At low ionic strength the heads are folded back toward the tail which also appears to be looped around the head (10S or folded form). At high ionic strength (0.4 to 0.5 M KCl), the heads are extended (6S or extended form). Phosphorylation shifts the equilibrium to the extended form at approximately physiological ionic strength.[8] HMM also appears to exist in two states in which the heads are extended or folded back towards the tail.[9]

HMM shows only small effects of phosphorylation on the rate constants of substrate binding and hydrolysis[10] (k_1' and k_2' in Scheme 1). However, the steady state rate of ATP hydrolysis for myosin and heavy meromyosin increases with ionic strength. Phosphorylation increases the steady state rate of ATP hydrolysis and for myosin the increase in rate is correlated with the transition from the folded to extended form.[8] The step in the mechanism which determines the steady state rate of hydrolysis is the rate of product dissociation (k_3'). Therefore the rate of opening of the substrate pocket is affected by the folding of the head. The site of phosphorylation on LC-2 of the myosin head is probably close to the head-tail junction. Since folding back of the head appears to depend on an electrostatic interaction, phosphorylation may act indirectly by altering the charge distribution, thereby favoring the transition to the extended form. Subfragment 1 has a relatively high rate of ATP hydrolysis at low ionic strength, which is independent of phosphorylation. Since phosphorylation of myosin increases the rate of ATP hydrolysis about ten fold to approximately the value for S1, the S1 rate corresponds to the activated form.

The rate constant for the dissociation of ADP from HMM and S1 shows a dependence on ionic strength and phosphorylation which is similar to the ATPase activity. At low ionic strength the rate constant is about 10 to 20 times larger for S1 than HMM and the rate for HMM increases with ionic strength and phosphorylation (unpublished observations). The results support the conclusion that the rate of opening of the active site is reduced by folding back of the head.

The binding of ATP or AMPPNP to myosin or HMM shifts the equilibrium towards the folded form.[9] This finding suggests that nucleoside phosphates are more strongly bound to the folded form which is consistent with a slower rate of product dissociation and therefore a smaller ATPase activity in the folded state.

Kinetic studies of the actomyosin ATPase are still incomplete. The rate constants for substrate binding and hydrolysis show only small differences for phosphorylated and dephosphorylated states[10] (k_1 and k_2 of Scheme 1). The binding of HMM to actin in the presence of ATP is increased five fold by phosphorylation of HMM[11,12] (K_b and K_c are increased by phosphorylation). Although in the regulation of striated muscle acto S1 there was very little effect of activation on the binding, some studies with acto HMM have shown that the binding is five to ten fold stronger in the presence of calcium.[13] Thus smooth and striated muscle acto HMM may show similar effects of activation on the binding of M.T and M.D.P. The effect does not account for the large activation of ATPase activity by actin, consequently product dissociation (k_3) must be increased by phosphorylation of the HMM. At present we do not have direct measurements of this rate constant.

The available evidence suggests that a primary step in activation is the increase in rate of opening of the substrate pocket of the phosphorylated HMM in the acto HMM.ADP.P complex. Part of the increase in rate could be attributed to the effect of phosphorylation of HMM itself and the remainder to a more effective interaction of actin with the phosphorylated enzyme. The two regulation mechanisms for striated and smooth muscle are similar kinetically in that the same steps are affected.

Some caution is necessary in drawing this conclusion since acto HMM of smooth muscle may be a poor model of the actin-myosin system as it occurs in the intact muscle. Myosin does not form thick filaments in the dephosphorylated state in solution while filaments are present in the relaxed muscle. If filaments could be formed from the folded state of myosin the actin binding sites of myosin would probably be inaccessible to F actin. The binding to actin could be greatly increased by phosphorylation which allows the heads to form the extended state. In solution the folded state of HMM could still bind to actin and since binding is stronger for the extended state, the interaction could displace the equilibrium toward the extended state in the absence of phosphorylation. Thus the relatively small effect of phosphorylation of HMM on its binding to actin in solution may greatly underestimate the effect in intact muscle. The structural mechanism of regulation of smooth muscle may be different from striated muscle even though kinetic mechanisms in solution are similar.

KINETICS AND CONTROL MECHANISMS

The actomyosin ATPase mechanism is an example of the coupling of ATP hydrolysis to perform mechanical work. The hydrolysis of ATP or GTP can also be used to control cellular processes. In some cases control mechanisms depend on binding reactions (equilibrium control mechanisms). Examples discussed at this meeting are myosin light chain kinase which is turned on by the binding of calcium calmodulin and cyclic AMP dependent kinase which is activated by dissociation of the regulatory subunit in response to cyclic AMP binding. The level of the effector, calcium or cyclic AMP is regulated while the activation process depends on a binding equilibrium of catalytic and regulatory subunits. The control of striated muscle actomyosin ATPase was originally thought to be an equilibrium mechanism in which the affinity for the actin subunit was regulated. However, the studies described here suggest that the control is on a rate process, the rate of dissociation of reaction products, ADP and phosphate. We will refer to this type of control as a dynamic control mechanism. A slow hydrolysis of ATP is necessary to keep the system switched off since myosin can bind to the relaxed state of actin and cooperatively activate the filament.

Two questions will be considered. First, do ATPases in which hydrolysis is coupled to work have similar kinetic mechanisms. Second, is there a similarity between kinetic mechanisms in which ATP hydrolysis is coupled to performance of work and dynamic control mechanisms in which a flux of ATP or GTP hydrolysis controls the state of the system.

Two mechano-chemical systems have been investigated in some detail, actomyosin and microtubule-dynein (reviewed by Johnson[14]). Both systems share properties which I consider to be necessary for mechanochemical coupling. The ATPase component rapidly hydrolyzes ATP to form a relatively stable enzyme-product intermediate. The complex of the two proteins is dissociated by ATP to reset the movement cycle. The binding of actin or the microtubule to the enzyme-product intermediate leads to an increased rate of product dissociation and formation of a strongly bound protein complex. Thus the system ossilates between weakly and strongly bound states and the transition between these states generates a force.

The ion transport ATPases perform osmotic work against an electro-chemical gradient. The sodium-potassium, calcium and hydrogen ion transport ATPases are related proteins. The mechanism of calcium transport[15] involves a rapid hydrolysis of ATP with formation of a relatively stable intermediate (in this case a phosphorylated enzyme), a transition to a state in which Ca is weakly bound ("inside" state), a fairly rapid dissociation of the reaction product (phosphate) and a return to the "outside" state to which calcium is strongly bound. Once again the ATPase cycle leads to an ossilation between states which bind the substance to be transported (ions or actin) strongly or weakly. Thus a similar strategy is employed by ATPases which couple hydrolysis to the performance of work.

Examples of dynamic control mechanisms which use the flux of ATP or GTP hydrolysis to control the state of the system are provided by the polymerization of actin and microtubules. Kirschner and Mitchison[16] refer to the mechanism as dynamic instability. In the case of microtubules the addition of a tubulin subunit is coupled to the hydrolysis of one molecule of GTP which is tightly bound to tubulin. Since the rate of addition of subunits depends on tubulin concentration while the hydrolysis of GTP is a first order reaction the growing microtubule can have a cap of tubulin GTP subunits. The rate of dissociation of a tubulin.GTP subunit from the end of the microtubule is slower than a tubulin.GDP subunit. Thus the microtubule end is stabilized by the cap of tubulin GTP subunits. A fluctuation in cap size leaving only GDP subunits on the end is followed by rapid depolymerization of the microtubule. Thus a flux of GTP hydrolysis is necessary to stabilize the microtubule end and the dynamic state permits large fluctuations in the microtubule length.

An example of a dynamic control system which is relevant to the topics presented at this meeting is the activation of adenylate cyclase by G protein.[17] The G protein-receptor-hormone system has properties which are formally similar to actomyosin. The G protein is a GTPase which forms a G.GDP state while myosin forms an M.ADP.P state. The activation cycles are similar.

$$
\begin{array}{ccc}
\text{R.H.G.GDP} \xrightarrow[+\text{GTP}]{-\text{GDP}} \text{RH.G.GTP} & \qquad & \text{A.M.ADP.P} \xrightarrow[+\text{ATP}]{-\text{ADP,P}} \text{AM.ATP} \\
+\text{R.H} \uparrow \qquad \qquad \downarrow -\text{R.H} & & +\text{A} \uparrow \qquad \qquad \downarrow -\text{A} \\
\text{G.GDP} \longleftarrow - - - \text{G.GTP} & & \text{M.ADP.P} \longleftarrow \underline{\qquad} \text{M.ATP}
\end{array}
$$

G.GDP binds to receptor-hormone (RH), GDP dissociates and GTP binds leading to the dissociation of G.GTP. Thus RH stimulates GTP hydrolysis even though the object of the activation mechanism is to generate G.GTP. The activation mechanism is essentially the same as for actomyosin, an increase in the rate of dissociation of the reaction product GDP versus ADP and P for myosin. The nature of the GDP intermediates has not been determined but stimulation by AlF_4^- , a possible analogue of phosphate, may implicate a GDP.P intermediate in the cycle.[17] The hormone has a function similar to that of calcium binding to actin. It increases the affinity for G.GDP and possibly increases the rate of GDP dissociation from the receptor-G.GDP complex.

The remaining steps in the mechanism are of course different from the actomyosin mechanism. G.GTP dissociates into α.GTP, β and γ subunits and the α.GTP binds to adenylate cyclase to activate this enzyme. The complex series of reactions by which hormone binding leads to formation of free G.GTP are in fact very familiar since the same kinetic mechanism is used by the cell to activate actomyosin. The flux of GTP hydrolysis is necessary to maintain the system in the switched on state.

The mechanisms of coupling of nucleoside triphosphate hydrolysis to the performance of work and to the dynamic control of cellular processes have some features in common. By comparing the mechanisms of systems which have been studied in detail one may gain insight into how other systems work.

REFERENCES

1. Adelstein, R. S. and Eisenberg, E. (1980) Annu. Rev. Biochem. 49, 921.
2. Huxley, H. E. (1972) Cold Spring Harbor Symp. Quant. Biol. 37, 361.
3. Cooke, R. (1986) CRC Crit. Rev. Biochem. 21, 53.
4. Botts, J., Takoshi, R., Torgerson, P., Hozumi, T., Muhlrad, A., Mornet, D. and Morales, M. (1984) Proc. Natl. Acad. Sci. U.S.A. 81, 2060.
5. Chalovich, J. M. and Eisenberg, E. (1982) J. Biol. Chem. 257, 2432.
6. Chalovich, J. M., Greene, L. E. and Eisenberg, E. (1983) Proc. Natl. Acad. Sci. U.S.A. 80, 4909.
7. Rosenfeld, S. S. and Taylor, E. W. (1987) J. Biol. Chem., in press.
8. Ikebe, M., Hinkins, S. and Hartshorne, D. J. (1983) Biochemistry 22, 4580.
9. Suzuki, H., Stafford III, W. F., Slayter, H. S. and Seidel, J. C. (1985) J. Biol. Chem. 260, 14810.
10. Rosenfeld, S. S. and Taylor, E. W. (1984) in Smooth Muscle Contraction, ed. by N. L. Stephens. M. Dekker, New York and Basel, 175.
11. Sellars, J. R. (1985) J. Biol. Chem. 260, 15815.
12. Ikebe, M. and Hartshorne, D. J. (1985) Biochemistry 24, 2380.
13. Wagner, P. D. (1984) Biochemistry 23, 5950.
14. Johnson, K. A. (1985) Annu. Rev. Biophys. Biophys. Chem. 14, 161.
15. Hasselbach, W. (1981) in Membrane Transport, ed. by S. L. Bonting and J. H. M. dePont. Elsevier-North Holland, New York, 183.
16. Mitchison, T. J. and Kirschner, M. W. (1984) Nature (London) 312, 232.
17. Ross, E. M. and Gillman, A. G. (1980) Annu. Rev. Biochem. 49, 533.

REGULATION OF MUSCLE CONTRACTION

S.V. Perry

Department of Physiology
Medical School
University of Birmingham, Birmingham B15 2TJ, U.K.

Contraction in muscle is associated with a high rate of hydrolysis of ATP by actomyosin. In all types of muscle the two protein components of the contractile system are located in different structures, the thin actin filaments and the thick myosin filaments. The substrate for contraction. MgATP, is a poor substrate for myosin alone but when it interacts with actin it is split at a high rate. Thus regulation involves controlling the interaction of actin with myosin that activates the MgATPase and leads to contraction.

In muscle activation occurs when the calcium concentration rises from the resting value of 10^{-7} M to about 10^{-5} M. As all actomyosin systems possess many common features the mechanism of the interactionm of actin and myosin that leads to the cross bridge cycle is probably very similar in all types of myscle. The characteristic physiological properties of different muscle types will depend on a number of factors. Of these the speed of the cross bridge cycle, largely determined by the enzymic properties of myosin, and the manner in which it is regulated are of particular importance. The rise in calcium concentration associated with contraction leads to changes in the of the proteins of the thick and thin filaments which, depending on the muscle type, may trigger or modulate the contractile response.

SKELETAL MUSCLE

Contraction in skeletal muscle is triggered by the binding of calcium, released from the sarcoplasmic reticulum, to the target protein, troponin C, a component of the I filament system. The I filament contains a linear polymer of tropomyosin in each of the two groves of the actin duplex arranged so that one tropomyosin molecule extends alongside seven actin monomers (Fig.1). The troponin complex is located with a periodicity of 380 A along the filament with the stioichiometry of one molecule of the complex for one of tropomyosin. The troponin complex contains one molecule each of troponin C (M_r 18000), troponin I (M_r 21000-24000) and troponin T (M_r 30000-35000).

Different isoforms of these proteins and of tropomyosin are present in different cell types. Usually a single isoform of each thin filament protein is present in a given cell type but in the cases of tropomyosin and troponin T at least two isoforms occur in some cells.

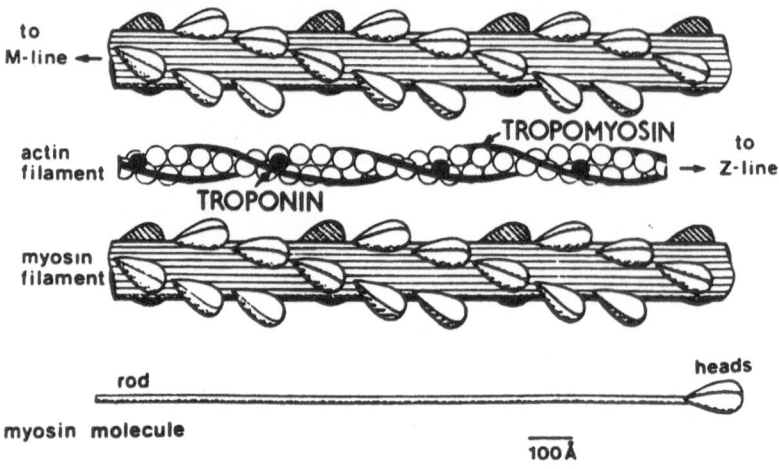

Fig. 1 Schematic representation of the A and I filaments and the myosin molecule of striated muscle.

The precise nature of the mode of action of the tropomyosin-troponin regulatory system is not completely understood. Nevertheless well defined functions can be ascribed to each of the components which must be related to their role in the regulatory process. In resting muscles actin is unable to interact with myosin in the manner required to activate the MgATPase due to the presence of the regulatory proteins. With in vitro systems in which tropomyosin and troponin are absent, activation occurs irrespective of whether calcium is present or not. Thus the two regulatory proteins confer calcium sensitivity on the MgATPase of actomyosin.

The inhibition of the MgATPase that occurs in resting muscle is lifted when calcium is bound to troponin C. This protein which is structurally very similar to calmodulin contains four calcium binding sites of high affinity that will contain no bound calcium in resting muscle but which would be expected to be filled on stimulation. Troponin C possesses two categories of binding sites. Sites I and II are specific for calcium whereas sites III and IV, with a slightly higher binding constant for calcium, also bind magnesium. As the free magnesium ion concentration in muscle is about 1-2mM, sites III and IV will be filled with magnesium in resting muscle. Therefore whereas sites I and II will be rapidly filled with calcium on stimulation, due to the slow off rate for magnesium, calcium will more slowly displace the former cation from sites III and IV. It is considered that the filling of sites I and II is the event that leads to the activation of the MgATPase.

Troponin I inhibits the MgATPase of actomyosin, a property that is much accentuated in the presence of tropomyosin. In effect this results in inhibition being obtained in vitro with actin: troponin I ratios of 3-4: 1 or higher i.e. approaching the 7:1 ratio found in the myofibril. The inhibition is neutralised by troponin C which interacts with troponin I to form a complex that is much strengthened in the presence of calcium. In the absence of other components of the regulatory system troponin C neutralises the inhibitory activity of troponin I both in the presence and absence of calcium. For calcium sensitivity of the MgATPase troponin T and tropomyosin must be present in addition to the other two components of the complex.

Two regions of each of the polypeptide chains of troponin I and C have been shown to be involved in the interaction between the two proteins. One of these on troponin I is very close to, or may even form part of, the region that interacts with actin. A 22 residue peptide can be isolated from troponin I by selective cleavage which contains this site and which possesses the inhibitory activity of the intact molecule (Syska et al., 1976). These findings suggest that when calcium is bound to troponin C it strengthens its interaction with troponin I and displaces actin from its binding site in the latter protein. Clearly as other changes take place both tropomyosin and troponin T are essential components for the MgATPase to be calcium sensitive. Troponin T forms complexes with troponin C and with tropomyosin, the latter interaction being of importance in positioning the troponin complex on the I filament and in some way transmitting its regulatory function to all the actin monomers along the filament. Indeed a major problem in the field is to understand how the troponin complex which at most extends along two actin monomers is able to extend its effect over seven which the stioichiometry demands.

On the basis of X-ray diffaction studies interpreted to indicate that the tropomyosin filament moves in the grove of the actin filament on contraction it has been suggested that it is the former protein that blocks the interaction of actin with myosin in resting muscle. On stimulation the conformational changes initiated by the binding of calcium are transmitted via the troponin complex probably through troponin T to tropomyosin. The latter protein is then considered to move in the grove and leave the interacting site on actin free to interact with myosin and participate in the cross bridge cycle (steric hypothesis). Despite its attractions there is increasing evidence that the mechanism is not quite so simple and a cycle of simple attachment followed by detachment is not compatible with an increasing amount of enzyme kinetic and physicochemical data. It is difficult also to incorporate what is known about the detailed mechanism of the interactions between actin troponin I and troponin C in the steric hypothesis (see Perry, 1986 for a review of some of the data).

The trigger for contraction is clearly the binding of calcium by troponin C but events occuring at the A filament and involving the myosin directly may modulate the cross bridge cycle. Lehman (1978) has reported evidence for calcium sensitivity of the MgATPase of myofibrils from which tropomyosin and troponin have been removed. Also the rise in calcium concentration on contraction will activate myosin light chain kinase which phosphorylates the P light chain on the myosin head and thus changes the charge distribution on the cross bridge itself. It would be surprising if this did not in some way affect the actomyosin interaction (see article on myosin light chain kinase in this volume).

CARDIAC MUSCLE

Cardiac muscle is also triggered by calcium binding at the troponin complex. The I filament is very similar in structure to that of skeletal muscle but is composed of tissue specific isoforms of the components of the troponin complex, tropomyosin and actin. The properties of these isoforms are very similar to their skeletal counterparts, the slight differences in amino acid sequences no doubt being responsible for minor differences in properties appropriate for cardiac muscle function.

The regions of troponin I that interact with troponin C and actin are virtually identical with the corresponding regions of the skeletal muscle isoforms of the protein. One striking difference of functional importance is at the N terminus where cardiac troponin I has an additional 26 residue peptide. This peptide contains a serine residue at position 20 which is readily phosphorylated by c-AMP dependent protein kinase. Phosphorylation of this site causes the calcium sensitivity of the MgATPase of cardiac actomyosin to fall i.e. higher calcium concentrations are required for 50% activation of the ATPase. In the normal beating rabbit heart for example serine 20 of the troponin I is 20-30% phosphorylated and on intervention with adrenaline rises to approach 100% due to the activation of the protein kinase. Thus phosphorylation of troponin I acts as a negative feedback control associated with the inotropic action of adrenaline. As the calcium level rises in response to the agonist the ATPase requires a higher calcium concentration for activation. This effect presumably explains the increase in speed of relaxation that occurs after treatment with β-adrenergic agonists.

Little can be said about A filament regulation in cardiac muscle. Myosin light chain kinase activity is lower in mammalian heart than in other muscle tissues. In the normal beating heart the P light chain is less than 50% phosphorylated and does not change significantly after intervention with adrenaline and other agents that increase the force of contraction. There are however, recent reports that phosphorylation of the light chain can increase the calcium sensitivity of isometric tension development in skinned fibres (see article on myosin light chain kinase).

SMOOTH MUSCLE

The regulatory systems of smooth muscle are not so well defined as those of striated muscle but nevertheless marked differences can be distinguished. Activation is a consequence of the triggering action of calcium in this muscle at the myosin filament. This can occur directly involving the regulatory light chain as is the case with molluscan adductor muscle in which tissue calcium is bound directly to the myosin without any evidence of covalent modification (Lehman et al., 1972). More usually as is the case in vertebrate smooth muscle the binding of calcium to calmodulin activates myosin light chain kinase resulting in P light chain phosphorylation and activation of the MgATPase (for a fuller discussion of this aspect see article on myosin light chain kinase).

Significant evidence also exists for thin filament regulation in vertebrate smooth muscle. To date no system corresponding to the troponin complex has been isolated from smooth muscle. Ebashi and collaborators (Nonomura & Ebashi, 1980), however, have reported the isolation of a system showing some similarities to the troponin complex consisting of two components, leiotonin A and C, which regulate contraction via the actin filament. Few other workers have been able to substantiate these findings but there is acccumulating evidence that the actin- binding protein, caldesmon, may play a role in smooth muscle regulation. Discovered by Kakiuchi & Sobue (1983), caldesmon binds strongly to calmodulin in the presence of calcium. In the absence of calcium or at the concentrations found in resting muscle caldesmon binds to actin and inhibits the MgATPase of actomyosin. Thus caldesmon could regulate the smooth muscle actomyosin ATPase by a so-called flip-flop mechanism. In resting smooth muscle the caldesmon will be bound to the actin preventing its interaction with myosin. The rise in calcium concentration on contraction will lead to activation of the ATPase by phosphorylation of the myosin and detachment of the caldesmon from actin leaving the latter protein free to participate in the cross-bridge cycle. Smooth muscle contains adequate amounts of caldesmon for such a role. Small amounts are present in striated muscle but its function is unknown.

REFERENCES

Kakiuchi, S. and Sobue, K., 1983, Control of the cytoskeleton by calmodulin and calmodulin binding proteins, Trends in Biochemical Sciences 8: 59.

Lehman, W., 1978, Thick filament calcium regulation in vertebrate striated muscle, Nature 274: 80.

Lehman, W., Kendrick-Jones, J. and Szent-Gyorgyi, A.G., 1972, Myosin linked regulatory systems : Comparative Studies, Cold Spring Harbor Symposium on Quantitative Biology, 37: 319.

Nonomura, Y. and Ebashi, S., 1980, Calcium regulatory mechanisms in vertebrate smooth muscle, Biomedical Research 1: 1.

Sysila, H., Wilkinson, J.M., Grand, R.J.A. and Perry, S.V., 1976, The relationship between biological activity and primary structure of troponin I from white skeletal muscle of the rabbit, Biochem. J., 153: 375.

Reviews

Bretscher, A.J., 1986, Thin filament regulatory proteins of smooth and non-muscle cells, Nature 321: 726.

England, P.J., 1983, Phosphorylation of cardiac muscle contractile proteins, In Cardiace Metabolism (Drake-Holland, A.J. and Noble, M.I.M., eds) John Wiley and Sons Ltd., 365.

Hartshorne, D.J. and Siemankowski, R.F., 1981, Regulation of smooth muscle actomyosin, Ann. Rev. Physiol., 43: 519.

Marston, S.B. and Smith, C.W.J., 1985, Thin filaments of smooth muscles, J. Muscle Research and Cell Mobility 6: 669.

Perry, S.V., 1979, The regulation of contractile activity in muscle, Biochemical Society Transactions 7: 593.

Perry, S.V., 1986, Activation of the contractile mechanism by calcium, In: Myology (Engel, A.G. and Banker, B.Q., eds) McGraw-Hill Book Company, New York, Part I, 613.

REFERENCES

DETERMINATION OF PHOSPHATES IN CARDIAC TROPONIN BY PHOSPHATE

ANALYSIS AND PHOSPHOSERINE MODIFICATION

Kristine Swiderek, Kornelia Jaquet, Helmut E.
Meyer, Ludwig M. G. Heilmeyer, jr.

Institut für Physiologische Chemie
Abt. Biochemie Supramolekularer Systeme
Ruhr-Universität Bochum, Universitätsstr. 150

Troponin (Tn) of high endogenous phosphate content (ca. 3 mol phosphate/mol protein) was isolated from beaf heart. Following separation of Tn into its subunits Tn T, Tn I and Tn C by reverse phase HPLC a phosphate content of about 1.4 mol phosphate/mol protein is found in the Tn T and the Tn I subunits, respectively. These results show, that cardiac Tn contains at least four endogenous phosphorylation sites.

As a model, phosphoserine present in phosphokemptide was transformed into S-ethyl-cysteine, which could be determined quantitavely by PTC-amino acid analysis. Phosphoserine, present in Tn I was modified analogously and its S-ethyl-cysteine content was determined subsequently as PTC-amino acid. The amount of S-ethyl-cysteine (1.7 mol/mol protein) was identical to the amount of phosphate in the protein.

INTRODUCTION

Cardiac troponin is a major component of thin filaments, regulating muscle contraction (for review see Perry, 1979). It consists of three subunits, Tn T of molecular weight 35000 (Burtnick et al., 1976), Tn I of molecular weight of 24000 (Grand et al., 1976) and Tn C of molecular weight 17500 (van Eerd & Takahahsi, 1975).

Tn can be phosphorylated by many enzymes (Perry, 1979), most of these phosphorylation sites are known. Moir & Perry (1977) have shown, that serine 20 of cardiac Tn I is phosphorylated in intact heart. However there are other phosphorylation sites, which are still unknown. Furthermore it is still unclear which function of Tn is affected by phosphorylation/dephosphorylation. To get information on phosphorylation of Tn in the intact organ, it is necessary to isolate troponin with high phosphate content. It is our goal to localize these phosphate groups in the primary sequence of the subunits.

MATERIALS AND METHODS

Cardiac Tn was prepared by a modified method (Beier, 1983) of Tsukui & Ebashi (1973). Tn could be separated into its subunits by HPLC on a reverse phase column (60x16 mm) filled with Organogen RP-gel 7um-S300, as described by Crabb & Heilmeyer (1984) for phosphorylase kinase. The Tn subunits were identified by SDS-gel electrophoresis. Phosphate content of holotroponin and of the subunits was determined according to Stull & Buss (1977). Transformation of phosphoserine to S- ethyl-cysteine and amino acid analysis was carried out as described by Meyer et al. (1986a, 1986b).

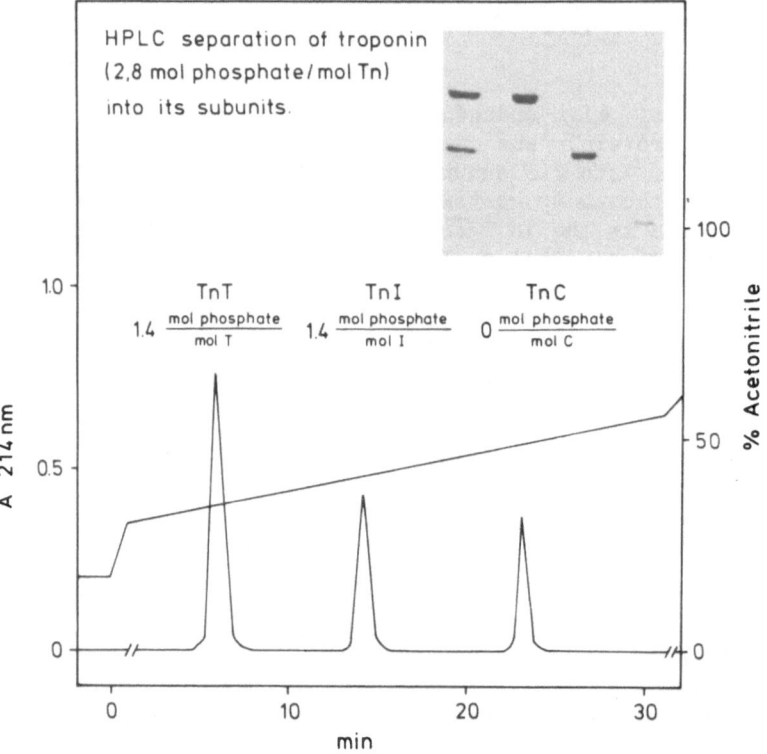

Figure 1   Phosphate content of Tn and its subunits after separation by reverse phase HPLC

Approximately 1 mg Tn, dissolved in 0.5M KCl, 20mM Tris, 15mM mercaptoethanol, pH 7.5 was injected on a Knauer column (60x16 mm), filled with HD-RP-gel 7um-S300 from Organogen (Heidelberg) at 20 % B. The gradient conditions are indicated, the flow rate was 7.8ml/min. Solvent A consists in 0.09 % trifluoracetic acid, solvent B was 84 % acetonitrile, containing 0.08 % trifluoracetic acid. The inset shows a SDS-gel pherogram of holotroponin and its isolated subunits.

RESULTS

Tn of high phosphate was isolated (ca. 3 mol phosphate/mol protein).
Tn was separated into its subunits by HPLC (Fig. 1) to avoid loss of
phosphate which is observed using classical separation methods. All
peaks which were eluted from a reverse phase column following

injection of the holotroponin were identified as the subunits by
using SDS-gel electrophoresis as indicated in Figure 1. The Tn
subunits were obtained in high yield (60-80 %). Furthermore, no loss
of phosphate was observed. A content of 1.1-1.8 mol phosphate/mol Tn
T and 1.2-1.7 mol phosphate/mol Tn I was determined.
As a first step towards localization of these phosphoserines in the
primary sequence we tried to transform this phosphoamino acid into
S-ethyl-cysteine which can be identified in the sequence. Fig. 2
shows the amino acid composition of Tn I before and after this
transformation. It can be seen that as a result of this treatment
the amount of serine decreases whereas a new amino acid, S-ethyl-
cysteine, appears (compare Fig. 2 a and b). Quantitaion yields 1.7
mols S-ethyl-cysteine/mol Tn I.

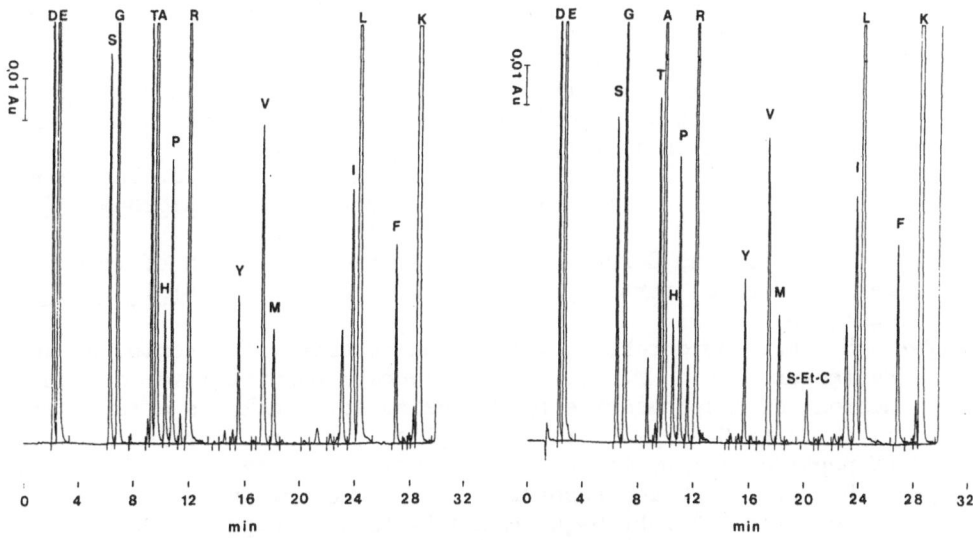

Figure 2   PTC-amino acid analysis of Tn I

500 pmols of unmodified (a) or modified (b) Tn I were
hydrolyzed and 40 % of the sample were analyzed for the
amino acid composition as described in the methods.

DISCUSSION

Several reported procedures for isolation of cardiac Tn yield a protein of low phosphate content (0.1-0.3 mols/mol protein), (Stull, 1980). In contrast our modification of the method of Tsukui & Ebashi which prevents dephosporylation allows to isolate cardiac Tn of high phosphate content (3 mol phosphate/mol protein. This phosphate content is distributed approximately equally between the Tn T and the Tn I subunit (compare Fig. 1). The HPLC separation avoids losses of phosphate, which occurs usually during the separation of the subunits employing ion exchange chromatography in presence of urea (Greaser & Gergely, 1971). Our results allow to postulate minimally four phosphorylation sites in the holotroponin, i. e. minimally two sites in Tn T and Tn I, respectively. Apparently phosphoserine in Tn I can be transformed quantitavely into S-ethyl-cysteine. The amount of S-ethyl-cysteine is identical to the phosphate content of Tn I. As an advantage S-ethyl-cysteine can be determined 100-fold more sensitively than phosphate.

This method shall be used to determine the phosphoserine content of peptides after enzymatic digestion of the Tn subunits. These phosphopeptides of the Tn subunits shall be separated by HPLC. The phosphoserine residues can be detected in the primary sequence by transformation to S-ethyl-cysteine.

REFERENCES

Beier, N., 1983, PHD.-thesis, Ruhr-University, Bochum

Burtnick, L. D., McCubbin, W. D. and Kay, C. M., 1975, The isolation and characterization of the tropomyosin binding component (Tn T) of bovine cardiac troponin, Can. J. Biochem., 54:546

Crabb, J. W. and Heilmeyer, L. M. G.,jr., 1984, High performance liquid chromatography purification and structural characterization of the subunits of rabbit muscle phosphorylase kinase, J. Biol. Chem., 259:6346

van Eerd, J.-P. and Takahashi, K., 1975, The amino acid sequence of bovine cardiac troponin C. Comparison with rabbit skeletal troponin C, Biochem. Biophys. Res. Commun., 64:122

Grand, R. J. A., Wilkinson, J. M. and Mole, L., 1976, The amino acid sequence of rabbit cardiac troponin I, Biochem. J., 159:633

Greaser, M. L. and Gergely, J., 1971, Reconstitution of troponin activity from three protein components, J. Biol. Chem., 246:4226

Meyer, H. E., Swiderek, K., Hoffmann-Posorske, E., Korte, H. and Heilmeyer, L. M. G., jr., 1986a, Quantitative determination of phosphoserine by PTC-amino acid analysis after modification to S-ethyl-cysteine. Application for picomolar amounts of peptides and proteins, J. Chromat., submitted

Meyer, H. E., Hoffmann-Posorske, E., Korte, H. and Heilmeyer, L. M. G., jr., 1986b, Sequence analysis of phosphoserine containing peptides. Modification for picomolar sensity, FEBS Lett., 204:61

Moir, A. J. G. and Perry, S. V., 1977, The sites of phosphorylation of rabbit cardiac troponin I by adenosine 3':5' cyclic monophosphate dependent protein kinase, Biochem. J., 167:333

Perry, S. V., 1979, The Twelfth CIBA Medal Lecture, <u>Biochem. Soc. Trans.</u>, 7:593

Stull, J. T., 1980, <u>in</u>:Adv. in Cyclic Nucl. Res., eds, P. Greengard and G. A. Robison, Raven Press, New York

Stull, J. T. and Buss, J. E., 1977, Phosphorylation of cardiac troponin by cyclic adenosine 3':5'-monophosphate dependent protein kinase, <u>J. Biol. Chem.</u>, 252:851

Tsukui, R. and Ebashi, S., 1973, Cardiac troponin, <u>J. Biol. Chem.</u>, 73:1119

MYOSIN LIGHT CHAIN KINASE

S.V. Perry

Department of Physiology
Medical School, University of Birmingham
Birmingham B15 2TJ, U.K.

Myosin light chain kinase is a widely distributed enzyme found in most tissues in which the actomyosin contractile system is present, the largest amounts occuring in skeletal and smooth muscles. To date the only muscle tissue in which the enzyme has been sought and not detected is molluscan adductor (Frearson et al., 1976). The natural substrate for the enzyme is the phosphorylatable (P) light chain in the intact myosin molecule (Fig. 1). This light chain has a M_r of 18000-20000 and is also known as light chain 2 (LC2), the dithiobisnitrobenzoic acid (DTNB) or regulatory light chain.

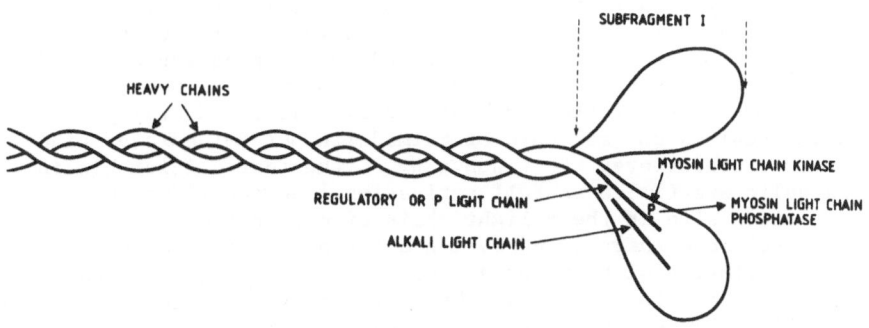

Fig. 1 Schematic representation of the myosin molecule.

Most of the enzymes isolated from striated muscle appear to have molecular weights in the range of 70000--90000 (Stull et al., 1985). From sequence studies the myosin light chain kinase of rabbit fast skeletal has been shown to have a molecular weight of 73000. The kinases isolated from smooth and non-muscle tissues are substantially larger with molecular weights of 130000-150000.

The activity of the enzyme can be studied either by following the incorporation of phosphorus into the substrate using [γ-32P]ATP or by determining the amount of phosphorylated P light chain by separating it from the non-phosphorylated form by electrophoresis in 6M urea at pH 7.6 (Perrie and Perry, 1970). In cases where the P light chain in a particular muscle type exists in isoforms (Westwood and Perry, 1982) or when it has more than one phosphorylation site (Cole et al., 1985) separation is best achieved by two-dimensional electrophoresis. Isolated P light chains can be used as substrate for assay and study of the enzyme but care must be exercised in extrapolating the results to the cell where the substrate is the intact myosin molecule in which a P light chain is attached to each of the two heads. It would not be surprising if the kinetics of phosphorylation were affected by association with the myosin heavy chain or by co-operative effects between the two heads. Phosphorylation of intact myosin is affected also by the state of aggregation i.e. whether it is in the monomeric or filament form. Proteolytic fragments of myosin such as heavy meromyosin or subfragment 1 can also be used as substrates providing care is taken to protect the P light chain from proteolysis to which it is very susceptible.

PROPERTIES OF THE STRIATED MUSCLE ENZYME

The enzyme most widely studied is that from rabbit fast skeletal muscle which occurs as a monomer. This kinase like the others is activated by calcium at concentrations equal to those found in contracting muscle. Magnesium is also essential for activity and ATP is the most active

$$\text{Myosin P light chain} + \text{ATP} \xrightarrow[\text{Kinase}]{Ca^{2+},\ Mg^{2+}} \text{Myosin P light chain monophosphate} + \text{ADP}$$

triphosphate donor. Published K_m values for the P light chain range from 5-40μM. The enzyme is highly specific for the P light chain but for a given enzyme the rates of phosphorylation vary according to the source of the light chains. In contrast P light chains are good substrates for a number of other protein kinases such as phosphorylase kinase, protein kinase C, cAMP-dependent protein kinase etc., It is generally agreed that activation requires one molecule of calmodulin per molecule of kinase but there is some controversy about how many of the four calcium binding sites ion calmodulin are filled on full activation. Serine 15 is the phosphorylation site in the P light chain of myosin from rabbit fast skeletal muscle and there is no evidence for other sites even after prolonged incubation with excess enzyme. Studies using isolated light chains as substrate indicate that the enzyme acts by a rapid equilibrium random bi bi mechanism (Geus et al., 1986).

By controlled proteolysis the kinase can be split into enzymically active fragments one of which of M_r about 30000 has been characterized. This fragment which is rich in α-helix has been described as the globular head of the molecular (Mayr and Heilmeyer 1983). The globular region contains the calmodulin-binding site which has been identified as a 27 residue peptide at the C-terminus of the molecule. The remainder of the molecular, the asymmetric tail, possesses less secondary structure and is rich in proline but as yet no function has been ascribed to it (Fig. 2). Recently the complete sequence of the kinase from rabbit fast skeletal muscle, a single chain of 670 amino acid residues has been reported (Takio et al., 1986).

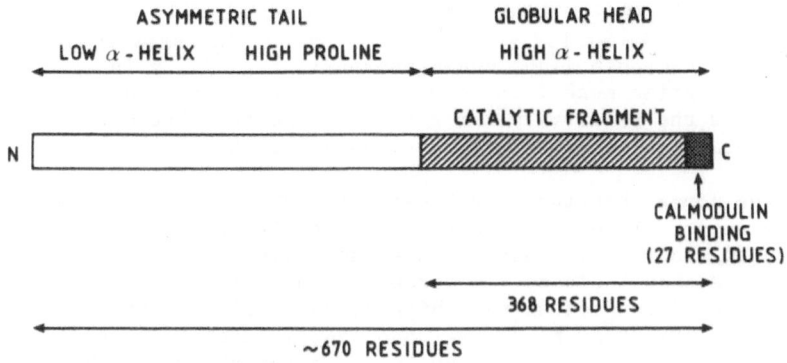

Fig. 2 Structural features of myosin light chain kinase from rabbit fast skeletal muscle.

BIOLOGICAL FUNCTION OF STRIATED MUSCLE MYOSIN LIGHT CHAIN KINASE

The kinase is a highly specific enzyme widely distributed with the amounts present varying between the different types of striated muscle. Although this suggests functional significance its role in striated muscle is still somewhat controversial. In striated muscle phosphorylation of the P light chain is not essential for contraction or activation of the MgATPase of actomyosin. Even in the muscles with the highest kinase activities the enzyme is not active enough to function synchronoously with the cross-bridge cycle although it is switched on by the same concentration of calcium as that which activates the ATPase. It is estimated that there is about one molecule of kinase for every two hundred molecules of myosin in rabbit fast skeletal muscle, which tissue has relatively high kinase activity. There is no universal agreement about the effects of phosphorylation on the MgATPase of actomyosin in vitro. With myosin from rabbit fast skeletal muscle phosphorylation was found to produce only slight activation of the enzyme under saturating conditions with actin (Morgan et al., 1976). Subsequently it has been reported in more detailed kinetic studies that a decrease in K_m but no change in V_{max} occurs after phosphorylation (Pemrick, 1980; Persechini and Stull, 1985). The results described in the latter investigation were only obtained with fresh myosin and disappeared after five days storage. On the other hand Cooke et al., (1982) reported a decrease in the MgATPase of myofibrils after phosphorylation. Variable findings have also been obtained in studies with intact muscle. The original claim that the fall in the maximum velocity of shortening that occurs after tetanus in fast skeletal muscle correlated with the rise in P light chain phosphorylation has not been substantiated. Stull and collaborators (Stull et al., 1985) have concluded that post tetanic potentiation is a consequence of light chain phosphorylation but not all the evidence available is in agreement with this view (see Westwood et al., 1984). One of the difficulties has been in determining accurately the level of phosphorylation of the light chains, particularly with in vivo experiments. This is reflected in the wide range of values reported for the extent of P light chain phosphorylation in resting muscle.

The activity of the kinase in homogenates of whole muscle is in the order fast skeletal > slow skeletal > cardiac. In striated muscle homogenates the myosin light chain phosphatase activity is much lower and in all cases varies much less between muscle types than does the kinase. This explains the relatively low rate of dephosphorylation of the P light chain in intact muscle after it has been stimulated.

In the normal beating rabbit heart the P light chain is 25-30% phosphorylated. Although the extent varies with the species, it is usually less than 50% and in all cases is not much affected by interventions such as adrenaline that increase the force of contraction. In the turtle heart which is unusual in possessing a high kinase activity, beat to beat changes have been observed (Sayers and Barany, 1983). Recently a frequency-dependent phosphorylation has been reported for the rabbit heart (Silver et al., 1986). With skinned fibres of both skeletal (Persechini et al., 1985) and cardiac muscle (Morona et al., 1986), in which preparations the calcium concentrations can be manipulated, there is evidence that increasing the extent of phosphorylation of the P ligh chain increases the calcium sensitivity i.e. increases the isometric tension at a given calcium concentration.

PROPERTIES OF SMOOTH MUSCLE MYOSIN LIGHT CHAIN KINASE

Despite the fact that the smooth muscle enzyme has a significantly higher molecular weight than the striated muscle enzyme it possesses a number of properties in common with the latter. Like the striated muscle enzyme it can be degraded by proteolytic enzymes to produce a fragment of lower molecular weight which is active in the absent of calcium and calmodulin. The enzyme will phosphorylate the P light chain from skeletal muscle myosin but less rapidly than light chains from smooth muscle myosin. The site on chicken gizzard P light chain that is rapidly phosphorylated is serine 19 which is the homologous site to that phosphorylated on the skeletal muscle light chain. Threonine 18 of chicken gizzard P light chain is phosphorylated in addition (Cole et al., 1985; Perry et al., 1985; Ikebe et al., 1984) but this is not a feature of smooth muscle kinase because skeletal muscle kinase will also phosphorylate two sites on gizzard P light chain.

The smooth muscle kinase possesses at least two sites that can be phosphorylated by cAMP dependent kinase. When one of these sites is phosphorylated the affinity of the kinase for calmodulin is reduced and under a given set of conditions the rate of light chain phosphorylation is reduced, thus providing a possible mechanism for β-adrenergic control of smooth muscle activity. In the presence of calmodulin the phosphorylation of one of the sites on the kinase is much reduced. This observation enabled Lukas et al., (1986) to isolate and sequence the calmodulin-binding site. It is homologous to the calmodulin-binding site of the skeletal kinase which is close to the C-terminus in the latter enzyme. Much less is known about the sequence of smooth muscle kinase but it would seem likely that it also has a globular head region containing the active centre and the calmodulin-binding domain.

BIOLOGICAL FUNCTION OF THE SMOOTH MUSCLE KINASE

In contrast to myosin from striated muscle, in the dephosphorylated state myosin from smooth muscle has very low MgATPase. Phosphorylation leads to activation of the MgATPase which is insensitive to calcium when phosphorylation is complete. Although there is general agreement that phosphorylation activates the MgATPase there are varying results on the precise relationship between it and the actin activation of the ATPase. In some instances significant activation can be obtained at low levels of phosphorylation and there are reports that phosphorylation can increase without change in ATPase activity. Further several investigators have described non-linear correlation in that activation of MgATPase expressed as percentage of the maximum is much less than the percentage of the light chains in the phosphorylated form. This has been interpreted to imply that only when the light chains in both heads of the molecule are phosphorylated is myosin activated by actin and that phosphorylation of the second head is more difficult than the first (Persechini and Hartshorne, 1981). Direct studies of the phosphorylation process, however, suggest that it is a random process (Trybus and Lowey, 1985).

Despite certain inconsistencies in the experimental data the general view is that for contraction to take place the P light chain must be phosphorylated. In resting smooth muscle the light chain is dephosphorylated and phosphorylation rises sharply as the tension increases on stimulation. After stimulation the level of phosphorylation falls but tension is maintained. This implies either that at lower levels of phosphorylation the activation of the MgATPase is adequate to maintain the cross-bridge cycle or that some other mechanism exists for maintaining tension. Murphy has suggested that there is a 'latch' mechanism that enables tension to be maintained at low cross-bridge cycling rates (Dillon et al., 1981).

FILAMENT FORMATION AND P LIGHT CHAIN PHOSPHORYLATION IN SMOOTH MUSCLE

Another consequence of the phosphorylation of myosin by the kinase that is confined to smooth and non-muscle systems is the effect on filament formation. It was noted by Watanabe and collaborators (Sukuki et al., 1982) that smooth myosin in solution exists in two forms with sedimentation velocities of 6S and 11S respectively. Both forms are monomers but examination in the electron microscope has revealed that the 11S form is a more compact molecule with tail folded over and interacting at a specific point with the neck region of the molecule. Smooth and non-muscle myosins adopt this conformation in the dephosphorylated form. On phosphorylation of the P light chain the molecule unfolds with its tail extended and sediments more slowly at 6S due to the increased frictional coefficient. The phosphorylated form readily aggregates into filaments at low ionic strength and hence phosphoryulation favours filament formation (see Smith et al., 1983 for review). Although the evidence would suggest that in resting smooth muscle the myosin is present as filaments this may not be the case in non-muscle cells where phosphorylation could be an important process for regulating filament formation and thus the assembly of contractile systems.

There can be little doubt that myosin light chain kinase has an important functional role in contraction. Every cross-bridge involving a single myosin head will acquire two negative charges (four if two heads are involved) as a consequence of phosphorylation at physiological pH values. This change in charge distribution would be expected to affect the inter-action between actin and myosin. In view of the similarity of the enzyme and substrate and the actomyosin system in all muscles it is surprising that the effect of phosphorylation appears to differ so markedly between striated and smooth muscles. Although this difference is apparent from the experimental findings in neither system do all workers report precisely the same results. This can only imply that the systems are not as well char-acterised as one might wish and much has still to be learnt to understand the function of the enzyme. Only when they are will it be possible to decide what are the common features of mechanism in all muscle types.

REFERENCES

Cole, H.A., Griffiths, H.S., Patchell, V.B. and Perry, S.V., 1985, Two site phosphorylation of the phosphorylatable light chain (20-KDa light chain) of chicken gizzard myosin, FEBS. Lett. 180: 165

Cooke, R., Franks, K. and Stull, J.T., 1982, Myosin phosphorylation regulates the ATPase activity of permeable skeletal muscle fibres, FEBS. Lett. 144: 33.

Dillon, P.F., Aksoy, M.O., Oriska, S. and Murphy, R.A., 1981, Myosin phosphorylation and the cross bridge cycle in arterial smooth muscle, Science. 211: 495.

Frearson, N., Focant, B.W.W., and Perry, S.V., 1976, Phosphorylation of a light chain component of myosin from smooth muscle, FEBS. Lett. 63: 27.

Geuss, U., Mayr, G.W. and Heilmeyer, L.M.G., 1985, Steady-state kinetics of skeletal muscle myosin light chain kinase indicate a strong down regulation by products, Eur. J. Biochem, 153: 327.

Ikebe, M., Hartshorne, D.J. and Eczinga, M., 1985, Indentification, phosphorylation and dephosphorylation of a second site for myosin light chain kinase on the 20000 dalton light chain of smooth muscle myosin.

Lukas, T.J., Burgess, W.H., Prendergast, F.G., Lau, W. and Watterson, D.M., 1986, Calmodulin binding domains; characterisation of a phosphorylation and calmodulin binding site from myosin light chain kinase, Biochemistry 25: 1458.

Mayr, G.W. and Heilmeyer, L.M.G., 1983, Shape and substructure of skeletal muscle myosin light chain kinase, Biochemistry 22: 4316.

Morano, I., Hofmann, F., Zimmer, M. and Ruegg, J.C., 1985, The influence of P light chain phosphorylation by myosin light chain kinase on the calcium sensitivity of chemically skinned fibres. FEBS. Lett. 189: 221.

Morgan, M., Perry, S.V. and Ottaway, J., 1976, Myosin light chain phosphatase, Biochem. J. 157: 687.

Pemrick, S., 1980, The phosphorylated L_2 light chain of skeletal myosin is a modifier of the actomyosin ATPase, J. Biol. Chem. 255: 8836.

Perrie, W.T. and Perry, S.V., 1970, An electrophoretic study of the low molecular weight components of myosin, Biochem. J. 119: 31.

Perry, S.V., Griffiths, H.S., Levine, B.A. and Patchell, V.B., 1985, The phosphorylation of the P light chains of gizzard myosin, Adv. Prot. Phosphatases. 2: 3.

Persechini, A. and Hartshorne, D.J., 1981, Phosphorylation of smooth muscle myosin: Evidence of cooperativity between myosin heads, Science 213: 1383.

Persechini, A. and Stull, J.T., 1986, Phosphorylation kinetics of skeletal muscle myosin and the effect of phosphorylation on actomyosin adenosinetriphosphatasc activity. Biochemistry 23: 4144.

Persechini, A., Stull, J.T. and Cooke, R., 1985, The effect of myosin phosphorylation on the contractile properties of skinned rabbit skeletal muscle fibres, J. Biol. Chem. 260: 7951.

Sayers, S.T. and Barany, KL., 1983, Myosin light chain phosphorylation during contraction in the turtle heart, FEBS. Lett. 154: 305.

Silver, P.J., Buja, L.M. and Stull, J.T., 1986, Frequency dependent myosin light chain phosphorylation isolated myocardium, J. Mol. Cell Cardiol. 18: 31.

Smith, R.C., Cande, W.Z., Craig, R., Tooth, P.J., Scholley, J.M. and Kendrick-Jones, J., 1983, Regulation of myosin filament assembly by light chain phosphorylation, Phil. Trans. Roy. Soc. Lond. B. 302: 73.

Stull, J.T., Nunnally, M.H., Moore, R.L. and Blumenthal, D.K., 1985, Myosin light chain kinases and myosin phosphorylation in skeletal muscle, Adv. Enzyme Regulation, 23: 123.

Suzuki, H., Kamata, T., Ohnishi, H. and Watanabe, S., 1982, ATP induced
     reversible changes in the comformation of chicken gizzard myosin and
     HMM, J. Biochem. Tokyo, 91: 1699.
Takio, K., Blumenthal, D.K., Walsh, K.A., Titani, K. and Krebs, E.G., 1986,
     Amino acid sequence of rabbit skeletal muscle myosin light chain
     kinase, Biochem. (In press).
Trybus, K.M. and Lowley, S., 1985, Mechanism of smooth myosin
     phosphorylation, J. Biol. Chem. 260: 19988.
Westwood, S.A. and Perry, S.V., 1982, Two forms of the P light chain of
     myosin in rabbit and bovine hearts, FEBS. Lett. 142 31.
Westwood, S.A., Hudlicka, O. and Perry, S.V., 1984, The effect of
     contractile activity on the phosphorylation of the P light chain of
     myosin of rabbit skeletal muscle in rats, Biochem. J. 218: 841.

CONTRACTILE PROTEIN ISOFORMS

Ute Gröschel-Stewart

Institute of Zoology
Technical University Darmstadt
D-6100 Darmstadt, FRG

All active cell movement originates in the interaction of only a few different classes of protein molecules. The major representatives are Actin and Tropomyosin, which can assemble into long and thin filamentous polymers, and Myosin, which can also aggregate to form thick filaments. They are found in varying proportions, in the three major categories: namely striated muscle, smooth muscle and non-muscle (or cytoplasmic) contractile system. Biochemical and ultrastructural studies have shown that the proteins from the different contractile systems have many characteristic features in common, and yet measurable differences have been found between them. Contractile proteins are present in numerous Isoforms, which can confer different regulatory and contractile properties to different types of muscle cells. As a consequence the three major categories can be subdivided further in the adult vertebrate systems; and, in addition, some of the components may be present as developmental stage specific (embryonic or neonatal) isoforms. The most frequently used methods, present and past, for the distinction of contractile protein isoforms, include:

1. Measurement of mechanical properties of isolated organs, organ strips and "skinned" muscle fibers.

2. Myosin ATPase activity in vitro and in situ (histochemistry).

3. Electrophoretic analysis of muscle proteins and extracts in the native state, in the denatured state, either alone or in combination with isoelectric focussing.

4. Peptide mapping, sequence analysis.

5. Immunological assays with antisera ± specific for isoforms.

6. Recombinant DNA techniques.

It was especially the latter method that suddenly opened up a previously unsuspected variation in muscle proteins. The physiological meaning of many of the isoforms is not yet clear, but the research in this area is progressing rapidly. Not all of the findings can get be coordinated, and to some readers the following account may therefore appear as a sheer enumeration.

I. Isoforms of Actin

This small assymetric molecule with a M_r of 42,000 (374-375 amino acid residues) is not only one of the most ubiquitous, but also most highly conserved proteins. It is encoded by a multigene family in all animals, protozoa and plants (as far as examined), but only by single gene in yeast (review see Hightower and Meagher, 1986). There are only 12 % nucleotide replacement substitution between the actin coding sequences of yeast and chicken skeletal muscle: Within vertebrate actins, the sequence differences are located primarily in the amino terminal region (Vandekerckhove and Weber, 1979). Six distinct actin isoform are present in mammals that seem to be expressed in pairs:

1. The striated muscle isoforms (α skeletal and α cardiac). In the human, cardiac actin mRNA accounts for 5 % of the actin in the adult skeletal muscle, and, vice versa, skeletal actin mRNA represents 50 % of the total mRNA in the adult human heart. The two genes are, however, not in close chromosomal proximity, but apparently on two separate autosomes (Gunning et al, 1983). Vascular smooth muscle α -actin and non-vasucular smooth muscle γ-actin are also coexpressed in varying proportions, and so are the two cytoplasmic β and γ isoforms. The muscle type actins are more closely related to each other than they are to cytoplasmic actins, and this may explain the immunological differences described earlier (Gröschel-Stewart, 1980). There is no account of developmental stage specific actin isoforms.

II. Isoforms of Tropomyosin

Tropomyosin appears to be always associated with actin. While its function as a regulatory protein in striated muscle can well be explained, the exact role of tropomyosin in smooth muscle and non-muscle systems is still rather speculative. Muscle tropomyosin are long rodshaped molecules, dimers of two largely α -helical polypeptides forming a coiled coil (subunit M_r of 33,000). A 19.7 residue periodicity is repeated 14 times. Non-muscle tropomyosins have only twelve repeats, and they do not have the tendency of muscular tropomyosins to aggregate with a head-tail overlap. In vertebrate muscle tissue, four major subunits are found: α and β striated, β and γ smooth. There is evidence for minor additional forms (γ and δ) in slow muscle (Perry, 1985). There is a correlation between speed of contraction and the distribution of tropomyosin isoforms in vertebrate striated muscle; the α isoform being predominant in fast muscle. $\alpha\alpha$ homodimer is found in the faster beating hearts of small animals, $\alpha \beta$ heterodimer in the slower beating hearts of larger animals; the $\beta\beta$ homodimer has not been described. In skeletal muscle, there is a predominance of in fast twitch muscles (or β missing altogether as in chicken pectoral muscle). Embryonic or neonatal isoforms have not been described so far; however, β predominates in the embryonic skeletal muscle, $\alpha\alpha$ in the cardiac muscle throughout development, the above mentioned transition to α , β occurring in the slower beating adult hearts. In chicken gizzard, the two isoforms β and γ are present at heterodimers (Sanders et al., 1986), other smooth muscles appear to have only one isoform, and most evidence speaks of one isoform in non-muscle systems. Inspite of their close structural relationship, antibodies to the various tropomyosin isoforms have not been helpful in establishing functional and evolutionary relationships. The major differences in primary structure between tropomyosin chains appear to be located in two restricted regions (residues 40 - 80 and the C-terminal residues 258-284). It has recently been shown for rat smooth and striated muscle tropomyosins that the isoforms are most probably encoded by the same gene, and that alternative exon splicing (around the above mentioned highly variable regions) is involved in the production of the mature corresponding mRNA (Ruiz-Apazo et al, 1985).

III. Isoforms of Myosin

Myosin is a highly asymmetric molecule with a long tail and two pear-shaped heads, consisting of two heavy chains forming an alpha-helical rod in the carboxy terminal part and N-terminal globular heads. Two pairs of light chains are associated with the head and the head-neck junction. The heads are involved in ATP hydrolysis and actin binding. The differences between contractile properties of the different striated muscles has been correlated with their enzymatic properties (Bárány, 1967). In addition to muscle type specificity found in both the heavy and light chain patterns, there are also developmental stage specific isoforms during muscle organogenesis (Whalen, 1985). Immunological studies have, in the past, been very useful to establish structural relationships between myosin isoforms, but more recently, molecular cloning techniques have advanced our knowledge enormously. The myosin heavy chains are encoded by a multigenic family of closely related members, some of them mapping on the same chromosome. Their expression seems to be largely regulated at the transitional level. The first complete amino acid sequence of a mammalian myosin heavy chain (rat embryonic skeletal) has recently been derived (Strehler et al, 1986). This gene comprises about 24 kb of genomic DNA and consists of 41 exons interrupted by 40 introns of variable size; not always interrupting at the known functional domain boundaries or the repetitive sequences in the rod portion. It is to be expected, that, in the near future, the sequences of all myosin heavy chains will become known. A short synopsis of our present knowledge of myosin isoforms will follow (for review, see Swynghedauw , 1986).

1. Skeletal Muscle Myosin

In chicken pectoral muscle the embryonic isoform (specific embryonic heavy chain, one embryonic specific light chain and one adult fast type light chain) will be replaced upon hatching by neonatal myosin (specific neonatal heavy chain with 3 adult fast type light chains). Neonatal myosin will disappear around day 60 of adult life, adult fast myosin begins to appear at day 20 (adult fast heavy chain, 3 adult fast type light chain). The transition from neonatal to adult type myosins depends on the presence of thyroid hormones. The transition to slow type myosin (specific slow muscle heavy chain and three adult type show light chains) can occur at all three stages of development upon specific stimuli (innervation).

2. Cardiac Muscle Myosin

In the heart, two heavy chain isoforms are found, α and β . In the atrium, two α heavy chains are found in combination with an atrium-specific light chain. The second light chain, being identical to the embryonic skeletal light chain. In the ventricle, the β chain homodimer predominates in the slower beating adult hearts and late in fetal life. In the faster beating hearts of smaller animals, The α chain homodimer predominates, the α β heterodimer is also found. The ventricular myosin heavy chain isoform is identical to the skeletal muscle slow myosin heavy chain. α and β heavy chain genes are arranged in tandem on the chromosome and they are differently expressed not only during development but also in response to thyroid hormone (Mahdawi et al, 1984). In ventricular myosin, one specific light chain is found, the second one being related or even identical to a skeletal slow one. The emerging evidence of how innervation and hormonal stimulation can influence and alternate the expression of developmental stage specific and contraction speed specific isoforms will be of great importance in the field of muscle pathology in general.

3. Smooth Muscle Myosin

Although smooth muscle myosins are distinct from striated muscle myosins by their regulation, their ATPase activity, and by their immunological properties, they have never been as thoroughly dealt with as the striated muscle myosins. Only recently, the presence of equimolar amounts of two myosin heavy chain isoforms in various smooth muscles have been des- cribed (Rooner et al, 1986).

Sofar, no evidence has been given for developmental stage specific isoforms, only the presence of an additional embryonic light chain has been described. Since smooth muscles have to perform so many various tasks, their uniformity in structure seems surprising. Possibly, these differences in performance may be controlled on a regulatory level, a field that is still far from clear.

4. Cytoplasmic myosin

According to our own findings (Gröschel-Stewart, 1985) there is a specific cytoplasmic myosin (non-crossreactive with smooth muscle myosin) in all cell dervided from the hematopoetic stemm cells, in endothelia and primary fibro- blasts. Other non-muscle cells, such as myoid and epithelial cells, seem to possess myosin either identical or closely related to smooth muscle myosin. Hopefully, recombinant DNA techniques so successfully applied in the analysis of striated muscle myosin will also applied in the near future to the more neglected smooth and cytoplasmic myosin systems.

REFERENCES

I. Overviews:

- Gröschel-Stewart, U. (1980) Intl. Rev. Cytology 65, 194-254.
- Gröschel-Stewart, U. & Drenckhahn, D. (1982) Collagen Rel. Res. 2, 381-463.
- Hightower, R.C. and Meagher, R.B. (1986) Genetics 114, 315-332.
- Perry, S.V. (1985) J. exp. Biol. 115, 31-42.
- Swynghedauw, B. (1986) Physiol. Rev. 66, 710-771.
- Whalen, R.G. (1985) J. exp. Biol. 115, 43-53.

II. Original Papers:

- Bárány, M. (1967) J. Gen. Physiol. 50, 197-218.
- Gröschel-Stewart, U., Rahousky, C. Franke, R., Peleg, I., Kahane, I., Eldor, A., Muhlrad, A. (1985) Cell Tissue Res. 241, 399-403.
- Gumming, G., Ponte, P., Blau, H., Kedes, L. (1983) Molec. Cell Biol. 3, 1985-1995.
- Mahdavi, V., Chambers, A.P., Nadal-Ginard, B. (1984) Proc. Natl. Acad. Sci. USA 81, 2626-2630.
- Rooner, A.S., Thompson, M.M., Murphy, R.A. (1986) Am. J. Physiology.
- Ruiz-Opazo, N., Weinberger, J., Nadal-Ginard, B. (1985) Nature 315, 67-70.
- Sanders, C., Burtnick, L.D., Smillie, L.B. (1986) J. Biol. Chem. 261, 12774-12778.
- Strehler, E.E., Strehler-Page, M.-A., Perriard, J.-C., Periasamy, M., Nadal-Ginard, B. (1986) J. Mol. Biol. 190, 291-317.
- Vandekerckhove, J., Weber, K. (1979) Differentiation 14, 123-133.

ACTIN POLYMERIZATION

Elke Schröer, Klaus Ruhnau, Norma Selve, and
Albrecht Wegner

Institut für Physiologische Chemie I
Ruhr-Universität Bochum
D-4630 Bochum, Federal Republic Germany

INTRODUCTION

Actin filaments and microtubules make up the principal dynamic
constituents of the cytoskeleton. Actin is capable of entering
into a variety of interactions with other proteins which
regulate its state. The simplest possible in vitro system -
that in which the actin monomers polymerize to form long linear
aggregates - has been found to embody a number of properties
reflecting the dynamics of the turnover of actin filaments in
cells. Actin filaments can quickly polymerize and depolymerize,
they can spontaneously break and associate end to end, and an
ATPase activity causes actin filaments to "treadmill", that is
to polymerize at one end and to depolymerize simultaneously at
the other end.

Nucleation and elongation of actin filaments

When actin monomers are polymerized by the addition of Mg, K or
Ca ions, polymerization proceeds in the initial phase slowly.
After some minutes the rate of polymerization is accelerated
and in the final stage the concentrations of monomeric and
polymeric actin approach their constant steady state values.
The polymerization curves (plot of the concentration of polymerized actin versus time) are sigmoidal (Fig. 1). The shape

of the polymerization curves can be explained by slow and un-
favored formation of actin filaments (nucleation). Once filaments
have been nucleated monomers quickly disappear by polymerization
onto filaments. At the final stage polymerization ceases because
the monomers have been consumed.[1,2] Quantitative analysis of the
polymerization curves showed that actin tetramers are the
smallest aggregates which polymerize and depolymerize like long

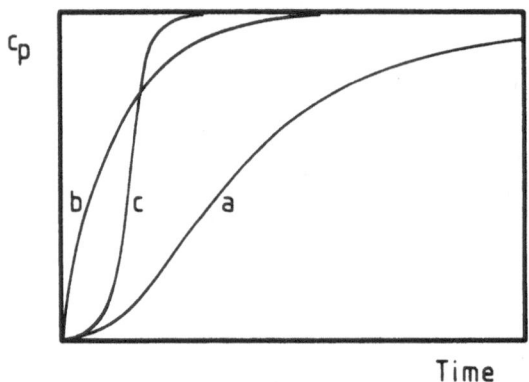

Fig. 1: Actin polymerization curves. c_p, concentration of polymerized actin. a, monomers polymerize by nucleation and elongation. b, monomers polymerize onto existing nuclei. c, monomers
polymerize by nucleation, elongation and fragmentation of
filaments.

actin filaments. Dimers and trimers are very unstable intermediates which readily dissociate.[3-5] These conclusions were
confirmed by more direct experiments in which the nucleating
activity of chemically crosslinked actin oligomers of defined
size was tested.[6] Further evidence for this interpretation of
the contour of the polymerization curves stems from investigations on nucleated polymerization. When monomers are mixed with
polymeric actin, no initial lag of polymerization is observed
because nuclei are present already at the beginning of the polymerization reaction (Fig. 1, curve b).

The rate of lengthening and shortening of actin filaments has
been determined using the electron microscope. In elegant

experiments actin filament bundles were isolated from microvilli. The isolated microvilli cores consist of parallel crosslinked actin filaments. The actin molecules are polarly arranged in actin filaments. In microvilli cores all actin filaments have the same polarity, leading to a structural polarity of the cores. The microvilli cores were incubated for various time intervals with different concentrations of monomeric actin. Strings of actin molecules polymerized onto the filament ends of the micro-villi cores were visualized by electron microscopy. One end of the actin filaments ("barbed end") was found to grow more than one order of magnitude faster than the other end ("pointed end). The structural polarity of actin filaments is reflected by the polarity of actin polymerization. The rate of binding of mono-mers to the barbed ends is so fast that it is almost diffusion controlled (rate constant $k^+ = 10^7$ M^{-1} s^{-1}).[7,8,6]

End to end association, fluctuation of the filament length and spontaneous fragmentation

End to end association of actin filaments has been demonstrated by electron microscopy. Short actin filaments were decorated with proteolytic fragments of myosin. Actin monomers were then polymerized onto these decorated filaments to procuce "arrow-heads". Following incubation of these arrowheads for some minutes "double arrowheads" appeared which were formed by end to end association (Fig. 2). This observation demonstrated unambigously that actin filaments associate end to end. [9,10] Recently, the rate of formation of long actin filaments from short fragments was analyzed quantitatively. According to this analysis the formation of long filaments can also be explained by fluctuation of filament length. Some short filaments disappear by accidental dissociation of actin molecules. Other filaments incorporate these actin molecules. Thus, some filaments grow at the expense of other filaments. Probably, both end to end association and fluctuation of filament length cause long fila-ments to be produced from short filaments.[11]

Slow spontaneous fragmentation of actin filaments has been detected by quantitative analysis of actin polymerization, and more directly by fractional centrifugation and electron micro-scopy. Under some conditions, actin polymerization curves reveal a shape typical for a system that reproduces itself, such as

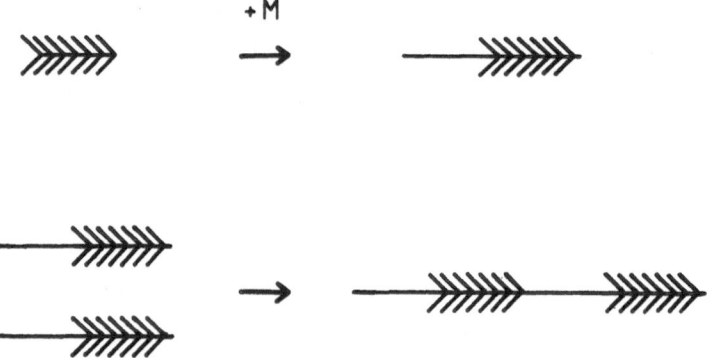

Fig. 2: Formation of arrowheads from decorated actin filaments and monomers (M)(upper line). End to end association of two arrowheads to form double arrowheads (lower line).

spontaneously breaking actin filaments. After a relatively long lag phase the polymerization rate increases strongly to reach the final concentration of polymerized actin quickly (Fig. 1, curve c).[3] Fragmentation has also been more directly demonstrated by fractional centrifugation and electron microscopy.[12]

ATP hydrolysis by actin: ATP caps and treadmilling

Actin filaments have an enzymic activity. Each actin subunit can hydrolyze ATP to yield ADP and inorganic phosphate.[13] Actin monomers bearing ATP associate with the ends of filaments. In the presence of millimolar Mg^{2+} concentrations ATP hydrolysis follows about 5 seconds later.[14,15] The association of actin monomers with elongating barbed ends of filaments is so fast that ATP hydrolysis lags behind association. The subunits at the elongating barbed ends carry a string of subunits with bound ATP, referred to as an "ATP cap". Subunits near the middle of filaments possess bound ADP as they assembled first and had sufficient time to allow ATP hydrolysis. The composition of shortening barbed filament ends is different. Filaments lose their ATP caps with dissociation of terminal subunits. Shortening filaments have ADP-containing subunits at their ends. The rate constant for dissociation of an ADP-bearing subunit from a filament end is about 5 times faster than that of an ATP-bearing subunit.[16] As a consequence, a plot of the rate of growth versus the actin monomer concentration reveals a discontinuity near the transition between polymerizing ATP-capped filaments and depolymerizing filaments containing terminal ADP-bearing subunits.[15]

Actin monomers associate with the pointed ends more than one order of magnitude more slowly than with the barbed ends. The association at lengthening pointed ends is so slow that the time between the association reactions of two monomers can be

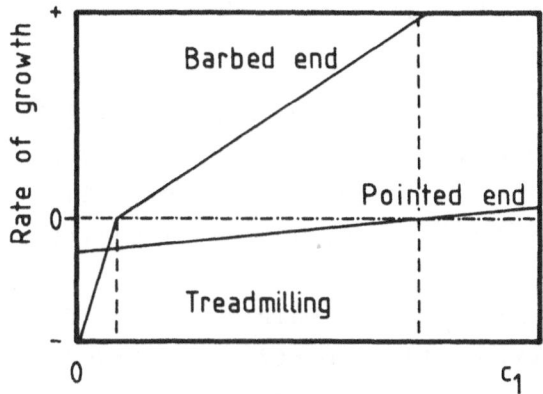

Fig. 3: Plot of the rate of growth versus the monomer concentration c_1.

sufficient for ATP hydrolysis.[17] Thus, actin filaments carry an ATP cap only at one end, namely the barbed end. ATP-capped ends have a higher affinity for actin monomers than ends without ATP caps. ATP-capped ends tend to polymerize while ends containing ADP-bearing subunits tend to depolymerize. Because of the different composition of the nucleotides at the two ends actin filaments "treadmill" that is they polymerize at the barbed ends and depolymerize simultaneously at the pointed ends.[18]

Treadmilling has been demonstrated by a number of experiments. The consumption of monomers at the barbed ends of filaments and the production of monomers at the pointed ends accelerates the exchange of monomers with filament subunits. Originally, this effect has been used to prove treadmilling.[18,19] Treadmilling has also been demonstrated by the increase of the actin monomer concentration caused by proteins which inhibit association and dissociation of actin selectively at the barbed ends by binding to these ends. Consumption of monomers at the barbed ends is blocked by these proteins while monomers are still produced at the shortening pointed ends.[20,21] Finally, treadmilling has been demonstrated by electron microscopy.[7,8] The rate at which actin filaments treadmill in vitro, has been determined to be about 2 micrometers per hour.[22] Recently, evidence has been provided suggesting that actin filaments treadmill also in vivo. The rate of treadmilling in vivo has been estimated to be even greater (40 micrometers per hour) than in vitro.[23]

References

1. Oosawa, F. & Kasai, M. (1962) J. Mol. Biol. 4, 10-21.

2. Wegner, A. & Engel, J. (1975) Biophys. Chem. 3, 215-225.

3. Wegner, A. & Savko, P. (1982) Biochemistry 21, 1909-1913.

4. Frieden, C. & Goddette, D. W. (1983) Biochemistry 22, 5836-5843.

5. Cooper, J. A., Buhle, E. C., Walker, S. B., Tsong, T. Y. & Pollard, T. D. (1983) Biochemistry 22, 2193-2202.

6. Lal, A. A., Korn, E. D. & Brenner, S. L. (1984) J. Biol. Chem. 259, 8794-8800.

7. Pollard, T. D. & Mooseker, M. S. (1981) J. Cell Biol. 88, 654-659.

8. Bonder, E. M., Fishkind, D. J. & Mooseker, M. S. (1983) Cell 34, 491-501.

9. Nakaoka, Y. & Kasai, M. (1969) J. Mol. Biol. 44, 319-332.

10. Kondo, H. & Ishiwata, S. (1976) J. Biochem. (Tokyo) 79, 159-171.

11. Carlier, M.-F., Pantaloni, D. & Korn, E. D. (1984) J. Biol. Chem. 259, 9987-9991.

12. Grazi, E. & Trombetta, G. (1985) Biochem. J. 232, 297-300.

13. Straub, F. B. & Feuer, G. (1950) Biochim. Biophys. Acta 4, 455-470.

14. Pollard, T. D. & Weeds, A. G. (1984) FEBS Lett. 170, 94-98.

15. Carlier, M.-F., Pantaloni, D. & Korn, E. D. (1984) J. Biol. Chem. 259, 9983-9986.

16. Pollard, T. D. (1984) J. Cell Biol. 99, 769-777.

17. Coué, M. & Korn, E. D. (1986) J. Biol. Chem. 261, 1588-1593.

18. Wegner, A. (1976) J. Mol. Biol. 108, 139-150.

19. Wegner, A. & Neuhaus, J.-M. (1981) J. Mol. Biol. 153, 681-693.

20. Brenner, S. L. & Korn, E. D. (1979) J. Biol. Chem. 254, 9982-9985.

21. Wegner, A. & Isenberg, G. (1983) Proc. Natl. Acad. Sci. USA 80, 4922-4925.

22. Selve, N. & Wegner, A. (1986) J. Mol. Biol. 187, 627-631.

23. Wang, Y. L. (1985) J. Cell Biol. 101, 597-602.

ADP-RIBOSYLATION OF ACTIN BY BOTULINUM C2 TOXIN

Klaus Aktories, Michael Bärmann*, Monika Laux,
Karl Heinrich Reuner, Peter Presek and Beate Schering

Rudolf-Buchheim-Institut and *Institut für Toxiko-
logie und Pharmakologie der Universität Gießen
Frankfurter Str. 107, D-6300 Gießen, FRG

INTRODUCTION

The ADP-ribosylation of proteins is an important pathophysiological mechanism by which various microbial toxins affect
eucaryotic cell functions. Well-studied examples of this group
of toxins are cholera-, pertussis- and diphtheria toxins (1,
2). Whereas cholera toxin and pertussis toxin interfere with
the hormon-sensitive adenylate cyclase by ADP-ribosylating the
regulatory N_s and N_i-proteins of the adenylate cyclase system,
respectively, diphtheria toxin is known to inhibit the protein
synthesis by ADP-ribosylation of elongation factor 2.

Recently, it has been shown that botulinum C2 toxin is
another bacterial toxin with ADP-ribosyltransferase activity
(3). This toxin, which is produced by certain strains of
Clostridium botulinum, induces an increased gut secretion,
vascular permeability, hypotonic effects, haemorrhaging in the
lungs and finally death (4). The toxin is binary in structure
and consists out of component I and II with relative molecular
weights of 55 and 100 kDa, respectively (5). Whereas component
II is apparently involved in the binding of the toxin to the
eucaryotic cell membrane, component I possesses the ADP-ribosyltransferase activity. Thus, both components are necessary
to elicit the toxic effects in vivo.

THE SUBSTRATE OF BOTULINUM C2 TOXIN

We have demonstrated that actin is ADP-ribosylated by
botulinum C2 toxin (6,7) (Figure 1). ADP-ribosylation of actin
by the toxin was found in the cytosolic fraction of cell homogenates from platelets, fibroblasts, liver cells, PC-12 cells,
mast cells, S49 lymphoma cells and neuroblastoma & glioma hybrid
cells and sperm. We identified actin as the target protein of
the ADP-ribosylation by means of peptide mapping, immunoprecipitation from whole cells and immunoblotting with actin antibodies. Finally, isolated actin from platelets and liver cells

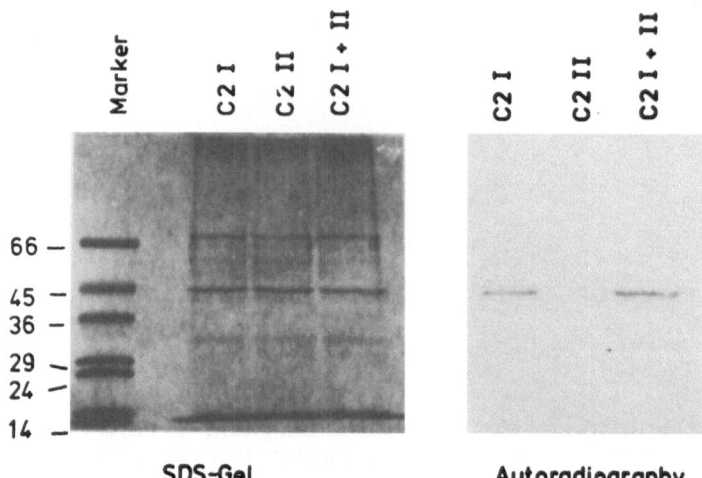

Figure 1: Botulinum C2 toxin-catalyzed ADP-ribosylation in platelet cytosol. Botulinum C2 toxin component I but not component II causes ADP-ribosylation of actin. Left panel: SDS-Gel of platelet cytosol treated for 1 h at 37° C without and with botulinum C2 toxin component I (C2I, 1 µg/ml), component II (C2II, 4 µg/ml) or components I and II (C2I + II) in the presence of ^{32}P-NAD as described in ref. 7. Right panel: Autoradiography of the SDS-Gel.

serves as a good substrate of botulinum C2 toxin. Thus, ADP-ribosylation of actin by botulinum C2 toxin is catalyzed in the absence of any additional cytosolic factor. Under our experimental conditions, isolated actin was not ADP-ribosylated by cholera toxin, pertussis toxin nor botulinum toxin A. Neither were the N-proteins of the adenylate cyclase system, transducin or tubulin substrates of botulinum C2 toxin. Thus, the substrate of botulinum C2 toxin differs from the eucaryotic substrates of cholera toxin or pertussis toxin, which have in common that they bind guanine nucleotides and possess GTPase activities (1).

Labelling of actin by botulinum C2 toxin shows an interesting specificity with respect to actin isoforms. Whereas purified actin from platelets and liver cells is a good substrate, skeletal muscle actin cannot serve as a substrate. Furthermore, botulinum C2 toxin ADP-ribosylates monomeric G-actin but not polymerized F-actin. This specificity is the likely explanation for the finding that phalloidin, which induces the polymerisation of actin (8), prevents the ADP-ribosylation of the microfilament protein in a dose-dependent manner (7). Most probably the covalent modification of actin is a mono-ADP-ribosylation, because the amount of label increased with the duration of the incubation time without an apparent increase in the molecular weight of the labelled protein. Furthermore, nicotinamide is able to reverse the toxin-induced ADP-ribosylation, which is typical for a mono-ADP-ribosylation. In con-

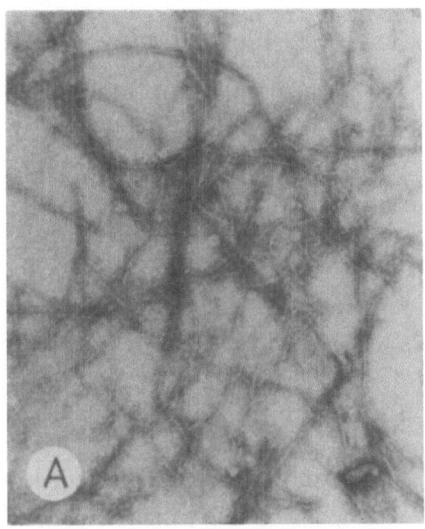 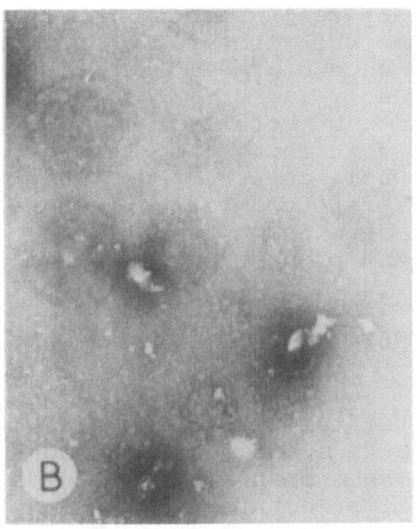

Figure 2: Electron microscopy of actin filaments. Isolated platelet G-actin was ADP-ribosylated in the presence of 1 mM NAD without (A) and with (B) botulinum C2 toxin component I (1 µg/ml) as described in ref. 7. Thereafter the polymerisation was induced by adding $MgCl_2$ (2 mM), ATP (0.5 mM), $CaCL_2$ (0.2 mM), DTT (0.1 mM), phalloidin (500 µM) and Tris/HCl (2 mM/pH 7.5, all final concentrations). After 10 min incubation at 25° C, aliquots of the sample were subjected to electron microscopy (Magnification 25 250 x).

trast, poly-ADP-ribosylation is irreversible (2).

ADP-ribosylation largely affects the functional properties of actin. Botulinum C2 toxin treatment dose dependently reduces the viscosity of an actin solution after induction of polymerisation by Mg^{2+} (6). The influence of the botulinum C2 toxin on the polymerisation of actin was further substantiated by means of electron microscopy. As shown in Figure 2, treatment of isolated platelet actin completely blocked the polymerisation of actin. Even in the presence of phalloidin ADP-ribosylated actin could not form the typical microfilament network.

ADP-RIBOSYLATION OF ACTIN IN INTACT CELLS

It has been shown that also other bacterial toxins can ADP-ribosylate cytoskeletal proteins (2). For instance tubulin serves as a substrate for cholera toxin and for pertussis toxin in vitro (2). But both toxins were not able to ADP-ribosylate actin in intact cells. In order to clarify whether actin is also a substrate of botulinum C2 toxin in intact cells two different approaches were chosen. When ^{32}P-prelabelled intact CEC were treated with botulinum C2 toxin, labelling of a 43 kDA protein was induced, which was identified as actin by means of proteolytic peptide maps. Furthermore, toxin pretreatment of intact cells reduced the labelling of actin in a second subsequent ADP-ribosylation performed with cell homogenates in the presence of ^{32}P-NAD and botulinum C2 toxin. Both findings strongly indicate that also in intact cells actin is a substrate of botulinum C2 toxin. ADP-ribosylation of actin was found in intact hepatocytes, mast cells, PC-12 cells, S49-lymphoma cells but not in platelets. Thus, similar as it has been reported for pertussis- and cholera toxin, platelets were not affected by the toxin, presumably, because the binding and/or the processing of the toxin is impaired in platelets.

BOTULINUM C2 TOXIN DISORGANIZES THE MICROFILAMENTS

Botulinum C2 toxin induces rounding up of cells, which correlates with the ADP-ribosylation of actin in intact cells in a time and dose-dependent manner (9). As rounding up of cells is a phenomenon which can be ascribed to changes in the cytoskeletal structure, these findings indicate that both processes are intimately connected. Further indications for the hypothesis that actin is also in intact cells the pathophysiological substrate of botulinum C2 toxin were obtained by studying the effects of the toxin on the microfilament network of cells. For this purpose the intracellular microfilament network was stained by means of fluorescein-labelled phalloidin. Toxin treatment of CEC largely affected the microfilament network in a time dependent manner. After 3 hrs of incubation the microfilament bundles appear more condensed and broken. Further incubation causes the almost complete destruction of the microfilament network and the appearance of diffuse staining distributed evenly over the entire cell. At this time most of the cells were round. Although the cytoskeleton is largely disorganized the cells are still adherent. A likely explanation for the disorganisation of the microfilament network may be a change in the dynamic intracellular G/F-actin ratio caused by toxin-induced ADP-ribosylation. Based on the finding that the ADP-ribosylation blocks the polymerisation of purified actin one can speculate that also in intact cells toxin treatment traps actin in its monomeric form. Thus, the modified actin is no longer available for formation of microfilaments. This hypothesis is supported by the findings that after toxin treatment of intact cells a decreased portion of actin can be isolated as Triton insoluble actin, which can be ascribed to cytoskeleton associated F-actin. In contrast, the portion of Triton soluble actin, which represents the cytoplasmic actin increases largely with toxin treatment. It should be pointed out that treatment of ^{32}P-preloaded cells causes exclusively labelling of cytoplasmic actin but not of that actin associated with cytoskeleton elements.

Taken together, botulinum C2 toxin mono-ADP-ribosylates isolated actin as well as actin in intact cells. ADP-ribosylation of actin drastically reduces its property to polymerize, which may cause a disorganization of the microfilament network of the cytoskeleton and a blockade of the physiological functions of actin. It is proposed that these properties of botulinum C2 toxin will be a powerful instrument for further studies on the physiological function of non-muscle actin.

REFERENCES

1. K. A. Wregett, Bacterial toxins and the role of ADP-ribosylation, J. Rec. Res. 6:95 (1986).
2. K. Ueda and O. Hayaishi, ADP-Ribosylation, Ann. Rev. Biochem. 54:73 (1985.
3. L. L. Simpson, Molecular basis for the pharmacological actions of clostridium botulinum C2 toxin, J. Pharmacol. Exp. Ther. 230:650 (1984).
4. L. L. Simpson, Molecular pharmacology of botulinum toxin and tetanus toxin, Ann. Rev. Pharmacol. Toxicol. 26:427 (1986).
5. I. Ohishi, M. Iwasaki, G. Sakaguchi, Purification and characterization of two components of botulinum C2 toxin, Infect. Immun. 30:668 (1980).
6. K. Aktories, M. Bärmann, I. Ohishi, S. Tsuyama, K.H. Jakobs and E. Habermann, Botulinum C2 toxin ADP-ribosylates actin, Nature 322:390 (1986).
7. K. Aktories, T. Ankenbauer, B. Schering and K.H. Jakobs, ADP-ribosylation of platelet actin by botulinum C2 toxin, Eur. J. Biochem. in press (1986).
8. T. Wieland, Modification of actins by phallotoxins, Naturwissenschaften 64:303 (1977).
9. K. H. Reuner, P. Presek, C.B. Boschek and K. Aktories, Botulinum C2 toxin ADP-ribosylates actin and disorganizes the microfilament network in intact cells, Europ. J. Cell Biol. in press

ACTIN-BINDING PROTEINS

Klaus Ruhnau, Norma Selve, Elke Schröer and
Albrecht Wegner

Institut für Physiologische Chemie I
Ruhr-Universität Bochum
D-4630 Bochum, Federal Republic Germany

INTRODUCTION

Actin filaments form networks and bundles within the cytoplasm
of cells. The actin filaments change continuously their organization by transition between the monomer and filament pool, by
fragmentation and elongation of filaments, by disruption of
filament bundles or by bundling of randomly oriented filaments.
There is good evidence that changes of the organization of actin
filaments are controlled by the action of transmembrane signals.
For example, actin polymerizes rapidly when platelets are stimulated with thrombin or ADP or pancreatic ß-cells with glucose.[1]

Numerous investigations have shown that proteins are associated
with actin monomers or filaments to regulate the arrangement
and turnover of actin. These proteins have been classified
according to their action on actin polymerization: 1. Actin-
depolymerizing proteins, 2. actin-filament-severing proteins,
3. actin-filament-crosslinking and -bundling proteins and other
proteins have been purified practically from all eukaryotic cells
and tissues. Some of these proteins turned out to be calcium-
sensitive, to be phosphorylated or to interact specifically with
intermediates of the phosphatidyl inositol cycle.[2-5]

1. Actin-depolymerizing proteins

In vitro experiments show that at physiological salt concentrations about 0.1 μM monomeric actin coexists with actin filaments. In many cells and tissues the concentration of unpolymerized actin has been found to be 100-fold greater (100 μM). This unexpectedly great concentration of unpolymerized actin could be explained by the occurrence of "profilin". This protein binds to monomeric actin to form the "profilactin" complex. The dissociation constant is about 10^{-8} M. The profilactin complex does not polymerize. Profilin causes actin filaments to depolymerize by removing actin monomers (Fig. 1).[6,7]

The concentration of unpolymerized actin often changes on stimulation of cells by extracellular signals. When platelets are activated by thrombin or ADP, intracellular actin polymerizes rapidly.[8] Before stimulation, amorphous aggregates of actin were detected in platelets. During stimulation the profilactin complex dissociates and actin polymerizes to form filament bundles. Today, one has developed ideas how extracellular stimulation leads to intracellular dissociation of the profilactin complex and assembly of actin filament bundles. It could be demonstrated that the profilactin complex dissociates at the surface of membranes containing phosphatidyl inositol-4,5-bisphosphate, an intermediate of the phosphatidyl inositol cycle. The resulting free actin can readily polymerize.[9] The dissociation of the profilactin complex during platelet stimulation is not yet understood in all details. The problem is that during platelet stimulation the phosphatidyl inositol-4,5-bisphosphate concentration decreases.

2. Actin-filament-severing proteins

This class of actin-binding proteins severs actin filaments. Also these proteins are ubiquitous. The majority of the severing proteins (gelsolin, villin, fragmin) is calcium-sensitive. Below micromolar Ca^{2+} they do not interact with actin filaments. In the presence of micromolar Ca^{2+} they sever actin filaments by insertion between actin filament subunits. Severing proceeds rapidly. A few seconds after incubation of actin filaments with severing proteins actin filaments are fragmented.[10-13] The severing protein remains bound to the barbed end to form "capped" actin filaments and inhibits association and dissociation of

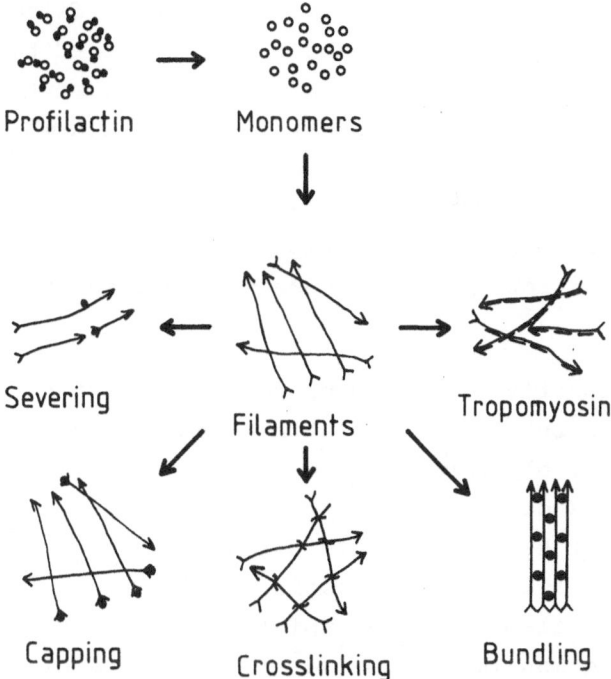

Fig. 1: Actin-binding proteins. The polar structure of actin filaments is indicated by arrows at the ends of actin filaments.

actin molecules at the barbed end and end to end association of actin filaments. Capping of the barbed ends causes the actin monomer concentration to increase about 5- to 10-fold because actin filaments treadmill, that is they polymerize at the barbed end and depolymerize simultaneously at the pointed end. The severing protein blocks monomer consumption at the barbed end while the pointed end still produces monomers.[14,15] The increase of the concentration of unpolymerized actin by severing proteins can be amplified by profilin.[7,15] Thus, severing proteins and profilin may regulate the distribution of actin between the monomer and filament pool in a calcium-sensitive manner.

Following removal of calcium from capped actin filaments long filaments are reformed. It has been proposed that the severing protein dissociates from the barbed end of filaments at low Ca^{2+} concentrations. The uncapped short filaments assemble then end to end.[10] Recently, evidence for another pathway of filament elongation has been provided. Some short capped filaments dis-integrate completely by fluctuation of the filament length. The other filaments grow at the expense of the disappearing fila-ments.[16]

The calcium-sensitive severing proteins interact not only with actin filaments but also with actin monomers. Gelsolin, one of the most intensively investigated severing proteins, forms a tight 1:1 or 1:2 complex with actin.[17] This gelsolin-actin complex does not sever filaments but acts as a calcium-sensitive capping protein. The gelsolin-actin complex associates with the barbed ends of filaments very rapidly in an almost diffusion controlled reaction ($k^+=10^7$ M^{-1} s^{-1}).[18] Also calcium-insensitive capping proteins have been purified.[19-21]

3. Actin filament-crosslinking and -bundling proteins

Crosslinking proteins connect actin filaments in any angle. These proteins confer rigidity and high viscosity on the actin cytoskeleton.[22] Bundling proteins arrange actin filaments in parallel alignment. Often bundled filaments form highly ordered crystal-like structures (Fig.1).[23] Most of the crosslinking and bundling proteins are not calcium-sensitive and are not regulated by phosphorylation. However, rigid crosslinked actin networks

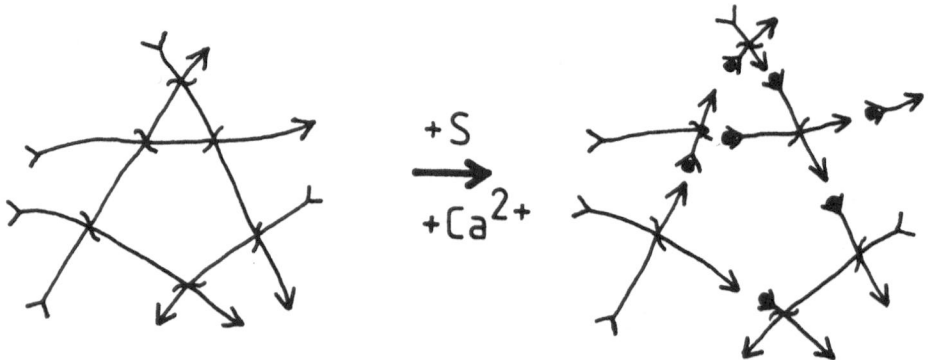

Fig. 2: Solation of a crosslinked actin gel by a severing protein in the presence of Ca^{2+}.

or filament bundles can be disassembled by the action of severing proteins which are calcium-sensitive. If the concentration of severing protein is sufficiently great, all actin filaments are severed between the binding sites of two crosslinking or bundling proteins. The networks or bundles are dissolved to yield actin solutions of low viscosity (Fig. 2).[10] Such calcium dependent transitions between gelation and solation may regulate locomotion, secretion and endocytosis of cells.

4. Other actin-binding proteins

Myosin and troponin are two other actin-binding proteins which are regulated by phosphorylation or by Ca^{2+}. These proteins are discussed in two other articles of this book. Tropomyosin is a further actin-binding protein which occurs in many types of cells and tissues.[24,25] Tropomyosin is a long rod-shaped protein that binds along actin filaments thereby covering six or seven actin filament subunits (Fig. 1). Tropomyosin stabilizes actin filaments. It inhibits spontaneous fragmentation of actin filaments and prevents severing by gelsolin and related proteins.[26,27]

The isolation of actin-binding proteins and the investigation on their action on actin polymerization has contributed to a better understanding of the turnover of actin filaments following stimulation of cells by extracellular signals.

REFERENCES

1. Gonnella, P. A. & Nachmias, V. T. (1981) J. Cell Biol. **89**, 146-151.

2. Weeds, A. (1982) Nature **296**, 811-816.

3. Craig, S. W. & Pollard, T. D. (1982) Trends Biochem. Sci. **7**, 88.

4. Jockusch, B. M. (1983) Mol. Cell Endocrinol. **29**, 1.

5. Hatano, S. (1984) Cell Struct. Funct. **9**, Suppl., s67-72.

6. Carlsson, L., Nyström, L.-E., Sundkvist, I., Markey, F. & Lindberg, U. (1977) J. Mol. Biol. **115**, 465-483.

7. Tobacman, L. S. & Korn, E. D. (1982) J. Biol. Chem. **257**, 4166-4170.

8. Carlsson, L., Markey, F., Blikstad, I., Persson, T. & Lindberg, U. (1979) Proc. Natl. Acad. Sci. USA **76**, 6376-638(

9. Lassing, I. & Lindberg, U. (1985) Nature **314**, 472-474.

10. Yin. H. L. & Stossel, T. P. (1979) Nature **281**, 583-586.

11. Hasegawa, T., Takahashi, S., Hayashi, H. & Hatano, S. (1980) Biochemistry 19, 2677-2683.

12. Bretscher, A. & Weber, K. (1980) Cell 20, 839-847.

13. Mooseker, M. S., Graves, T. A., Wharton, K. A., Falco, N. & Howe, C. L. (1980) J. Cell Biol. 87, 809-822.

14. Brenner, S. L. & Korn, E. D. (1979) J. Biol. Chem. 9982-9985.

15. Wegner, A. & Isenberg, G. (1983) Proc. Natl. Acad. Sci. 80, 4922-4925.

16. Janmey, P. A., Chaponnier, C., Lind, S. E., Zaner, K. S., Stossel, T. P. & Yin, H. L. (1985) Biochemistry 24, 3714-3723.

17. Nishida, E., Kuwaki, T., Maekawa, S. & Sakai, H. (1981) J. Biochem. (Tokyo) 89, 1655-1658.

18. Selve, N. & Wegner, A. (1986) Eur. J. Biochem. 155, 397-401.

19. Isenberg, G., Aebi, U. & Pollard, T. D. (1980) Nature 288, 455-459.

20. Kilimann, M. W. & Isenberg, G. (1982) EMBO J. 1, 889-894.

21. Schröer, E. & Wegner, A. (1985) Eur. J. Biochem. 153, 515-520.

22. Wang, K. (1977) Biochemistry 16, 1857-1865.

23. Bryan, J. & Kane, R. E. (1978) J. Mol. Biol. 125, 207-224.

24. Bailey, K. (1948) Biochem. J. 43, 271-278.

25. Cohen, I. & Cohen, C. (1972) J. Mol. Biol. 68, 383-387.

26. Wegner, A. (1982) J. Mol. Biol. 161, 217-227.

27. Hinssen, H. (1981) Eur. J. Cell Biol. 23, 234-240.

MODULATION OF CARDIAC Ca CHANNELS BY PHOSPHORYLATION

W. Trautwein

II. Physiologisches Institut
Universität des Saarlandes
6650 Homburg/Saar, FRG

Ca ions play an important role in signal transmission across the cell membrane and within the cell. In heart muscle they control the force of contraction and participate in the spontaneous impulse generation in the sinus node, the primary pacemaker, as well as in conduction across the atrio-ventricular node. Ca ions enter the cell through voltage-controlled Ca channels which open upon depolarization of the cell membrane giving rise to a Ca current into the cell. Two types of Ca channels have been found in heart cells. The T-type (T-for transient) is activated and rapidly inactivated at voltages far negative to those regulating the L-type (L-for low threshold) channel (1, 34, 35). The Ca current through the T-type channel is small in amplitude as compared to the current through the L-type channel and nothing is known as yet about its pharmacology and physiological control (1, 34). By contrast, the control of L-type Ca channels by ß-adrenergic stimulation or mucarinic inhibition has been quantitatively studied in recent years (20, 26, 47, for review). Evidence has been presented that phosphorylation of the L-type Ca channel glycoprotein is involved in this control.

The Ca current (whole cell and single channel currents)
Single ventricular cardiac cells from the guinea pig heart isolated by a treatment with collagenase are suitable for the measurement of Ca currents (Fig. 1A). For such measurements, a voltage clamp technique developed by Neher, Sakmann and collaborators (18) is used. A suction glass pipette (internal tip diameter about 2.5 μm) is brought into contact with the cell surface (Fig. 1B) and a tight seal between the cell membrane and the pipette tip is obtained by mild suction (Fig. 1C). This configuration (cell-attached patch) allows to record the current through a single (Fig. 1E) or several (Fig. 1F) ion channels in the patch. The current of the whole cell (Fig. 1D) can be recorded when the patch membrane is destroyed by a larger negative pressure, so that the pipette interior is in free communication with the interior of the cell (Fig. 1D). Under this condition various agents,

including enzymes, can be applied intracellularly by
diffusion from the electrode tip into the cytoplasm. The
solution in the tip can be changed through a fine
polyethylene tube which is moved into the tip of the patch
pipette(Fig. 1D). The techniques are well described in the
literature (26, 42, see also Hescheler et al. this volume).

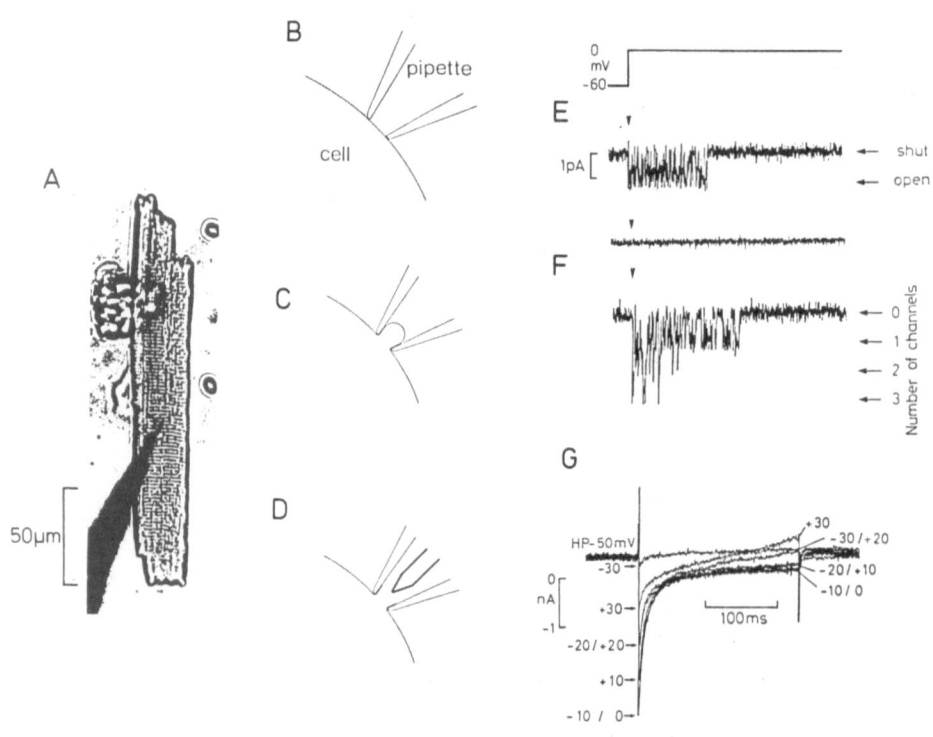

Fig. 1

A) Ventricular cell from the adult guinea pig isolated by a
collagenase treatment. The shadow of the suction pipette is
seen. B) Suction pipette in close contact with the sarcolemma
before Gigaohm seal formation between sarcolemma and
glasswall C). D) After rupture of the membrane patch. E)
Upper panel voltage clamp pulse across the membrane patch
(configuration as in C) and single channel current (middle
trace) with activity or without activity (blank, lower
trace). F) was recorded from another patch which contained at
least 3 single channels as seen by the superposition of their
activity. G) Superimposed whole-cell currents (recorded in
the configuration D) in response to different test potentials
as indicated in the figure. For recording the currents in G)
pipettes solution contained (in mM) KCl 50, K-aspartate 100,
$MgCl_2$ 1,0, HEPES 5, K_2ATP 3, EGTA 0.1, and several protease
inhibitors. For recording single channel currents the
pipettes were filled with $BaCl_2$ 90, $NaCl_2$ 2, KCl 4, HEPES 5,
TTX 0.2. Ba ions were used as charge carriers through the Ca
channel because their permeability is about twice as large as
that for Ca ions. The extracellular solution in all
experiments contained NaCl 140, KCl 5.4, CaCl 3.6, $MgCl_2$ 1,

glucose 10, HEPES 5. For technical details see 18, 24, 26.
Whole-cell Ca currents in response to step depolarization of
sufficient amplitude (threshold -35 mV) rise within a few ms
to a peak (activation) from which they decay in a much slower
time course (inactivation). The amplitude of the Ca current
depends among other factors on the membrane potential (see 46
for review). Fig. 1G shows superimposed Ca curents in
response to depolarizations to different test potentials.
There is a bell-shaped relation between the amplitude of the
voltage-clamp pulse and the amplitude of the Ca current: the
current is small at -30 mV, the maximal amplitude is reached
between 0 and 10 mV and at potentials more positive the
current amplitude decreases. Such Ca currents are the sum of
the activity of a large number of channels in the surface of
the cell (see 46, for numbers 33). The current through a
single channel recorded with a patch electrode in the 'cell
attached' configuration (Fig. 1C) is characterized by a rapid
flickering between the closed and open states (Fig. 1E, upper
pannel). If only a single channel is present in the patch
membrane this channel does not always open on depolarization,
i.e. blank traces occur (Fig. 1E, lower, pannel). During a
large series of depolarizations both types of traces, with
activity and blanks are grouped (9). When the patch contains
at least 3 channels up to 3 current levels of 3 times the
unitary amplitude are observed. These amplitudes correspond
to superpositions of openings of the 3 channels (Fig. 1F). In
line with the time course of the macroscopic current recorded
from the whole cell, the opening probability of the channel
is high early on depolarization and decreases during
maintained depolarization (7, 9). This decrease
(inactivation) is seen either as a sudden closure of the
channel (Fig. 1E) or a decrease in the number of
superpositions prior to closure (Fig. 1F).

Effect of neurotransmitters on the Ca channel

Adrenaline increases the amplitude of the whole-cell Ca
current (Fig. 2A, B, 38, 47). The larger Ca uptake on ß-
adrenergic stimulation is causally related to the increase in
both rate of beating and force of contraction of the cardiac
cell. At the level of the single channel, the following
changes could account for the increase in the amplitude of
the Ca current. 1) A larger single channel current (i) 2)
recruitement of channels, i.e. an increase in the total
number of channels (N_t) 3) a prolongation of individual open
times and a shortening of closed times leading to an
increased open probability (p_o) in traces with activity or 4)
an increased probability (p_f) of the channel to open on
depolarization leading to a reduction in empty traces
(blanks) in a series of depolarizations. The 4 quantities
determine the macroscopic current (I) as follows:
$I = N_T \times i \times p_o \times p_f$ (Cavalié et al. 1986; Tsien et al. 1986).

On ß-adrenergic stimulation a smaller number of blanks in an
ensemble of current traces is observed (increased p_f, compare
Fig. 2Ca with Da). In addition the fraction of time the
channel spends in the open state (p_o) during a depolarization
step is increased due to prolongation of open channel
lifetime and abbreviation of short times between channel
openings (5, 6). The dots mark the blanks (p_f) in a sequence
of about 190 depolarizations whereas the horizontal lines

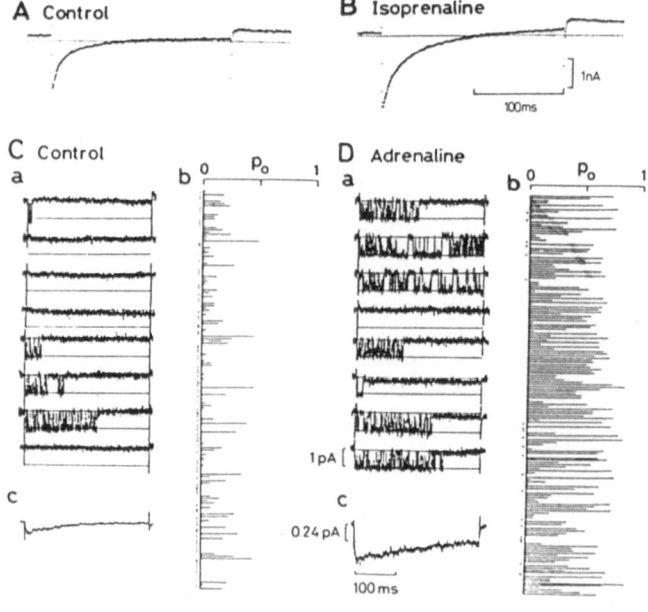

Fig. 2

A) and B) Whole-cell current elicited by a pulse from -38 mV
to +1 mV (duration 300 ms) in the absence A) and in presence
B) of 5×10^{-8} M ISP. C) and D) Activity of a single Ca channel
in response to pulses from -65 mV to +5 mV before Ca) and
after Da) application of 10^{-6} M adrenaline. The pipette
solution contained 90 mM $BaCl_2$. Open and closed level of the
current are shown by the solid lines. From 180 consecutive
traces open probability (p_o) during the pulse is calculated
and shown by horizontal bars in Cb) and Db). Dots indicate
traces without activity. The ensemble average current of each
180 traces are shown in Cc) and Dc).

indicate the fraction of open time (p_o) during each
depolarization. It is obvious that the average over the
series of the single channel currents with the smaller number
of blanks is of larger amplitude than the average current of
the control traces (Fig. 2Dc compared with Fig. 2Dc). It
should be noted that the amplitude of the single channel
current is not altered by adrenaline nor is any evidence for
an increase in the total number of channels. The increase in
p_o alone could maximally increase the average current by a
factor of 1.5 (47). Thus, the 3-4-fold increase in the
average (Fig. 2Cc and Dc) and whole-cell current (Fig. 4A) is
primarily caused by the increase in channel availability
(p_f).

Mechanism of ß-adrenergic stimulation
ß-adrenergic stimulation does not directly affect the channel
buth rather acts through a cascade of reactions with a final
phosphorylation of the channel protein (12, 38, for review
see 45) (see Fig. 3).

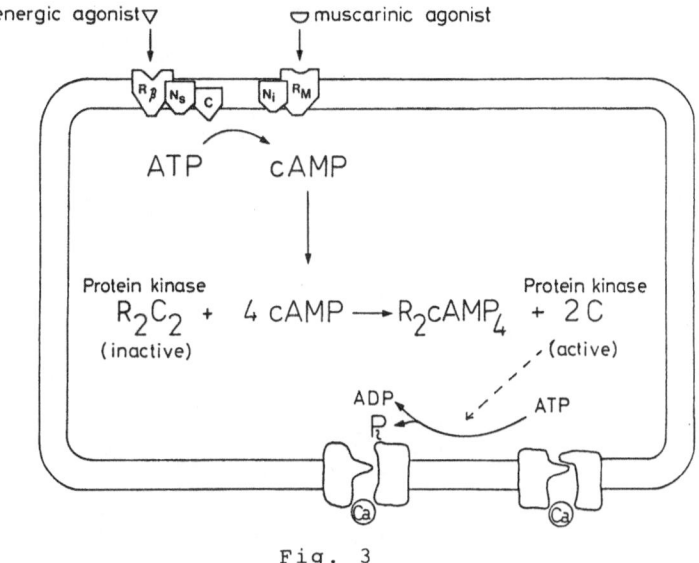

$$R_2C_2 + 4\,cAMP \longrightarrow R_2cAMP_4 + 2\,C$$

Fig. 3

Schematic diagramm of the sequence of reactions between the agonist binding and phosphorylation of the channel protein. Explanation in the text.

Binding of the agonist to the receptor activates through an N protein the membrane-bound adenylate cyclase resulting in an increase of the intracellular cAMP concentration. cAMP dissociates the inactive cAMP-dependent protein kinase (cAMP-PK) into 2 regulatory and 2 active catalytic subunits which phosphorylate the channel protein. Evidence for the operation of the cascade in the control of the Ca current came from experiments in which heart cells were loaded with defined concentration of either cAMP or catalytic subunit of cAMP-PK by diffusion from the pipette tip into the cell. Intracellular application of both cAMP or catalytic subunit increased the Ca current in much the same way as did adrenaline or ISP when applied from outside (26, see insets in Fig. 4CD).

The dose response curve in Fig. 4B suggests that an increase of cAMP above 1 µM enhances the Ca current amplitude which reaches the maximum value at an intracellular concentration of about 15 µM. The threshold concentration for the catalytic subunit in Fig. 4C is 0.3 µM and the maximum effect on the Ca current amplitude is seen with 3 µM catalytic subunit in the pipette. The 3 interventions, ISP application, and infusion of cAMP or catalytic subunit increased the density of the Ca current from a control value around 10 µA/cm^2 to a maximum value of around 30 µA/cm^2. The effects of the 3 agents are not additive at maximal concentrations as it should be expected when they converge on the same site of the Ca channel.

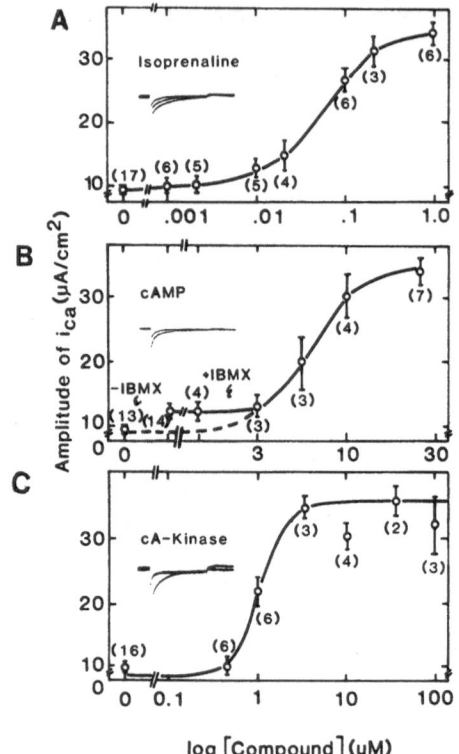

Fig. 4

Relation between the density of the Ca current and the bath
concentration of isoprenaline A), the cAMP concentrations in
the pipette B) and the C subunit concentration in the pipette
C). Current density equals the current divided by the
membrane surface. The surface was obtained by measuring the
membrane capacity; specific capacity 1 μF/cm^2. All
concentrations in logarithmic scale. Data expressed as
mean±SEM (number of experiments). In B) the pipette
contained, in addition to cAMP, 10 μM IBMX
(Isobutylmethylxantin) in order to avoid fluctuations in the
current amplitude at cAMP concentrations below 3 μM. IBMX
increased the Ca current in absence of cAMP. Above 3 μM cAMP
the Ca current was only little affected by 10 μM IBMX. Broken
line indicates the estimated relation between the Ca current
and low concentrations of cAMP in the absence of IBMX. The
insets in A), B) and C) show original current traces (from
26).

The effect of ATP analogues and protein kinase inhibitor on the Ca current.

Since phosphorylation of a protein by the cAMP-PK requires
ATP as the phosphate donor, one might expect that
intracellular application of a non-hydrolysable ATP-analogue,
AMP-PNP (adenylyl-imidodiphosphate), diminishes the effect of
ISP on the Ca current. Such a depression of the ISP-response
is shown in Fig. 5A where 10^{-7} M ISP was applied prior and

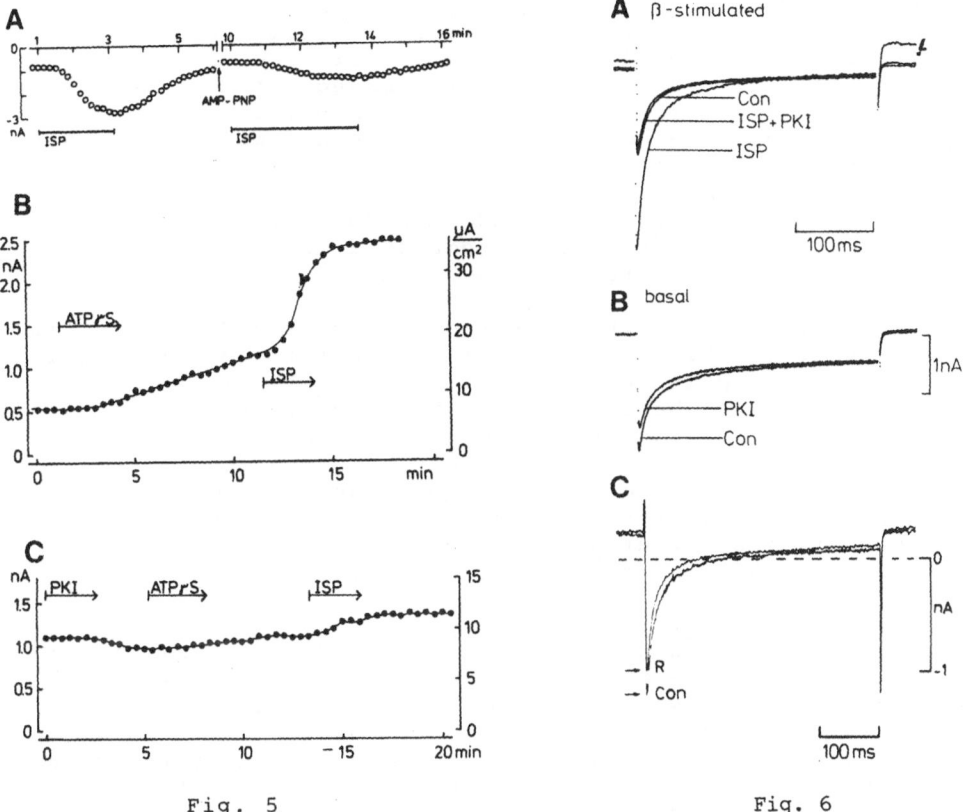

Fig. 5

Fig. 6

Fig. 5. Effect of intracellularly applied ATP analogues on the aplitude of the CA current (left ordinate in A) C)). In B) and C) the Ca current in response to 10^{-7}M ISP (indicated by bars) before and after cell loading with AMP-PMP. B) Time course of the amplitude of the Ca current before and during loading the cell with ATP S. The ISP concentration which produced a maximum increase in Ca current was 10^{-8}M. C) Same experiment as in B), but during continuous diffusion of the protein kinase inhibitor (PKI) from the pipette tip into the cell. A) from 26; B) and C) from 27.

Fig. 6. A) Effect of intracellularly applied inhibitor of the cAMP-dependent protein kinase on the ISP-enhanced Ca current and B) on the control current in the absence of ISP. The ISP concentration ws 10×10^{8}M, the pipette concentration of PKI was 1 μM C) Effect of regulatory subunit on cAMP-PK on the control current. Pipette concentration of R subunit was 40 μM.

after insusion of AMP-PNP into the myocyte. This result in consistent with the idea that the hydrolysis of the phosphate of ATP is involved in the actionof ISP on the Ca channel since AMP-PNP has been reported to be catalysed by adenylate cyclase to form cAMP and imidodiphosphate (51). A similar result was obtained, when insted of ISP application, the catalytic subunit of cAMP-PK was intracellularly applied to a cell loaded with 5 mM AMP-PNP. Just the opposite effects, that is, an enhancement of the ISP response, was observed when cells were loaded with the dephosphorylation-resistent ATP S (26). Diffusion of ATP-S from a pipette filled with 5 mM into the cell resulted in a slow steady increase in the Ca current (Fig. 5B). Under this condition, a threshold concentration of ISP produced a maximal response i.e., an increase in the amplitude of the Ca current by a factor of 3.5.

Complete suppression of the ISP-response on the Ca current is seen in cells in which a protein kinase inhibitor was internally applied (PKI, 1 μM in the pipette, Fig. 6A). A similar, but weaker effect, was observed on internal application of the regulatory subunit of cAMP-PK in concentrations of 10-14 μM. The regulatory subunit of the cAMP-PK was reported to bind and inactivate the catalytic subunit (30). These results support the view that the increase in the amplitude of the Ca current by β-adrenergic stimulation is brought about by a phosphorylation of the channel catalysed by cAMP-PK.

Phosphorylation and the amplitude of the basal Ca current
AMP and PNP, PKI, and R subunit did not reduce the amplitude of the basal Ca current by more than 20% (see Fig. 5A, 6B,C). This observation raises 2 interesting points: 1) There is a basal cAMP-PK activity which can contribute to the regulation of the channel. The approximately 20% depression of the basal Ca current may reflect a basal adenylate cyclase activity (basal cAMP level) in the absence of β-adrenergic stimulation (2, 10). 2) For the remaining 80% of basal Ca current, phosphorylation by cAMP-PK is not required. This would indicate that phosphorylation by cAMP-PK is not a prerequisite for maintaining a large part of basal Ca current. Contribution of other protein kinases (the activity of which cannot be suppressed by either PKI or R) to the regulation of the Ca channel activity is not likely, since ATP S did not increase the Ca current significantly in the presence of PKI (Fig. 5C compared to the control experiment in 6B). This notion is in line with the observation that protein phosphatases do reverse the enhancement by ISP, but, do not eliminate the Ca current completely (see below). The results support the view that Ca channels in the dephosphorylated state have a certain probability to open, i.e., they are still available to respond on depolarization (38,27).

Downregulation of the phosphorylated Ca channels
The Ca current enhanced by ISP is reversed to the control level by the catalytic subunit of protein phosphatase 1 (PPase-1; c.f. 22, 23, 28, see also Hescheler et al., this volume). This is shown in the experiment of Fig. 7, where 5x10^{-8}M ISP increased the Ca current 2.7-fold to a steady level. When the cell was dialysed with 6.400 U/ml (2.0 μM) PPhase-1 in the presence of ISP, the Ca current amplitude returned to the control within 7-10 min. A smaller concentration of PPase-1 (700 U/ml 0.22 μM) reduced the ISP-induced increment in the amplitude of the Ca current by about 50% suggesting a dose-dependent nature of the PPase-1 action. PPase-1 is an intrinsic enzyme of heart muscle

(23). The concentration which in myocyte is probably similar to that in skeletal muscle (about 0.5 µM, see 23) is in the same range as used in the experiment of Fig. 7. It should be noticed that the basal amplitude of the CA current in Fig. 7 was only little affected by PPase-1, corroberating the hypothesis that even in the dephosphorylated state Ca channels are available to open on depolarization.

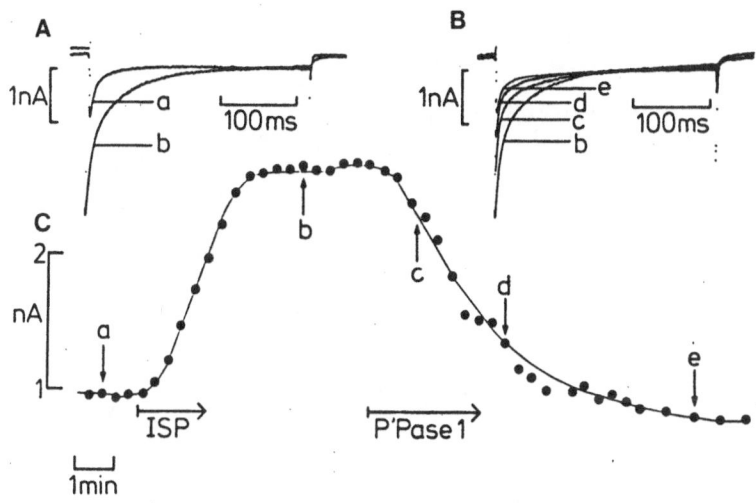

Fig. 7

Intracellular application of PPase-1 suppresses the ISP-enhanced Ca current (left ordinate, right ordinate Ca current density). The superimposed current traces on top were recorded during the experiment at the times indicated below (a-e). Voltage pulses from -40 mV to 0 mV. (From 28).

Phosphorylation of Ca channels reconstituted in lipid planar bilayers

In the living cell Ca channels can be blocked by organic compounds like dihydropyridines and phenylalkylamines (8, 21, 32). The tritiated congeners of these blockers bind in vitro to sites which have the same pharmacological characteristics as those observed in vivo (stereospecific and allosteric interaction between distinct sites) (17, 40). For reconstitution of the Ca channel into lipid bilayers, dihydropyridine-binding sites were purified from T-tubules of skeletal muscle (for method see 14). In this structure, the density of binding sites is about 10 times larger than in heart and hence, a larger yield of channel protein after purification is obtained (14, 17, 41, 44). Ca currents which are blocked by dihydropyridines can be recorded in intact skeletal muscle fibers (37). The nitrendipine receptor was purified about 500-fold to an apparent density of 2 nmol/mg protein. Gel electrophoresis of the purified receptor yielded 3 major peptides of an apparent M_r 142 KD, 56 KD and 31 KD. In the presence of a reducing agent, 3 additional peptides of

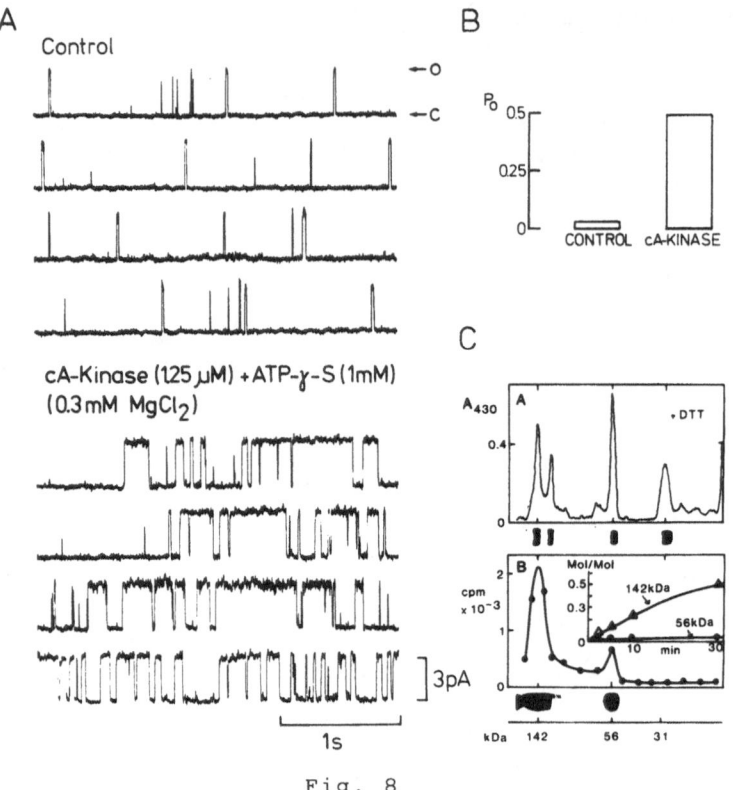

Fig. 8

A) Steady-state recordings of spontaneous single channel
openings, control and after addition of catalytic subunit and
ATP S to the bath. B) Increase in open probability by the
catalytic subunit. C) Analysis of the purified nitrendipine-
binding site on an SDS Gel. Densitometric scan of a silver-
stained Gel. The areas under the 142, 122, 56 and 31 KD
peptides have a ratio of 1.7x1x2.1x1. B) Phosphorylation of
150 ng purified receptor by 7 nMol pure catalytic subunit of
cAMP kinase in the presence of 3 μM ^{32}P-ATP (86 c.p.m.
fMol^{-1}. For inset see text. (From 15).

M_r 122 KD, 26 KD and 22 KD of unknown function were observed
(see Fig. 9CA, 14, 15). The complete receptor complex was
reconstituted into liposomes prior to incorporation into
lipid bilayers formed at the tip of patch pipettes (15, 19).

Channels open spontaneously (Fig. 8, upper panel), when a
potential gradient is applied across the bilayer. The
conductance of the channel is about 20 pS as it is for the Ca
channel in vivo for comparable ionic conditions (7). These
channels can be phosphorylated by adding 1 μM catalytic
subunit of cAMP protein kinase and 1 mM ATP S to the bath.
Under this condition, the channel activity is greatly
increased (Fig. 8A) The time, the channel spends in the open
state is strongly prolonged, and the closed times are
shortened (Fig. 8A,B). This result is very similar to that
obtained in single Ca channel measurements in vivo after
application of adrenaline (see Fig. 2).

170

The 142 KD and 56 KD peptides can be phosphorylated by cAMP kinase (Fig. 8CB, for ref. see 15). The apparent specificity of the phosphorylation was tested in the presence of low concentrations of catalytic subunit of cAMP protein kinase (7 nM) and ATP (3.1 μM). The peptides had different initial phosphorylation rates (Fig. 9CB inset). The cAMP kinase incorporates 1 and 0.1 mol phosphate per mol 142 KD and 56 KD peptide, respectively. This suggests that the phosphorylation of the 142 KD peptide results in the effect on the single channel shown in Fig. 9A and 3.

Muscarinic inhibition of the Ca current
Acetylcholine has little effect, if any, on the control Ca current of ventricular cardiac cells. If, however, the current is enhanced by ß-adrenergic stimulation, this enhancement is reduced by ACh. The dose-response curves in Fig. 9 show that ACh does neither depress the control Ca current nor the maximally stimulated current. However, 10^{-5}M ACh shifts the relation between the half maximal ISP concentration and the Ca current density by roughly one decade to a lower concentration (20, 36). In biochemical studies, ACh was found to decrease ISP or epinephrine-enhanced cAMP levels (3, 13, 31) and it was assumed that ACh inhibits the generation of cAMP and consequently the concentration of active catalytic subunit of the cAMP-dependent kinase (see scheme in Fig. 3). In line with this assumption, ACh cannot depress the Ca current enhanced by intracellular application of catalytic subunit or cAMP (20). The lack of an ACh-effect on the Ca current under this condition does exclude the possibility that ACh does antagonize the effects rather than the generation of cAMP, or does directly inhibit the cAMP-PK. In fact, in rat hearts, a significant lowering of the cAMP-dependent protein kinase activity by ACh was not observed (29). Thus, the bulk of biochemical and electrophysiological findings suggests that ACh depresses the ISP-enhanced Ca current by an effect on or prior to the adenylate cyclase (see scheme Fig. 3).

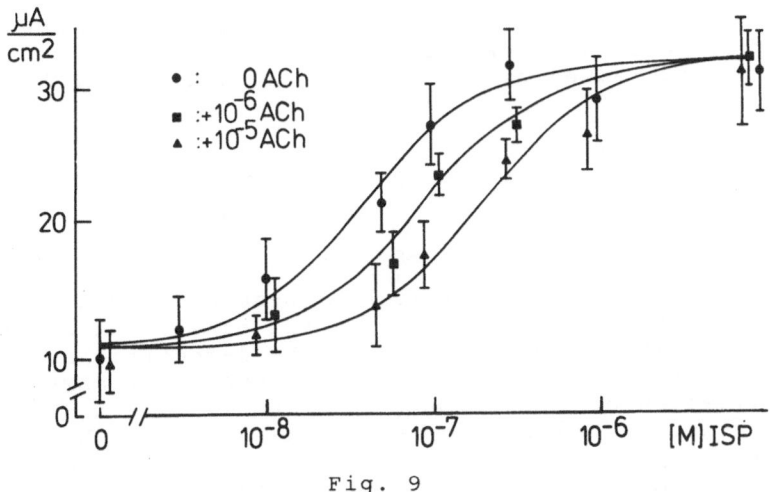

Fig. 9

Relationship between ISP-concentration (abscissa and Ca current density (ordinate) at different ACh concentrations.

171

The adenylate cyclase is controlled by stimulatory (N_s) and inhibitory (N_i) transducer proteins which couple the stimulatory (ß-adrenergic) and inhibitory (muscarinic) signals from the receptors to the catalytic subunit of the adenylate cyclase. Both proteins have GTPase activity (11, 16, 25, 39), i.e., they hydrolyse GTP during the turnover between activated and inactivated states (39).

When GTP is replaced intracellularly by the nonhydrolysed GTP analogue GMP-PNP, both transducer proteins should stay persistently activated and the signal transmission from both, ß-adrenergic and muscarinic receptors to the catalytic subunit of the adenylate cyclase should be interrupted (43). In our experiments (20), after cell dialysis with GMP-PNP, both, ISP and ACh failed to depress the Ca current. The latter observation is in line with the hypothesis that in ventricular cells the muscarinic action is mediated by a protein with GTPase activity, presumably the N_i protein. A test of this hypothesis is to incubate cells in pertussis toxin (PT) which was found to block specifically the inhibitory transducer protein by ADP ribosylation (48, 49, 50). In the pertussis-treated cell, an increase in amplitude of the Ca current by ISP was found (suggesting an intact N_s protein), but ACh failed to antagonize the enhancement (Fig. 10, 20, see also 4).

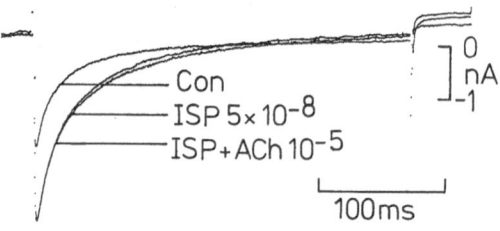

Fig. 10

Superimposed current traces recorded from a cell incubated in 5×10^{-7} g/ml pertussis toxin for 5 h at 23°C. CON, control Ca current is increased by ISP but not depressed by ACh. (From 20).

Conclusion

The modulation of the Ca channel by ß-adrenergic stimulation has been elaborated quantitatively and can be summarized as follows. Binding of the agonist to the ß-receptor adenylate cyclase activity cAMP-free catalytic subunit of cAMP-PK phosphorylation of the Ca channel protein Ca flux. The final phosphorylation step of the cascade has been shown at the level of the purified channel protein reconstituted in an artificial membrane and correlated to functional changes, i.e. to an increase of Ca channel activity. The muscarinic inhibition of the CA current was shown to be due to stimulation of the transducer protein N_i linking the muscarinic receptors to the adenylate cyclase.

REFERENCES

1. Bean, B.P. (1985) J. Gen. Physiol. 86, 1-30.

2. Beavo, J.A., Bechtel, B.J., Krebs, EG. (1974b) Proc. Natl. Acad. Sci. USA. 71, 3580-3583.

3. Biegon, R., Epstein, P., Pappano, A. (1980) J. Pharmacol Exp. Ther. 215, 348-356.

4. Breitwieser, G., Szabo, G. (1985) Nature, 317, 538-540.

5. Brum, G., Osterrieder, W., Trautwein, W. (1984) Pflügers Arch. 401, 111-118.

6. Cachelin, A.B., De Peyer, J.E:, Kokubun, S., Reuter, H. (1983). Nature, 304, 462-464.

7. Cavalié, A., Ochi, R., Pelzer, D., Trautwein, W. (1983) Pflügers Arch. 398, 284-297.

8. Cavalié, A., Pelzer, D., Trautwein, W. (1985) J. Physiol. Lond. 358, 59 P.

9. Cavalié, A., Pelzer, D., Trautwein, W. (1986) Pflügers Arch. 406, 241-258.

10. Corbin, J.D., Keely, S.L. (1977) J. Biol. Chem. 252, 910-918.

11. Codina, J., Hildebrandt, J., Sunyer, T., Sekura, R., Manclark, C., Iengar, R., Birnbaumer, L. (1984). Adv. Cyclic. Nucleotide Res. 17, 111-125.

12. Drummond, G.I., Severson, D.L. (1979)Circ. Res. 44, 145-153.

13. Endoh, M. (1980) Naunyn-Schmiedeberg's Arch. Pharmacol. 312, 175-182.

14. Flockerzi, V., Oeken, H.J., Hofmann F. (1986) Eur. J. Biochem. 155, 613-620.

15. Flockerzi, V., Oeken, H.J., Hofmann, F., Pelzer, D., Cavalié, A., Trautwein, W. (1986) Nature, 323, 66-68.

16. Gilman, A. (1984) Cell 36, 577-57917.Glossmann, H., Ferry, D.R., Goll, A., Striessnig, J., Zernig, G. (1985) Arzneim-Forschung/Drug. Res. 35, 1917-1935.

18. Hamill, O.P., Marty, A., Neher, E., Sakmann, B., Sigworth, J. (1981) Pflügers Arch. 391, 85-100.

19. Hanke, W., Methfessel, C., Wilmsen, U., Boheim, GC. (1984) Biolectrochem. Bioenerget. 12, 329-339.

20. Hescheler, J., Kameyama, M., Trautwein, W. (1986) Pflügers Arch. 407, 182-189.

21.Hess, P., Lansman, J.B., Tsien, R.W. (1984) Nature, 311, 538-544

22.Ingebritzen, T.S., Cohen, P. (1983) Eur. J. Biochem. 132, 255-261.

23.Ingebritzen, T.S., Stewart, A.A., Cohen, P. (1983) Eur. J. Biochem. 132, 297-307.

24.Isenberg, G., Klöckner, U. (1982) Pflügers Arch. 395, 30-41.

25.Jacobs, K., Aktories, K., Schultz, G. (1979) Naunyn-Schmiedeberg's Arch. Pharmacol. 310, 113-119.

26.Kameyama, M., Hofmann, F., Trautwein, W. (1985) Pflügers Arch. 405, 285-293.

27.Kameyama, M., Hescheler, J., Hofmann, F., Trautwein, W. (1986) Pflügers Arch. 407, 123-128.

28.Kameyama, M., Hescheler, J., Mieskes, G., Trautwein, W. (1986) Pflügers Arch. 407, 461-463.

29.Keely, S., Lincoln, T., Corbin, J. (1978) Am. J. Physiol. 234(4), H432-H438.

30.Krebs, E.G., Beavo, J.A. (1979) Ann. Rev. Biochem. 48, 923-959.

31.Linden, J., Hollen, C., Patel, A. (1985) Circ. Res. 56, 728-735.

32.McDonald, T.F., Pelzer, D., Trautwein, W. (1984) J. Physiol. Lond. 352, 217-241.

33.McDonald, T.F., Cavalié, A., Trautwein, W., Pelzer, D. (1986) Pflügers Arch. 406, 437-448.

34.Mitra, R., Morad, M. (1986) Proc. Natl. Acad. Sci. USA, 83, 5340-5344.

35.Nilius, B., Hess, P., Lansmann, J., Tsien, R.W. (1985) Nature (London) 316, 443-446.

36.Ochi, R. (1981) Academic Press New York, pp 79-86.

37.Palade, P.T., Almers, W. (1985) Pflügers Arch. 405, 91-101.

38.Reuter, H., (1983) Nature, 301, 569-574.

39.Rodbell, M. (1980) Nature 284, 151-159.

40.Ruth, P., Flockerzi, V., v. Nettelbladt, E., Oeken, J., Hofmann, F. (1985) Eur. J. Biochem. 150, 313-322.

41.Ruth, P., Flockerzi, V., Oeken, H.J., Hofmann, F. (1986) Eur. J. Biochem. 155, 613-620.

42.Sakmann B., Neher, E. (eds) Single Channel Recording, Plenum (New York).

43.Selinger, Z., Cassel, D. (1981) Adv. Cyclic Nucleotide Res. 14, 15-22.

44.Trautwein, W. (1987) Circ. Res. (in press).

45.Tsien, R.W. (1977) Adv. Cyclic. Nucleotide Res. 8, 363-420.

46. Tsien, R.W. (1983) Ann. Rev. Physiol. 45, 341-358.

47. Tsien, R.W., Bean, B.P., Hess, P., Lansman, J.B., Nilius, B., Novicky, M.C. (1986) J. Mol. Cell Cardiol. 18, 691-710.

48. Ui, M. (1984) TIPS (1984) 277-279.

49. West, R., Moss, J., Vaughan, M., Liu, T., Liu, T.Y. (1986) J. Biol. Chem. 260, 15718-15722.

50. Wolff, J., Hope Cook, G., Goldhammer, A., Londons, C., Hewlett, E. (1984) Adv. Cyclic Nucleotide Res. 17, 161-172.

51. Yount, R.G. (1975) Adv. Enzymol. 34, 1-56.

REGULATION OF THE Ca-CHANNEL BY PHOSPHORYLATION-DEPHOSPHORYLATION

J. Hescheler[1], M. Kameyama[1], W. Trautwein[1],
G. Mieskes[3], and F. Hofmann[2]

[1]II. Physiologisches Institut
[2]Physiologische Chemie, Universität des Saarlandes
 D-6650 Homburg/Saar, SFB 246
[3]Klinische Biochemie, Universität Gottingen
 D-3400 Gottingen, SFB 238

INTRODUCTION

Similar as other second messengers, Ca^{++} ions can trigger many intracellular enzymatic reactions (e.g. activation of the contractile filaments, 3,4). One of the strategic points for regulation of the intracellular Ca concentration are Ca channels, transmembrane proteins, which have -under physiological conditions- a high selectivity for conducting Ca ions. Beside their voltage dependency, it was found that the Ca channels are also substrates for different enzymatic reactions (13,14). Here we report that cAMP-dependent protein kinase (cAMP-PK) can increase the Ca current (I_{Ca}) amplitude by phosphorylation of a channel related protein. The phosphorylation is antagonized by several phosphatases.

PREPARATIONS AND ELECTROPHYSIOLOGICAL METHODS

Experiments were carried out on single cardiac myocytes prepared from guinea pig heart by collagenase treatment (for detail see 6,9). As shown in the microscopic picture (Fig. 1B) the myocytes were typically rod shaped and striated. Cells were voltage clamped and the membrane currents recorded with the 'whole-cell-clamp' technique (15, see Fig. 2 in Trautwein, this book). Because of the free exchange between pipette solution and the cytoplasm after disruption of the membrane patch under the pipette's tip, this technique also allowed to 'infuse' proteins and other agents into the cell. From a computer simulation using a simplified compartment model (for detail see 10,12), we would expect that small molecules like cAMP or ATP-derivatives (MW around 500, diffusion coefficient around $5x10^{-6}$ cm^2/s) reach 90% of the pipette's concentration after about 4 min, while large molecules (e.g. catalytic subunits of the cAMP dependent kinase or phosphatases, MW around 50,000, diffusion coefficient around $1.5x10^{-6}$ cm^2/s) need more than 15 min to enter the cytoplasm (see Fig 1C).

RESULTS AND DISCUSSION

Voltage clamp pulses from -40 to 0 mV evoked a Ca inward current of about 1 nA amplitude (measured as peak inward current) and a relatively fast (100 ms) inactivation (1.8 mM Ca^{++}). When 3.6 mM Ba^{++} was used as charge

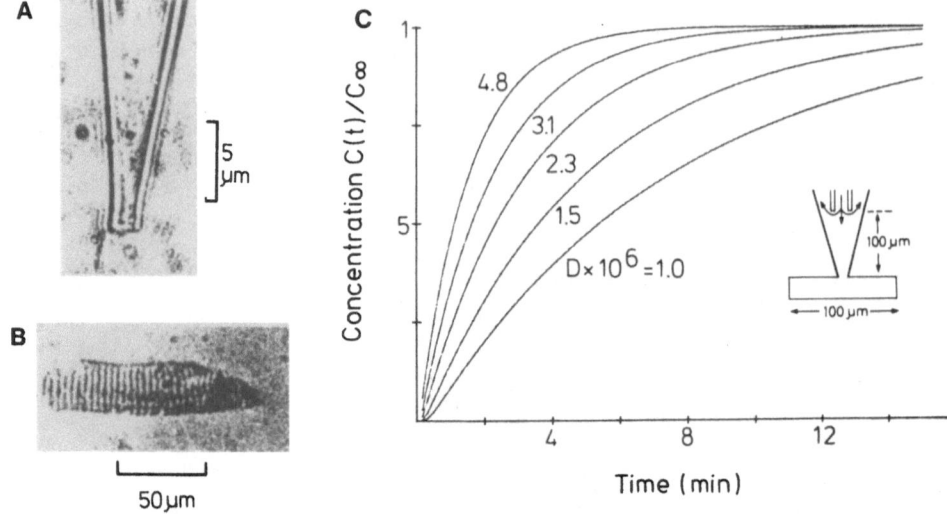

Figure 1: **A)** microscopic picture of a glass pipette, opening diameter approximately 1.3 μm. The pipettes were filled with a cytoplasma resembling solution: (concentrations in mM) K-aspartate 100, KCl 50, $MgCl_2$ 2, EGTA 0.1, K_2-ATP 3, creatine phosphate 5, creatine phosphokinase 20 U/ml, HEPES 5; pH was adjusted to 7.4 by KOH (36°C). In some experiments, several protease inhibitors were added to prevent proteolytic digestion of the enzymes in the pipette solution.
B) microscopic picture of an isolated myocyte of the ventricle of the guinea pig heart. The cell was on the bottom of a small chamber which was continuously perfused with a modified Tyrode's solution containing (in mM): NaCl 112, $NaHCO_3$ 24, KCl 5.4, $CaCl_2$ 1.8, $MgCl_2$ 1, glucose 10, HEPES 5. The solution was gassed with 95% O_2 and 5% CO_2; the pH was 7.4 (36°C).
C) computer simulation of the diffusion of molecules from the pipette into the cytoplasm. Shown is the time course of the rise of the intracellular concentration (for detail see 10,12)

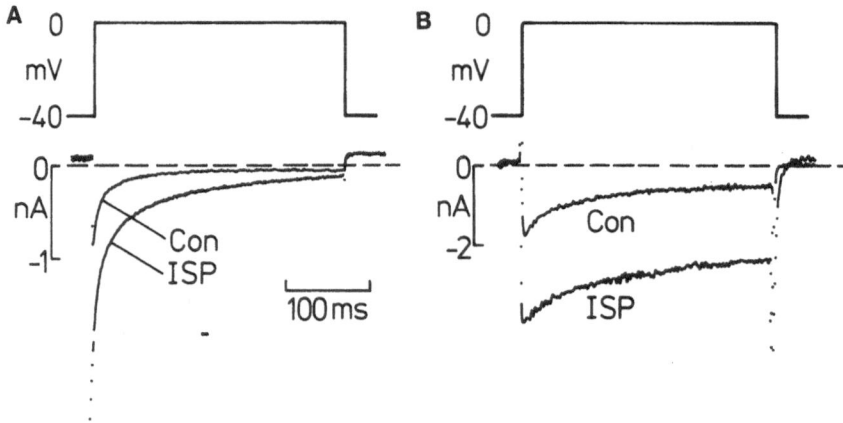

Figure 2: **A)** ISP (100 nM) was extracellularly applied and increased the peak Ca^{++} current. The extracellular Ca^{++} concentration was 1.8 mM.
B) same experimental protocol as in A, but Ca^{++} was replaced by 3.6 mM Ba^{++}.

carrier through the channel, a larger amplitude and a slower time course of inactivation were observed. Extracellularly applied isoprenaline (ISP) greatly increased the peak amplitude, up to 3-4 fold at maximal concentrations. A slowing of the inactivation, which was better seen under the Ba^{++} condition, correlates with the prolonged open times in single channel recordings (see Trautwein, this volume).

By analogy with the ß-adrenergic regulation of glycogen metabolism, we hypothetised that these ISP effects on the Ca current are also mediated by a cAMP dependent phosphorylation. This hypothesis was supported by the finding that the ISP effects were mimicked by intracellularly applied cAMP, by the catalytic subunit of the cAMP-PK (C-SU) and by ATP-v-S (17). As shown in Fig. 3A all compounds led to a maximal enhanced I_{Ca}, which was about 3 fold the control amplitude. ISP, cAMP, C-SU and ATP-v-S effects were non additive at high concentrations but additive at submaximal concentrations. A slowing of I_{Ca} inactivation course by ATP-v-S was observed by Chad and Eckert (2). Taken together, these data suggest a

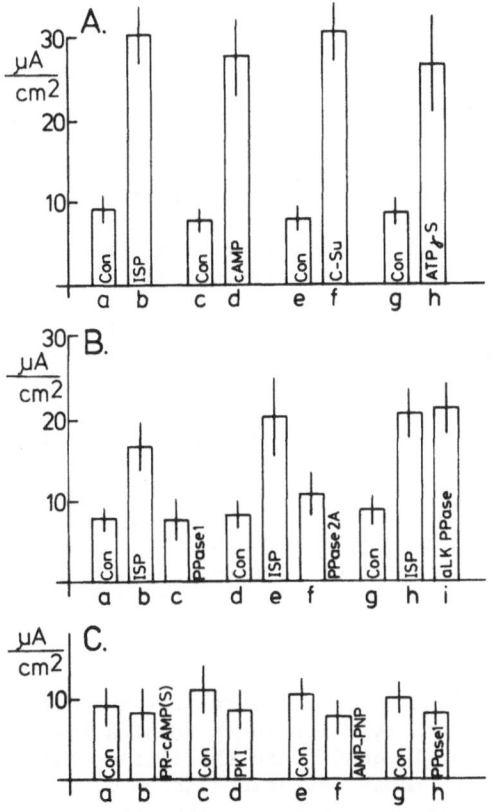

Figure 3: (A), (B), (C) mean Ca current densities ($\pm$SD). The current density was obtained by division of the I_{Ca} amplitude by the cellular surface ($_2$= cell capacity times $1cm^2/\mu F$)
A) ISP (1 µM), cAMP (33 µM), C-SU (10 µM) and ATP-v-S (3 mM) increased the I_{Ca} density about 3 fold.
B) PPase 1 (2 µM) and 2A (2.3 µM) antagonized the ISP (50 nM) effect on I_{Ca}. Alkaline PPase was ineffective.
C) PR-cAMP(S) (100 µM), PKI (2 µM), AMP-PNP (5 mM) as well as PPase 1 (2 µM) depressed the basal I_{Ca} by only about 20%.

cascade mechanism of the components, which ends up at a final reaction step on the Ca channel. Beta-adrenergic receptors stimulate the adenylyl cyclase leading to an increased intracellular cAMP level. The increased cAMP activates the cAMP-dependent protein kinase which phosphorylates a protein related to the Ca channel (10, 13, 14). Recent experiments with purified channels suggest that this 'related protein' is identical to a subunit of the Ca channel protein (5).

The ISP effect on I_{Ca} is rapidly reversed when the ß-adrenergic agonist is washed out of the bath (10). If the phosphorylation theory is correct, Ca channels have to be dephosphorylated during that phase (7,11). We

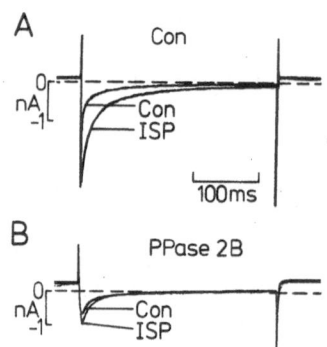

Figure 4: A) Control experiment, 50 nM ISP increased I_{Ca} about 2-fold.
B) The cell was infused with approximately 0.1 μM PPase 2B for 15 min. ISP (same concentration as in A), which was thereafter superfused, was markedly less effective than in the control experiment.

therefore looked for phosphatases (PPases) which can antagonize the cAMP dependent phosphorylation. Fig. 3B shows that the catalytic subunits of the protein specific PPase 1 and 2A (Cohen's classification, 7,8) can decrease I_{Ca} of a ß-adrenergically stimulated myocyte. In other words, the intracellularly applied PPases (in μM range) outweighted the cAMP dependent phosphorylation. Alkaline and acid PPase, as well as heat inactivated PPase 1 and 2A had no effect on ß-adrenergically stimulated I_{Ca} (11).
In a recent experiment we found that the Ca dependent PPase 2B (identical with calcineurin) might also dephosphorylate the Ca channel. Fig. 4 shows in A the increase of I_{Ca} by 50 nM ISP (about 2 fold). Another cell (Fig. 4B) was first infused with PPase 2B (approximately 0.1 μM). Then the same ISP concentration was superfused, but the effect was much smaller than in the control cell suggesting that PPase 2B is also involved in the Ca channels dephosphorylation. Because of the Ca dependency of this PPase, there might be a negative feedback mechanism, i.e. increased ß-adrenergic stimulation leads to an increased Ca influx into the cell and thereby to an increased dephosphorylation, which again antagonizes the ß-adrenergic stimulation. Similar results with PPase 2B have been reported by Chad and Eckert (2), who found in addition a faster time course of inactivation under PPase 2B.

Is the phosphorylation of the Ca channel a prerequisite for its activity or does the channel has a basal, phosphorylation-independent activity? To answer this question we applied several agents into the intracellular space, which are known to depress the phosphorylation (see Fig. 3C). PR-cAMP(S)(18) as well as the heat stable protein kinase inhibitor (PKI) suppress the activation of the cAMP-dependent protein kinase (1,16); the non hydrolysable ATP analogue, AMP-PNP, prevents phosphorylation and PPase 1 accelerates the dephosphorylation. All compounds decreased I_{Ca} by only 20%, but never abolished it completely. From this observation we concluded that the Ca channel has a certain 'basal' activity in the non-phosphorylated state, which can be increased (about 4 fold) by cAMP dependent phosphorylation(11,12). Whether other enzymatic reactions on the Ca channel can be activated, e.g. by other hormones or neurotransmitters, has to be examined in the future.

REFERENCES

1. Ashby CD., Walsh DA. (1972) J. Biol. Chem. 247:6637
2. Chad J., Eckert R. (1986) J.Physiol. 378:31
3. Cohen P. (1982) Nature 296:613
4. Fabiato A., Fabiato F. (1979) Ann. Rev. Physiol. 41:473
5. Flockerzi V., Oeken HJ., Hofmann F., Pelzer D., Cavalie A.,Trautwein W. (1986) Nature 323:66
6. Hescheler J., Kameyama M., Trautwein W. (1986) Pflügers Arch 407:182
7. Ingebritzen TS., Cohen P (1983) Science 221:331
8. Ingebritzen TS., Cohen P (1983) Eur. J. Biochem. 132:255
9. Isenberg G., Klöckner U. (1982) Pflügers Arch 395:6
10. Kameyama M., Hofmann F., Trautwein w. (1985) Pflügers Arch. 405:285
11. Kameyama M., Hescheler J., Mieskes G., Trautwein W. (1986) Pflügers Arch 407:461
12. Kameyama M., Hescheler J., Hofmann F., Trautwein W. (1986) Pflügers Arch 407:123
13. Osterrieder W., Brum G., Hescheler J., Trautwein W., Flockerzi V., Hofmann F. (1982) Nature, 293:576
14. Reuter H. (1983) Nature 301:569
15. Sakmann B., Neher E. (eds) Plenum Press (New York, 1983)
16. De Wit R., Hoppe J., Stec W., Baraniak J., Jastorff B. (1982) Eur. J. Biochem. 122:95
17. the thiophosphate group donated by this nucleotide is resistant to dephosphorylation
18. PR-cAMP(S) suppresses the dissoziation of the regulatory and catalytic subunit of the cAMP dependent protein kinase and thereby inactivates the enzyme.

III. STRUCTURE FUNCTION RELATIONSHIP OF CASCADE ENZYMES

THE PROTEIN KINASE FAMILY

Kenneth A. Walsh

Department of Biochemistry, SJ-70
University of Washington
Seattle, Washington 98195

An ever-expanding list of protein kinases is now known to serve diverse physiological roles in cellular systems (Table I). By phosphorylating serine, threonine or tyrosine residues in target proteins, these enzymes transduce metabolic or hormonal signals with profound cellular consequences (Krebs & Beavo, 1979; Flockhart & Corbin, 1982; Nishizuka, 1984; Stull et al., 1985). Serine- or threonine-specific protein kinases were first found to exercise control of diverse metabolic processes. Tyrosine-specific protein kinases are now found both as domains of trans-membrane growth factor receptors and encoded by viral oncogenes (Sefton & Hunter, 1985). Thus, some protein kinases are integral membrane proteins that are directly involved in the transduction of extracellular signals for intracellular purposes. Others act as amplifiers of intracellular signals, as regulators of structural or enzymatic components of cells, and as the unregulated products of certain oncogenic lesions.

The protein kinases vary in their specificity. Perhaps the most thoroughly studied is cAMP-dependent protein kinase, which specifically phosphorylates serine or threonine residues that follow an Arg-Arg-X sequence. Others are known to favor acidic regions but many remain uncharacterized in their sequence specificity. Identification of the protein kinase responsible for a physiological response is complicated by the overlapping specificities of the kinases and by the existence of multiple phosphorylation sites in individual target proteins. Perhaps the most striking effects and the clearest roles of protein kinases are their actions as on/off switches for target proteins. However, when multiple phosphorylation sites exist on a single target protein (e.g. glycogen synthase, Kuret et al., 1985), the interaction of primary and secondary phosphorylation sites complicates interpretation of the regulatory process, particularly toward target proteins carrying endogenous phosphoryl groups.

The regulatory action of the protein kinases may be regulated in turn by extracellular and intracellular effectors. These may be mitogens as in the case of epidermal growth factor (EGF), a second messenger such as cAMP, calcium/calmodulin or diacylglycerol, or an intracellular phosphorylation event involving another protein kinase, as in the regulation of phosphorylase kinase (Table I). In each case the protein kinase

Table I

**Members of the Protein Kinase Family
Containing Homologous Segments**

Regulator	Ser/Thr-target	Tyr-target
Cyclic NMP	cAMP-dependent protein kinase	
	cGMP-dependent protein kinase	
Calcium/calmodulin	Phosphorylase kinase	
	Myosin light chain kinase	
Diacylglycerol	Protein kinase C	
Growth factor		Epidermal growth factor receptor
		Insulin receptor
		Platelet-derived growth factor receptor
Not known	Casein kinase	pp60^{src}
		v-erb-B and related onco-proteins

This list is representative only. For example, many other oncogene products are also homologous proteins. Many other protein kinases are of unknown structure at this time. It now appears likely that most protein kinases will prove to be evolutionarily related.

is acting as a molecular amplifier transducing a chemical signal into a physiological response.

Although the actions of certain Ser/Thr-directed protein kinases have been recognized for twenty years and the analogous actions of tyrosine-directed protein kinases for less than ten, it remained for amino acid (or DNA) sequence analyses of the protein kinases to demonstrate that all of the known protein kinases bore evidence of a common ancestral relationship, implying that analogous mechanisms underlie their function (Takio et al., 1984). In 1982 the amino acid sequence of the catalytic subunit of cAMP-dependent protein kinase was compared with that of pp60[src], a tyrosine-specific protein kinase translated from the Rous sarcoma oncogene (Barker & Dayhoff, 1982). Their similarity presented clear evidence of homology of the corresponding genes and prompted an exploration of the detailed sequences of other protein kinases that is continuing today. Several generalities have become clear about the relationships among the protein kinases, and indeed about relationships within other families of homologous proteins. Specifically it appears that such complex proteins are best thought of as alignments of individual substructural domains that may have been assembled by primordial splicing mechanisms. As a result, similarities or differences in molecular weight are very poor guides to the likelihood of homologous relationships. In these respects the protein kinases present an unusually interesting family of proteins to study because homologous catalytic domains are regulated either by separate domains fused to a single polypeptide chain or by different subunits (gene products) associated non-covalently in an oligomeric protein. The homologous catalytic subunits (or domains) respond to quite different modulators (cAMP, calmodulin, diacylglycerol, growth factors, etc.), apparently by virtue of differences in the corresponding regulatory subunits (or domains). In these cases, the chimeric nature of the multidomain kinases facilitates their individualized responses to specific regulators.

Before discussing the details of these relationships within the family of protein kinases, it is important to consider the tools that are used to recognize the homologous relationships between any two amino acid sequences.

DETECTION OF HOMOLOGOUS RELATIONSHIPS AMONG PROTEINS

The word homology has been used in different senses in different disciplines. To a biologist, a homologous relationship denotes the divergence of genotypes from a common ancestor in the course of evolution. To a chemist, ethane and propane are homologous, but of course no evolutionary relationship is suggested. Protein chemists refer to similar, but not identical, amino acid sequences as homologous on the basis of chemical similarity; implicit in that recognition are their origins by divergent evolution from a common ancestral gene. There are now many examples of homologous proteins serving analogous functions in a wide range of organisms (e.g. the hemoglobins). Of even more interest are the findings of single families of homologous proteins serving diverse physiological needs in a single organism (e.g. homologous serine proteases in digestion, blood coagulation, fertilization, fibrinolysis, complement activation, etc.). In these cases the ancestral genetic material must have duplicated to allow the evolution and expression of these related genes in a single cell or organism. From the functional point of view, a demonstration of homology is sufficient for confident prediction of similarity in three-dimensional structures at least at the level of chain folding. Thus it has become accepted that details of the three dimensional structure of one member of a homologous set provides a working model of all members of that set, e.g. the structure of the pancreatic serine proteases provides a working model of the active sites of blood coagulation factors.

If a comparison of two proteins reveals more than 50% of the amino acids in corresponding loci, the similarity of the two proteins is evident and there is little reason to doubt a common ancestral relationship. However, if the percent identity between the two proteins is of the order of 20% or less, it becomes more difficult to assess whether the possibility of an evolutionary relationship is greater than might have been achieved by mere chance alignment of amino acids (Doolittle, 1981). If proteins serve the same function in quite different organisms (for example, cytochrome C in yeast and in humans), conclusions of homology can be reasonably based on sequences in spite of extensive dissimilarity. However, when proteins serve quite different functions, as in the EGF receptor (a trans-membrane protein that phosphorylates tyrosines in target proteins in response to a growth factor) and phosphorylase kinase (a soluble, serine-directed kinase, controlled by calcium and itself responsive to phosphorylation), one must evaluate statistically the similarity to ensure that deduction of homology is well-founded. "Alignment scores" are calculated using algorithms devised for this purpose (Dayhoff et al., 1983; Lipman & Pearson, 1985). Some take into consideration similarities as well as identities (e.g. of valine and isoleucine, or arginine and lysine) that tend to be conserved during evolution. Comparisons of two sequences are evaluated with respect to control scores derived after artificially scrambling the amino acid compositions. The real comparisons are then expressed as alignment scores in units of standard deviation from the mean of the scrambled scores. In general, alignment scores of 5 standard deviation units or greater are taken as evidence of a homologous relationship.

Comparisons of two proteins by these methods are relatively simple using available algorithms; however, it is becoming more difficult to seek homologous counterparts to a newly derived amino acid sequence because of the growing size of the databanks. It is tempting to ascribe significance to short segments of identical or similar sequence and it is essential to evaluate whether such similarities are of sufficient length and identity to imply homologous relationships (Doolittle, 1981).

THE PROTEIN KINASE FAMILY

In the case of the protein kinases, homologous relationships were found among segments of 200 amino acid residues or more in proteins ranging in molecular weight from 40,000 to 75,000. In addition, as reported for many other proteins, evidence of internal gene duplication was indicated by the occurrence of tandem homologous segments in the same protein. For example, the regulatory subunit of cAMP-dependent protein kinase has two tandem segments that appear to be homologous (Takio et al., 1982) and account for the binding of two cAMP molecules to a single chain. Interestingly, each of these repeated domains is homologous with a cAMP-binding domain in CAP, a protein that binds a promoter in E. coli in response to a "hunger" signal of cAMP (Weber et al., 1982). In the same polypeptide chain, the C-terminal domain of CAP binds DNA, but it bears no relationship structurally or functionally to any part of the cAMP-dependent protein kinase molecule. Since the three-dimensional structure of CAP has been solved by X-ray crystallographic analysis, that structure provides some basis for visualizing each of the cAMP-binding domains in the regulatory subunit of protein kinase.

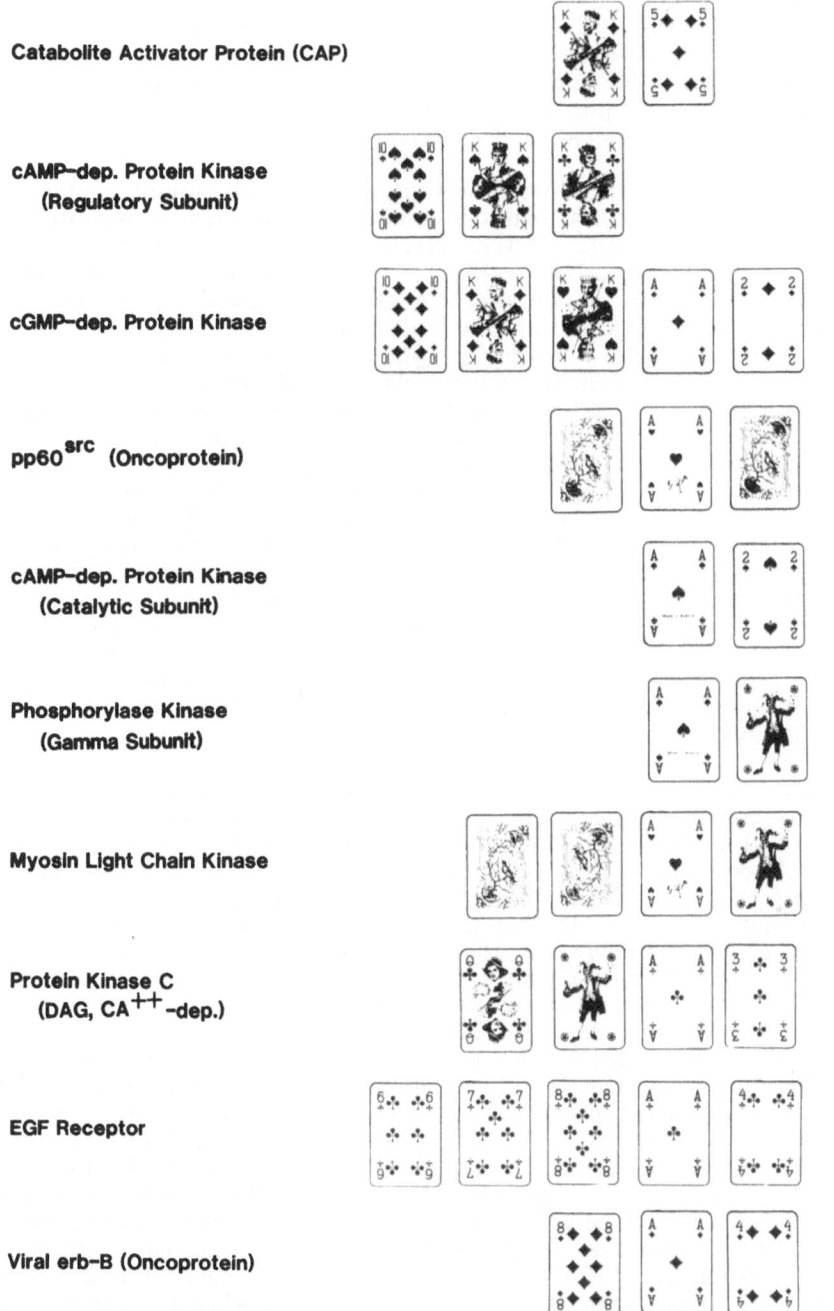

Figure 1. Diagrammatic summary of structural relationships among the protein kinases. Playing cards are used to denote domains or segments. The same card in a different suit indicates homology. The length of the protein is suggested by the number of cards. Ace, catalytic domain; King, binds cyclic nucleotide; Queen, binds diacylglycerol (DAG); Joker, calmodulin interaction site. Segment 5 in CAP interacts with DNA. Segments 6-7 in EGFR include the receptor site; segment 8 is a transmembrane segment. Segment 10 is involved in subunit interactions.

But one is left in this case with the realization that splicing or excision events must have been involved in the evolutionary relationship between the regulatory subunit and CAP (Figure 1). It might be anticipated that other proteins binding cyclic nucleotides for other purposes might also have homologous cAMP-binding segments and it will be interesting to see if such domains are spliced to yet other proteins to couple cAMP-binding signals to other physiological functions.

Turning to the catalytic subunit of cAMP-dependent protein kinase, one finds a homologous segment in virtually every other protein kinase that has been examined, whether directed toward serine, threonine, or tyrosine in target proteins (Table 1). For example, the gamma subunit of phosphorylase kinase contains a segment of about 260 residues that resembles the catalytic subunit of cAMP-dependent protein kinase (Reimann et al., 1984). However, phosphorylase kinase does not respond to cAMP, nor is its calcium-dependence found in cAMP-dependent protein kinase. The gamma subunit of phosphorylase kinase is known to bind calmodulin tightly in the presence of calcium and it is assumed that the C-terminal 100 residues of the gamma subunit, which has no homologous counterpart in cAMP-dependent protein kinase, is responsible for the interaction with and regulation by calmodulin. Similarly, the 65,000 dalton, calcium-dependent, serine-directed enzyme, myosin light chain kinase (MLCK) has a segment homologous with the catalytic domains of both other protein kinases and a C-terminal segment with a tentative relationship to the putative calmodulin-binding segment of phosphorylase kinase (Takio et al., 1985). On statistical grounds it is not possible to state that the C-terminal segments of the two proteins are homologous, but in the case of MLCK it has now been shown that that segment does bind calmodulin at nanomolar concentrations (Blumenthal et al., 1985).

The familial relationship among the protein kinases is best illustrated by considering the 670 residues of cGMP-dependent protein kinase (Takio et al., 1984), of which the amino-terminal half is homologous to the regulatory subunit of cAMP-dependent protein kinase and the C-terminal half to the catalytic domains of the various protein kinases, however regulated. Thus, in this case, as predicted some years before by Lincoln and Corbin (1977) on the one hand and by Gill (1977) on the other, the regulatory and the catalytic domains appear to be fused into one chimeric protein. The relationships among these and other proteins are illustrated in Figure 1 and Table II. These lists can now be expanded to include more than ten oncogene translation products, several receptors and most recently, the protein kinases C (Parker et al., 1986). As captured segments of genes, some oncogenes appear to be derived from truncated versions of receptors, as indicated by the very high alignment score in the comparison of the oncoprotein erb-B and the EGF receptor (Table II) (Downward et al., 1980). The protein kinases C constitute a family of calcium/diacylglycerol-dependent serine/threonine-directed protein kinases that appear to be involved in mitogenic signal pathways (Nishizuka, 1984). One segment within each of these proteins appears to be homologous with the other protein kinases, but the remainder of each 80,000 dalton protein appears to be unrelated, raising questions again about the splicing mechanisms that have given rise to this branch of the family of chimeric proteins.

OTHER MULTI-FUNCTIONAL PROTEINS

The protein kinases are not alone in showing chimeric character and bearing evidence of their origin by complex evolutionary pathways. It has been known for many years that serine proteases in digestion,

Table II

Alignment Scores[a] of Catalytic Domains of Various Protein Kinases[b]

		Tyr-specific Kinases			Ser/Thr-specific Kinases				
		src	erb-B	EGFR	MLCK	cAK	cGK	PK-C	Pb-K
(Tyr)	src	--	21	20	13	13	10	13	9
	erb-B		--	53	13	10	10	12	9
	EGFR			--	10	11	11	12	10
(Ser/Thr)									
	MLCK				--	17	15	15	18
	cAK					--	28	28	18
	cGK						--	22	18
	PK-C							--	16
	Pb-K								--

[a]Expressed in units of standard deviation from the mean of randomly generated sequences of the same composition (Dayhoff et al., 1983).

[b]Abbreviations used: src and erb-B are oncoproteins from Rous sarcoma virus and avian erythroblastosis virus, respectively. EGFR, epidermal growth factor receptor; MLCK, myosin light chain kinase; cAK and cGK, cAMP or cGMP-dependent protein kinases; PK-C, protein kinase C; Pb-k, phosphorylase b kinase.

coagulation, fibrinolysis, etc., had homologous segments among polypeptide chains that varied from 23,000 to 80,000 in molecular weight (e.g. Katayama et al., 1979). Similarly, it has recently been shown that an amino-terminal domain in calpain is homologous with papain and the cathepsins, whereas a C-terminal segment is homologous with the calmodulin family (Ohno et al., 1984). A striking example is the distribution of seven proteins involved in successive steps of fatty acid synthesis among only two polypeptide chains in both rabbit and yeast (McCarthy & Hardie, 1984). Remarkably, the arrangement of the seven proteins is in a different order in the two species. These findings lend strong support to the idea that contemporary multi-functional proteins have been assembled from integral domains during evolution to satisfy biological demands for multi-enzyme complexes, for multi-ligand capacity and, within the protein kinase family, for the transduction of chemical signals into phosphorylation events that control cellular processes.

Recognition of the domain substructures of the protein kinases and of the homologous relationships among segments of these chimeric proteins has given us a basis for interpreting studies of one member of the family in terms of the control and function of the others. As detailed three-dimensional information on any member of this family is derived, it will immediately provide a working hypothesis for the interaction of the others with their regulators on the one hand and their targets on the other.

REFERENCES

Barker, W.C. & Dayhoff, M.O. (1982), Proc. Natl. Acad. Sci. USA 79, 2836

Blumenthal, D.K., Takio, K., Edelman, A.M., Charbonneau, H., Titani, K., Walsh, K.A. & Krebs, E.G. (1985) Proc. Natl. Acad. Sci. USA 82, 3187

Dayhoff, M.O, Barker, W.C. & Hunt, L.T. (1983) Methods Enzymol. 91, 524

Doolittle, R.F. (1981) Science 214, 149

Downward, J., Yarden, Y., Mayes, E., Scrace, G., Totty, N., Stockwell, P., Ullrich, A., Schlessinger, J. & Waterfield, M.D. (1984) Nature (London) 307, 521

Flockhart, D.A. & Corbin, J.D. (1982) CRC Crit. Rev. Biochem. 12, 133

Gill, G.N. (1977) J. Cyclic Nucleotide Res. 3, 153

Katayama, K., Ericsson, L.H., Enfield, D.L., Walsh, K.A., Neurath, H., Davie, E.W. & Titani, K. (1979) Proc. Natl. Acad. Sci. USA 76, 4990

Krebs, E.G. & Beavo, J.A. (1979) Annu. Rev. Biochem. 48, 923

Kuret, J., Woodgett, J.R., Cohen, P. (1985) Europ. J. Biochem. 151, 39

Lincoln, T.M. & Corbin, J.D. (1977) Proc. Natl. Acad. Sci. USA 74, 3239

Lipman, D.J. & Pearson, W.R. (1985) Science 227, 1435

McCarthy, A.D. & Hardie, D.G. (1984) Trends Biochem. Sci. (Pers. Ed.) 9, 60

Nishizuka, Y. (1984) Nature (London) 308, 693

Ohno, S., Emori, Y., Imajoh, S., Kawasaki, H., Kisaragi, M. & Suzuki, K. (1984) Nature (London) 312, 565

Parker, P.J., Coussens, L., Totty, N., Rhee, L., Young, S., Chen, E., Stabel, S., Waterfield, M.D. & Ullrich, A. (1986) Science 233, 853

Reimann, E.M., Titani, K., Ericsson, L.H., Wade, R.D., Fischer, E.H., & Walsh, K.A. (1984) Biochemistry 23, 4185

Sefton, B.M. & Hunter, T. (1984) Adv. Cyclic Nucleotide Protein Phosphorylation Res. 18, 195

Stull, J.T., Nunnally, M.H., Moore, R.L. & Blumenthal, D.K. (1985) Adv. Enzyme Regul. 23, 123

Takio, K., Blumenthal, D.K., Edelman, A.M., Walsh, K.A., Krebs, E.G. & Titani, K. (1985) Biochemistry 24, 6028

Takio, K., Smith, S.B., Walsh, K.A., Krebs, E.G., & Titani, K. (1982) Proc. Natl. Acad. Sci. USA 79, 2544

Takio, K., Wade, R.D. Smith, S.B., Krebs, E.G., Walsh, K.A. & Titani, K. (1984) Biochemistry 23, 4207

Weber, I.T., Takio, K., Titani, K. & Steitz, T.A. (1982) Proc. Natl. Acad. Sci. USA 79, 7679

PROTEIN KINASE C OF INTESTINAL EPITHELIUM:

ITS ROLE IN THE CONTROL OF IONIC TRANSPORT

Pedro S. Lazo and Gloria Velasco

Departamento de Bioquímica, Universidad de Oviedo
33071, Oviedo, Spain

SUMMARY

Protein kinase C activity has been identified in rat and rabbit small intestine epithelial cells. When the crude cytosolic fraction was assayed, a Ca activated protein kinase activity which was independent of any exogenously added phospholipid was identified. Protein kinase C activity was resolved from other protein kinase activities by ion exchange chromatography. Phosphatidylserine was required for protein kinase C to be active. The activation of protein kinase C was a function of phosphatidylserine concentrations up to 50 μg/ml. In the presence of phosphatidylserine and diolein, the Ka for activation by Ca was 10^{-7} M. The phorbol ester TPA , used at 1-5 ng/ml concentrations, substituted for diacylglycerol in activating protein kinase C. The results are dicussed within a body of evidence which indicates that protein kinase C is implicated in the control of the ionic transport in the intestinal epithelium.

INTRODUCTION

Protein kinase C (*) is a part of a signal transduction mechanism, controlling a variety of cellular processes. This signal transduction mechanism controls both short-term and long-term processes, such as secretion, smooth muscle contraction and cell proliferation[1,2] . PKC was first found in a cytosolic fraction from rat brain and subsequently in other mammalian tissues[3-5]. The knowledge about the target proteins for the PKC is scarce, although it is known that receptors for growth factors[6,7] and the β-adrenergic receptor are substrates for the enzyme and it has been sugested that Na/H exchanger migth be a substrate for the enzyme[8,9].

Intestinal secretion is considered to be under dual control of cyclic AMP and Ca, through changes in their intracellular concentrations[10,11], which, according to the current hypothesis of the biological effects of these second messengers, would act to regulate the activity of protein kinases. Knowing that agonists which activate the PI-PKC transduction mechanism in other systems stimulate intestinal secretion[12] we have investigated the presence of this enzyme in intestinal epithelial cells.

METHODS

Preparation of enterocyte cytosolic fraction. Enterocytes were obtained from jejunum and ileum by treatment of the everted intestine with hyaluronidase as described[13] . The cells were resuspended in 20 mM Tris buffer, pH 7.5, containing 1 mM EDTA, 1 mM EGTA and 50 mM β-mercaptoethanol and were homogenized in a Potter-Elvehjem with 20 strokes at 2,400 rpm. The cell extract was centrifuged for 60 min at 100,000 x g. The supernatant of this centrifugation is referred to as cytosolic fraction.

DEAE Cellulose chromatography. PKC was eluted from DE52 (Whatman) columns with a linear gradient of NaCl or in a stepwise chromatography as follows: 1-2 ml of cytosolic fraction was applied onto 0.5-1 ml columns of DE52 which had been previously equilibrated with homogeneization buffer. After washing it with 8 ml of the same buffer, it was eluted with 1 ml each of 0.1, 0.2 and 0.4 M NaCl.The eluates were immediately used for assay of PKC activity.

Protein kinase C assay. PKC was assayed at pH 6.8 by measuring the incorporation of ^{32}P from $|\gamma-^{32}P|$ATP into H1 histone as described[14] .

RESULTS

Identification of protein kinase C. During the preliminary studies to identify PKC in rat enterocytes we used a cytosolic fraction as a source of enzyme. The results indicated that this fraction contained a Ca dependent protein kinase which was activated up to 4-fold in the presence of Ca. However, the activity was not enhanced when, in addition to Ca, phospholipids were present in the assay mixture. On the other hand, when eluates from DE52 columns were assayed, very little or no stimulation was obtained by Ca alone, while in the presence of Ca and phospholipids the activity was stimulated 4 to 10 fold depending on the preparation used (Table 1).

DEAE cellulose (DE52) chromatography of the cytosolic fraction resolved protein kinase activity into two fractions(Fig. 1). A major peak, representing up to 80 % total activity eluted at 110-140 mM salt. This activity appeared to be Ca and phospholipid independent; in fact, in some preparations it was observed that in the presence of Ca alone, or Ca and phospholipid, the activity was inhibited as much as 50% respecting to the control assayed in the presence of EDTA/EGTA. Another fraction representing 20-40 % of the total activity eluted at 40-60 mM salt and appeared as PKC(i.e. Ca and phospholipid dependent). For rutine assays of PKC, cytosolic fractions were chromatographied on small columns (0.5-1.0 ml) of DE52. After washing as described in Methods, a stepwise elution with 1 ml each of 0.1, 0.2 and 0.4 M NaCl was performed.Conductivity measurements of the eluates gave salt concentra-

Table 1
Requirements for activation of protein kinase C in homogenates of epithelial tissue and a partially purified fraction

| | Protein kinase activity(pmol\|min) | |
Addition	Homogenate	Purified fraction
None	1.68	2.6
0.2 mM Ca	4.38	2.7
0.2 mM Ca + PS + Diolein	4.42	10.7

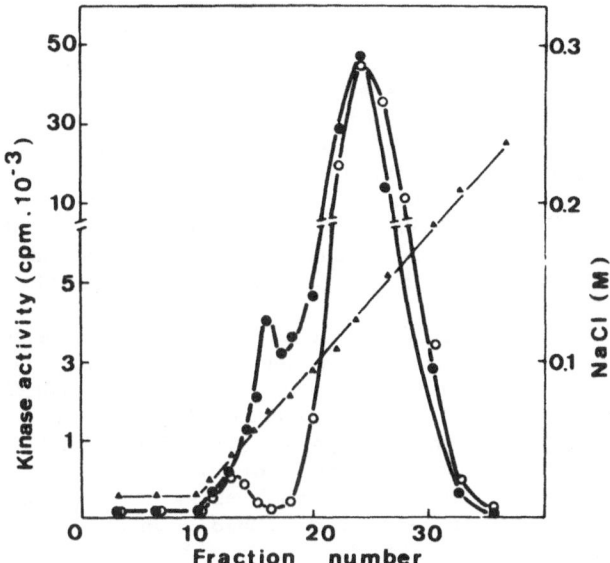

Figure 1. DEAE Cellulose fractionation of protein kinase C. A cytosolic fraction was loaded onto a DE52 column equilibrated with 20 mM Tris buffer,PH 7.5, 1 MM EDTA, 1 mM EGTA, 50 mM β-mercaptoethanol. The column was washed with the same buffer and a Nacl gradient (▲) applied. Protein kinase activity was assayed in the absence of Ca and phospholipids (o) or in thr presence of 25 μg/ml of PS, 10 μg/ml of diolein and 0.2 mM free Ca (●).

tions of 15–20 mM, 40–50 mM and 110–140 mM respectively in various experiments. Protein kinase activity was assayed in the three eluates and the results showed that PKC was confined to the 40–50 mM salt eluate.

Properties of protein kinase C.Table 2 shows that PS activates PKC from intestinal epithelium. In the presence of this phospholipid, the enzyme was activated in a concentration dependent manner for concentrations up to 50 μg/ml.A maximum of six fold stimulation was observed with 80 to 100 μg/ml. In the presence of PS, the presence of a low concentration of a diacylglycerol caused a greater activation of the enzyme. The Ka for Ca in the presence of phospholipid and diolein was 10^{-7} M. Phospholipids other than PS showed little or non activation of PKC. Of the diacylglycerols tested, both saturated and unsaturated diacylglycerols were effective in stimulating PKC.

The tumour promoting phorbol ester TPA can activate PKC from intestinal epithelium in the presence of micromolar Ca and PS, that is, the agent can substitute for diacylglycerol in activating the enzyme. The activation was concentration dependent at concentrations ranging from 1 to 5 ng/ml (not shown).

Table 2
Lipid requirements for activation of protein kinase C

Addition	PKC activity (pmol/min)	Activation (fold)
None	0.6	–
0.2 mM Ca + 40 μg/ml PS	2.0	3.3
0.2 mM Ca + 40 μg/ml PS + 10 μg/ml Diolein	8.0	13.3
0.2 mM Ca + 40 μg/ml PC + 10 μg/ml Diolein	0.9	1.5
0.2 mM Ca + 40 μg/ml PS + 10 μg/ml Distearin	7.5	12.5

DISCUSSION

We have identified a Ca dependent protein kinase in enterocytes which showed the activation characteristics of PKC. The enzyme activity was dependent upon the presence of PS. It was found that diacylglycerols containig either saturated of unsaturated fatty acids in their hydrophobic moieties were equally effective. The concept that the identified activity is PKC is also substantiated by the fact that tumour promoting phorbol ester TPA activated the enzyme at very low concentrations in the presence of phospholipids and Ca thus substituting for diacylgycerol but having a much higher affinity for the enzyme.

Intestinal secretion is known to be regulated by the intracellular concentration of cyclic AMP and Ca and cyclic nucleotides (cAMP and cGMP) and calcium-calmodulin dependent protein kinases have been reported in this tissue[15-17]. The presence of PKC in intestinal cells, together with other evidences suggest that the PI-PKC transduction system might operate in the control of intestinal ionic transport. Thus, it has been shown that the phorbol ester TPA produces intestinal hypersecretion[18] ; dantrolene and TMB-8, which trap Ca within intracellular stores increased Na and Cl absorption[19] and in our laboratory, using a preparation of enterocytes with a permeabilised plasma membrane[20] , we have shown that IP3 causes release of Ca from the endoplasmic reticulum[21]. It is therefore conceivable that secretatogues such as serotonin or acetylcholine which utilise Ca as second messenger stimulate the production of IP3 and diacylclycerol from phosphatidylinositol 4,5, bisphosphate which would lead in turn to the activation of PKC (Fig. 2). PKC, as well as other protein kinases, would regulate ionic transport by phsophorylating specific membrane proteins. It is unknown at present which proteins are substrates for these protein kinases although it is expectable that they are related to the secretion of Cl, which is enhanced during intestinal secretion, and/or the cotransport of Na and Cl which is inhibited during intestinal secretion. Fig. 2 also depicts a possible interaction between cAMP and Ca in the control of ionic transport since it has been suggested that cAMP can induce a Ca signal. In our hands, however, cAMP did not cause Ca release

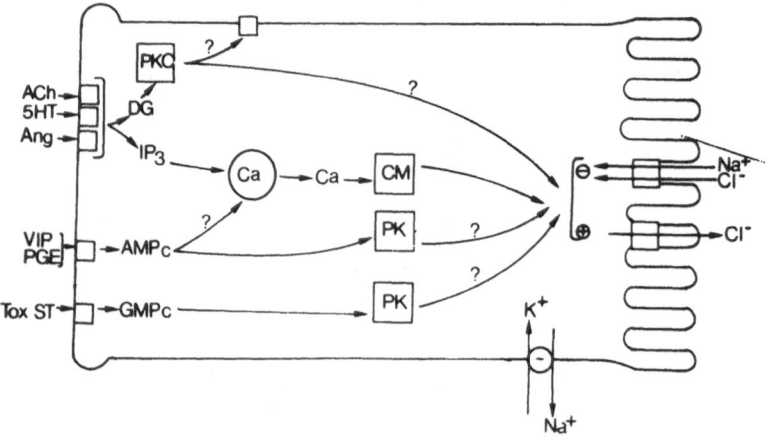

Figure 2 A model on the involvement of Ca and PKC in the control of ionic transport in the intestinal epithelium.

from any intracellular pool in conditions in which IP3 caused the release of Ca from the endoplasmic reticulum[21].

Acknowledgements. This work was supported in part by grants from the F.I.S. and the C.A.I.C.Y.T. G. Velasco is recipient of a fellowship from the F.I.S.

REFERENCES
1. Berridge, M.J. and Irvine, R.F. (1984) Nature 312, 315-321.
2. Nishizuka, Y. (1984) Nature 308, 693-698.
3. Takai, Y., Kishimoto, A., Kawahara, Y., Mori, T. and Nishizuka, Y. (1979) J. Biol. Chem. 254, 3692-3695.
4. Takai, Y., Kishimoto, A., Iwasa, Y., Kawahara, Y., Mori, T. Nishizuka, Y, Tamura, A. and Fujii, T. (1979). J. Biochem. (Tokio) 86, 575-578.
5. Minakuchi, R., Takai, Y., Yu, B. and Nishizuka, Y. (1981) J. Biochem. (Tokio) 89, 1651-1654.
6. Cochet, C., Gill, G.N., Meisenhelder, J., Cooper, J.A. and Hunter, T. (1984) J. Biol. Chem. 259, 2553-2558.
7. Hunter, T., Ling, N. and Cooper, J.A. (1984) Nature 311, 480-483.
8. Moolenar, W.H., Tertoolen, L.G.J. and DeLaat, S.W. (1984) Nature 312, 371-374.
9. Kelleger, D.J., Pessin, J.E., Ruoho, A.E. and Johnson, G.L. (1984) Proc. Natl. Acad. Sci. U.S.A. 81, 4316-4320.
10. Field, M. (1979) in Mechanisms of Intestinal Secretion (Binder, H.J. ed.) pp 83-91. Alan R. Liss Inc. New York.
11. Ilundain, A. and Naftalin, R. (1979) Nature 279, 446-448.
12. Turnberg, L.A. (1983) Scan. J. Gastroent. 18, suppl. 87,85-89.
13. Kimmich, G. (1970) Biochemistry 13, 3659-3668.
14. Velasco,G., Iglesias,C.F., Domínguez,P., Barros, F., Gascón, S. and Lazo, P.S. (1986) Biochem. Biophys. Res. Commun. 139, 875-882.
15. Kimberg,D.V., Shlatz, L.J. and Cattieu, K.A. (1979) in Mechanisms of Intestinal Secretion (Binder, H.J. ed.) pp 131-146. Alan R. Liss Inc. New York.
16. De Jonge, H.R. (1981) Adv. Cyclic Nucleot. Res. (Dumont, J.E., Greengard, P. and Robinson, G.A. eds.) pp 315-333. Raven Press. N.Y.
17. Donowitz, M. (1983) Am. J. Physiol. 245, G165-G177.
18. Fondacaro, F.D. and Henderson, L.S. (1985) Amer. J. Physiol. 249, G422-G426.
19. Donowitz, M., Cusolito, S. and Sharp, G.W.G. (1986) Am. J. Physiol. 250, G691-G698.
20. Velasco, G., Domínguez, P., Shears, S.B. and Lazo, P.S. (1986) Biochim. Biophys. Acta. In press.
21. Velasco, G., Shears, S.B., Michell, R.H. and Lazo, P.S. (1986) Biochem. Biophys. Res. Commun. 139, 612-618.

(*) Abbreviations. PKC: protein kinase C; IP3: inositol 1,4,5 trisphosphate; PS: phosphatidylserine; PI: phosphatidylinositol; PC: phosphatidylcholine; TPA: 12-tetradecanoylphorbol-13 acetate; EGTA: ethylene glycol bis (β-aminoethylether)-N,N,N',N' tetraacetic acid; EDTA: ethylenediaminetetracetic acid.

PROSTAGLANDIN-E$_1$- AND SODIUM NITROPRUSSIDE-REGULATED PROTEIN PHOSPHORYLATION IN PLATELETS

Maria Nieberding, Rainer Waldmann and Ulrich Walter

Medizinische Univ.-Klinik Würzburg, Labor für Klinische Biochemie, Josef-Schneider-Straße 2, D - 8700 Würzburg, FRG

A variety of agents like ADP, collagen, thrombin, platelet-activating-factor (PAF)* and others activate platelets with a concomitant increase in the phosphorylation of myosin light chain (Mr 20 K) and of a soluble protein with apparent Mr of 44 K [1,2]. Certain vasodilators such as cGMP-elevating agents (SNP, nitroglycerin) and cAMP-elevating agents (PGE$_1$, PGI$_2$) are able to inhibit platelet activation and the activation-associated protein phosphorylation [1,2]. Since cAK and cGK may mediate the effects of cyclic nucleotide-regulating agents in platelets, we have now compared the vasodilator-regulated protein phosphorylation observed in intact platelets with the cyclic nucleotide stimulated protein phosphorylation found in platelet membranes [3,4]. The results show for at least one phosphoprotein that the SNP-and PGE$_1$ -stimulated protein phosphorylation is mediated by cGK and cAK, respectively.

In intact platelets, PGE$_1$ stimulated the phosphorylation of four proteins (Mr of 240 K, 68 K, 50 K, 22 K), which was also observed with dibutyryl-cAMP in intact platelets (not shown) and in response to cAMP in platelet membranes (Figure 1 A). In contrast, SNP (Figure 1 A) and 8-Br-cGMP (not shown) increased the phosphorylation of a protein with Mr of 50 K, which was also phosphorylated in response to cGMP in platelet membranes.

---

* Abbreviations: PAF:platelet-activating-factor; SNP:sodium nitroprusside; PGE$_1$: prostaglandin-E$_1$; PGI$_2$:prostaglandin I$_2$; cAK:cAMP-dependent protein kinase; cGK:cGMP-dependent protein kinase; cG:cGMP; cA:cAMP; IP$_3$:inositol triphosphate; DG:1, 2 diacylglycerol; PKC:protein kinase C; PI, Phosphatidylinositol

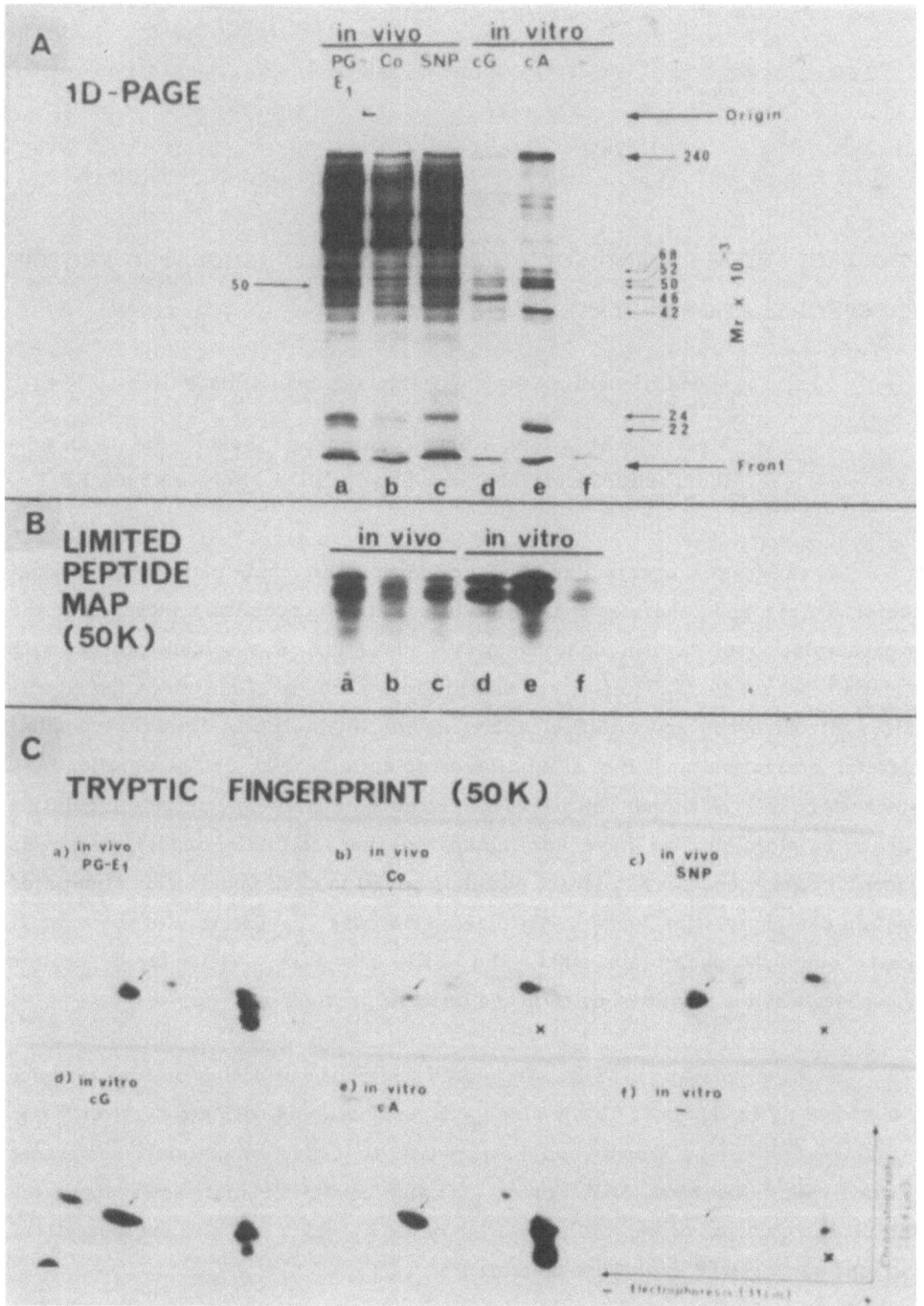

Figure 1:

Autoradiograph showing the effects of PGE$_1$, SNP and cyclic nucleotides on protein phosphorylation in platelets.

The phosphorylation pattern of intact platelets (in vivo) and of platelet membranes (in vitro) regulated by cyclic nucleotide-elevating agents (PGE$_1$, SNP) and by cyclic nucleotides was analyzed by one dimensional SDS-polyacrylamide gel electrophoresis (A), by limited peptide mapping (B) and by tryptic fingerprinting (C).

(A): The phosphorylation pattern induced by SNP or PGE$_1$ in vivo (lane a–c) and by cGMP or cAMP in vitro (lane d–f) was compared. In vivo, PGE$_1$ stimulated the phosphorylation of at least four major proteins (Mr 240 K, 68 K, 50 K, 22 K) as indicated by arrows. The proteins were also phosphorylated in response to cAMP in vitro. SNP (in vivo) and cGMP (in vitro) enhanced the phosphorylation of one protein with Mr of 50 K. In membranes, cGMP and cAMP also stimulated the phosphorylation of proteins with apparent Mr of 46 K and 42 K, respectively. With intact platelets, PGE$_1$ and SNP caused a moderate increase in the phosphorylation of two proteins with Mr of 52 K and 24 K.

(B): The 50 K phosphoprotein used in this analysis was obtained from the phosphorylation experiments (shown in A) and digested by Staphylococcus aureus V 8 protease. Limited digestion of the 50 K protein produced a triplet of closely migrating phosphopeptides, which appeared to be identical for all conditions examined. Counting of the peptides obtained after V 8 digestions indicated that SNP and PGE$_1$ caused a 2 – 3 fold increase in the phosphorylation of the 50 K-protein.

(C): The peptides obtained from the 50 K phosphoprotein by limited digestion (shown in B) were further analyzed by tryptic fingerprinting. Extensive digestion by trypsin and seperation by electrophoresis and ascending chromatography produced one peptide, whose phosphorylation was stimulated by the agents indicated.

Consistently, SNP caused a small increase in the phosphorylation of a 24 K-protein and a decrease in the phosphorylation of a 22 K-protein. In membranes, the cAMP- and cGMP-stimulated phosphorylation was specifically blocked by a synthetic fragment of the Walsh-inhibitor or by an inhibitory anti-cGK antiserum, respectively (not shown).

To identify, whether identical proteins were phosphorylated in intact platelets and platelet membranes, the 50 K phosphoprotein was analyzed also by limited peptide mapping and tryptic fingerprinting. This particular protein was chosen, because it was phosphorylated in response to cyclic nucleotide-elevating agents and cyclic nucleotide analogs in intact platelets as well as in response to cyclic nucleotides in platelet membranes. For limited proteolysis the gel piece containing the 50 K-protein was digested by Staphylococcus aureus V 8 protease. For all the conditions examined (SNP and PGE$_1$ in vivo; cGMP and cAMP in vitro) an identical pattern of phosphopeptides was observed (Figure 1 B). These phosphopeptides were further analyzed by tryptic fingerprinting. Again for all the conditions examined enhanced phosphorylation of a single major phosphopeptide (indicated by the arrow in Figure 1 C) was observed.

These results indicate that in platelets cAK mediates the phosphorylation of the 50 K-protein induced by PGE$_1$ whereas cGK mediates the phosphorylation induced by SNP. Increased phosphorylation of the 50 K-protein is common for both cAMP-und cGMP-elevating agents and may be of central importance for the mechanism of action of these two classes of vasodilators. Recently it has been

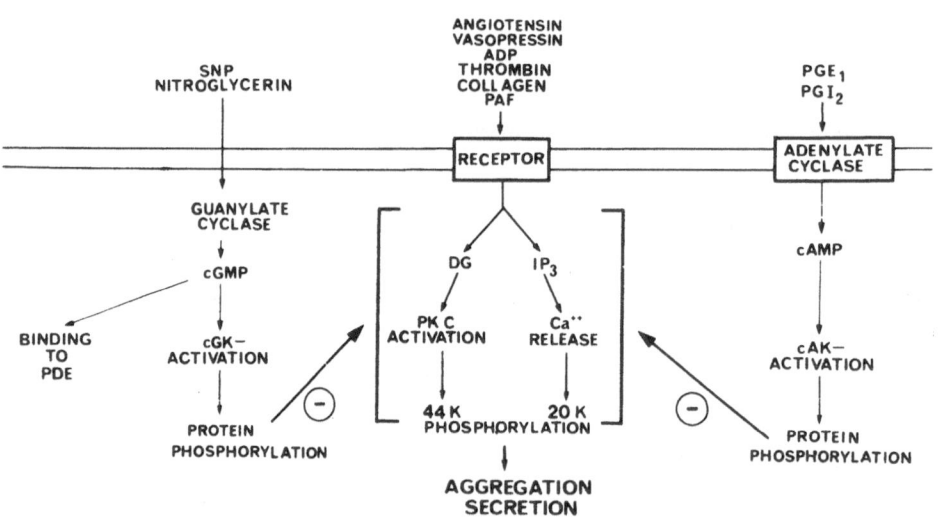

Figure 2:
Possible roles of cyclic nucleotide regulated protein kinases for the mechanism of action of platelet inhibitors such as PGE$_1$ and SNP.

shown with platelets that cAMP- and cGMP-elevating agents not only inhibit platelet activation but also inhibit the activator induced increase of 1, 2 diacylglycerol, IP_3 and cytosolic calcium [1,2,5,6]. It is our working hypothesis (Figure 2) that SNP and PGE_1 via cGK- and cAK-regulated phosphorylation of the 50 K-protein and perhaps of additional proteins inhibit the platelet activator-induced increase of DG, IP_3 and Ca^{2+}. This would result in the inhibition of both PKC- and Ca^{2+}/Calmodulin-induced protein phosphorylation as well as in the inhibition of platelet aggregation. Currently, we investigate whether the substrate proteins of platelet cAK and cGK regulate phospholipase C, other enzymes of the PI-turnover and Ca^{2+} removal or Ca^{2+} mobilization.

Acknowledgements: M.N. is supported by a grant from the Fritz-Thyssen-Stiftung. R.W. was a recipient of a postgraduate fellowship from the Boehringer Ingelheim Fond.

References

1. R. J. Haslam, S.E. Salama, J.E.B. Fox, J.A. Lynham and M.M.L. Davidson, Roles of cyclic nucleotides and of protein phosphorylation in the regulation of platelet function, in "Platelets: Cellular response mechanism and their biological significance", A. Rotman, F.A. Meyer, C. Gitler and A. Silberberg, eds. Wiley, New York (1980)

2. Y. Takai, U. Kikkawa, K. Kaibuchi and Y. Nishizuka, Membrane phospholipid metabolism and signal transduction for protein phorphorylation, Adv. Cyclic Nucleotide and Protein Phosphorylation Research 18: 119 - 158 (1984)

3. R. Waldmann, S. Bauer, C. Göbel, F. Hofmann, K.H. Jakobs and U. Walter, Demonstration of cGMP-dependent protein kinase and cGMP-dependent phosphorylation in cell-free extracts of platelets, Eur. J. Biochem. 158: 203 - 210 (1986)

4. R. Waldmann, M. Nieberding and U. Walter, Vasodilator-stimulated protein phosphorylation in platelets is mediated by cAMP- und cGMP-dependent protein kinases, Eur. J. Biochem., submitted for publication.

5. D. E. Mac Intyre, M. Bushfield and A.M. Shaw, Regulation of platelet cytosolic free calcium by cyclic nucleotides and protein kinase C, FEBS Lett. 188, 383 - 388 (1985)

6. S. Nakashima, T. Tohmatsu, H. Hattori, Y. Okano and Y. Nozawa, Inhibitory action of cGMP on secretion, polyphosphoinositide hydrolysis and calcium mobilization in thrombin stimulated human platelets, Biochem. Biophys. Res. Commun. 135, 1099 -1104 (1986)

STRUCTURE, FUNCTION AND REGULATION OF PROTEIN PHOSPHATASE

Edmond H. Fischer, Steven J. McNall, and Nicholas Tonks

Department of Biochemistry
University of Washington
Seattle, Washington 98195

In the opening lecture, the events surrounding the discovery that glycogen phosphorylase activity is regulated by phosphorylation-dephosphorylation were presented. Since that time, this type of protein modification has proved to be a major mechanism by which many eukaryotic cellular processes, including their metabolism, growth, differentiation, transformation and gene expression, can be regulated. We have already heard about the many advances that have been made in our understanding of the subunit structure, mechanism of action, functional sites and regulation of the cAMP-dependent protein kinase, of phosphorylase kinase, and several other kinases, including protein tyrosine kinases. Until recently, however, much less was known about the protein phosphatases that catalyse protein dephosphorylation. Reasons for this include: 1) difficulty in obtaining substrates since an appropriate phosphoprotein must first be isolated and then phosphorylated with a specific protein kinase, and 2) general instability, low intracellular concentrations, and broad specificity of the phosphatases in addition to their tendency to be affected by many allosteric agents. This has made it difficult to determine if one were working with a single phosphatase or a mixture of enzymes. Despite these problems, several phosphatases have now been purified and extensively studied and some of their properties will be discussed.

II. Classification of Protein Phosphatase

The central questions in the study of protein phosphatases concern the number of phosphatases, their substrate specificities (are there relatively few enzymes of broad specificity or a large number of phosphatases of very narrow specificity?) and how they are regulated in vivo.

Cohen and Ingerbertsen have proposed a classification system based on a study of the phosphatases that dephosphorylate phosphorylase kinase. Four enzymes have been identified that appear to account for most, if not all, of the cytosolic protein phosphatases that have been described thus far. These activities are divided into 2 categories (Table 1).

Type	Protein Phosphatase	Inhibited by Inhibitor-1 & Inhibitor-2	Specificity for Phosphorylase Kinase	Substrate Specificity	Regulators
1	Protein Phosphatase-1	Yes	β-subunit	Broad	Inhibitor-1 Inhibitor-2 GSK-3
2	Protein Phosphatase-2A	No	α-subunit	Broad	Polyamines
2	Protein Phosphatase-2B	No	α-subunit	Narrow	Ca^{2+} Calmodulin
2	Protein Phosphatases-2C	No	α-subunit	Broad	Mg^{2+}

Abbreviation: GSK3 - glycogen synthase kinase-3.

Type 1 protein phosphatases dephosphorylate the β subunit of phosphorylase kinase more rapidly than the α subunit and are potently inhibited by nanamolar concentrations of 2 thermostable inhibitor proteins, inhibitor 1 and 2. Type 2 phosphatases, of which there are 3 distinct catalytic subunits (2A, 2B and 2C) dephosphorylate the α subunit more rapidly than the β subunit and are essentially unaffected by the inhibitor proteins.

The four phosphatases have been identified in all tissues examined thus far, although their relative concentration and intracellular distributions vary considerably. The properties of these enzymes will be briefly considered in the following sections.

III. Type 1 Protein Phosphatase (Phosphorylase Phosphatase)

Several forms of this enzyme are known to exist. In all cases, the enzyme appears to consist of a Mr = 37,000 catalytic subunit whose activity is regulated through interactions with various regulatory subunits. For example, in glycogen particles, the catalytic subunit is associated with a Mr = 103,000 protein known as the "G-protein". This G-protein, in addition to containing a glycogen binding site, can also be phosphorylated. Upon phosphorylation of G, inhibitor-1 becomes a much more potent inhibitor of this phosphatase. Similar proteins have also been described which bind the catalytic subunit to microsomes ("M" proteins) and myosin.

What will be discussed today is the structure and regulation of another form of this enzyme which can be isolated from the muscle cytosol in a M_r 70,000 inactive complex. This work was carried out in collaboration with three of the participants of this course, namely, Lisa Ballou, Emma Villa Moruzzi and Steve McNall. David Brautigan, now at Brown University, was also involved in its earlier phases.

This enzyme is probably the one responsible for the inactivation of phosphorylase a, originally described by the Cori's some 30 years ago and which they called "prosthetic-group removing" enzyme "PR enzyme" and which I spoke of in my first lecture. Subsequently, it has been referred to as phosphorylase phosphatase. Since it is one of the major phosphatases involved in the inactivation of phosphorylase and activation of glycogen synthase, it might well be under insulin control.

The enzyme can be isolated in pure form as a M_r 70,000 complex showing two bands on SDS gel electrophoresis, a 38k band corresponding to the catalytic subunit and a 31k band corresponding to the regulatory subunit identified as inhibitor-2. While this form of the enzyme is totally inactive, it can be activated in two ways. The first involves Mn^{2+} ions: while this process is probably non-physiological, it has been most useful in providing some insight into the structure of the protein. The second requires Mg-ATP (as discovered just about twenty years ago by Wilfried Merlevede working on an adrenal cortex enzyme) and an activating factor F_A, subsequently identified as glycogen synthase kinase 3 (GSK 3). But, let's turn first to the Mn^{2+} activation system.

1. Activation by trypsin/Mn^{2+}

It was known for a very long time - in fact, since the early days of the Cori's - that the enzyme could be activated by Mn^{2+} under certain circumstances, but the extent of this reaction varied greatly from prep to prep - and even with the same preparation, during the course of purification. It eventually became clear that two separate processes were required: first, a proteolysis of the enzyme, and then the metal-ion effect. Either of these alone wouldn't do anything.

SDS gel analysis showed that trypsin rapidly destroys the regulatory subunit and leaves a slightly altered catalytic subunit from which approximately 5,000 daltons (or approximately 40 residues) have been split from the C-terminus. This catalytic core is totally inactive until exposed to Mn^{2+}.

There were three possibilities to account for these observations:

a) that limited proteolysis had inactivated the enzyme;

b) that a fragment of the inhibitory subunit still remained attached to - and blocked - the catalytic site; it would be displaced by Mn^{2+}.

c) that the catalytic subunit already existed in an inactive conformation in the inactive complex.

To resolve this matter, ways had to be devised to separate the two subunits under non-denaturing conditions; and this was achieved by ion exchange FPLC. The free catalytic subunit that emerged was totally inactive when measured in the absence of Mn^{2+}, and active with Mn^{2+}. It is interesting to note that this cytosolic enzyme is subjected to two levels of inhibition: first, it is bound to inhibitor 2, which can block its activity; second, the catalytic subunit exists in an inactive conformation. Even if the enzyme underwent dissociation, it would still be inactive. In that sense, the enzyme is unlike the cAMP-dependent protein kinase, in which dissociation of the inactive complex brings about the liberation of active catalytic subunits. Why are these two degrees of inhibition superimposed on one another? One possibility is that it could provide a fail-safe mechanism to prevent activation of the enzyme, for instance, after the accidental destruction of the inhibitory subunit by intracellular proteases. The phosphatase is a hydrolase that might wreak havoc within the cell if allowed to operate unchecked. Alternatively, this double level of inhibition could be part of a higher order of control in which hormones such as insulin could participate.

Mn^{2+} activation is not simple and occurs in at least two steps: a) an immediate activation to produce a form sensitive to EDTA - as if one were dealing with an ordinary "metalloenzyme". b) Then, in a slower reaction that takes an hour or more, a conversion to a form no longer sensitive to EDTA. Use of $^{54}Mn^{2+}$ shows no incorporation of radioactivity. Therefore, it was concluded that the enzyme either underwent a change in conformation from an inactive to an active state, or some kind of covalent modification such as an SH/SS interchange. Only Mn^{2+} or Co^{2+} works, not Mg^{2+} or Ca^{2+}.

The first alternative, i.e., a simple change in conformation, doesn't seem very probable: one would be dealing with a "stamping mechanism" in which metal ion goes in, does its thing, and comes out. The active conformation would have to be thermodynamically favored. The other possibility, i.e., that SH groups might be involved is suggested by the following experiments: if FPLC is carried out in the absence of SH compounds, the enzyme that emerges is recovered in an inactive state that no longer responds to Mn^{2+}. The preparation must first be exposed to reducing agents in order to regain its sensitivity to the metal ions. The reaction is time, temperature and concentration dependent and has an optimum pH of around 9. The assumption that SH groups might be involved also comes from the inhibition of the enzyme by disulfides such as oxidized glutathione (GSSG) that forms a mixed disulfide with the protein. The inactive species can be reactivated by high concentrations of DTT. A number of other SH compounds are effective in reactivating the enzyme (DTT > 2 mercapto-β-ethylamine > 2 Me > GSH > Cys, while ascorbic acid, NADH or NADPH are without effect). These data might indicate that there are two sets of SH groups present in the enzyme; while they seem to differ in chemical reactivity, they both appear to contribute to the expression of enzymatic activity.

2. Activation by F_A/GSK-3 and Mg-ATP

While the activation of the enzyme by Mn^{2+} is interesting from a mechanistic point of view, that is, for what it can teach us on the structures of the active and inactive species, it is probably not of physiological importance. By contrast, the second system of activation, i.e. the one involving F_A and Mg-ATP, is most probably physiologically significant. The reaction results in an immediate incorporation of ^{32}P from γ-labeled ATP that accompanies the activation of the enzyme, but no more than 0.1 to 0.2 equivalent of phosphate is incorporated. If EDTA were added in the course of the reaction to block the kinase, the counts would decrease because the enzyme undergoes autodephosphorylation. If a phosphatase inhibitor such as PPi were added, far more counts would be incorporated. Incorporation occurs exclusively in the regulatory subunit, none in the catalytic moiety.

A model was proposed to account for these observations:

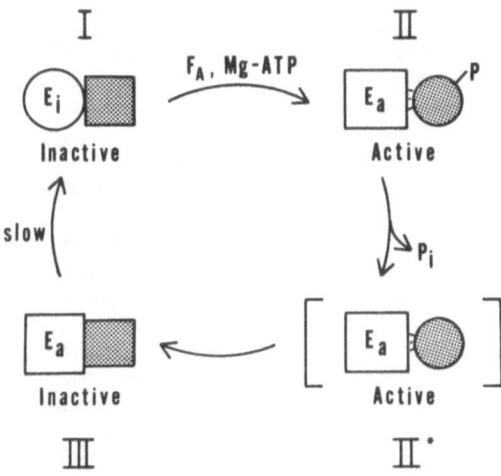

Figure 1.

a) during phosphorylation of the R subunit, there is a synchronous change in conformation of the catalytic subunit, switching it from its inactive to active state. At the same time the regulatory subunit (inhibitor 2), also undergoes changes in conformation, i.e., a slow switch from its inhibitory to a non-inhibitory state. The enzyme ungergoes dephosphorylation, but the non-inhibitory state of the regulatory subunit is maintained for a while even after all the phosphate has been released. After a while the R subunit regains its inhibitory power and the enzyme becomes inactive, even though the catalytic subunit remains in its active conformation. Finally, in a slow reaction ($t_{\frac{1}{2}}$ = 12.5 min) the catalytic subunit returns to its original inactive state. This interconversion seems to have an absolute requirement for inhibitor 2. In other words, we are dealing with a hysteretic system.

Since the regulatory subunit is required for both the activation and inactivation processes, it is clear that its function extends beyond that of a simple inhibitor and that it should be considered as a modulator of enzyme activity as already proposed by Merlevede.

3. Substrate specificity of the various active forms

One might ask whether the 3 active forms of phosphorylase phosphatase (the enzyme activated by F_A, Mn^{2+} alone or trypsin-Mn^{2+}), are identical in terms of their enzymatic activities and substrate specificities. Data obtained indicate that they are not. For example, while the phosphorylated regulatory subunit (RII) of type 2 cAMP-dependent protein kinase proved to be a good substrate for all activated forms of the phosphatase, synthetic peptides patterned according to the phosphorylated site were dephosphorylated only by E_a^{Mn} and E_a^{Tr-Mn}, but hardly at all by $E_a^{F_A}$. Similar data were obtained with peptides patterned according to the phosphorylated sites of the α and β subunits of phosphorylase kinase, and of phosphorylase $\underline{a}$.

4. Hormonal regulation of phosphorylase phosphatase

Glycogen degradation is initiated when a number of enzymes become phosphorylated following hormonal stimulation of adenylate cyclase. To bring about glycogen synthesis and the arrest of glycogenolysis, as observed under the influence of insulin, either protein kinases must be inhibited or protein phosphatases activated. The M_r = 70,000 phosphatase could be a candidate as a target for insulin activation since it acts on glycogen synthase, phosphorylase, the β subunit of phosphorylase kinase and the regulatory subunit of type II cAMP-dependent protein kinase.

How could insulin act on this system? It is known that when insulin binds to its receptor, it triggers the activation of a protein-tyrosine kinase residing in the β subunit. In turn, this tyrosine kinase could act on other kinases or the phosphatase itself. Alternatively, it could elicit the formation of a specific second messenger. On the activation cycle of phosphorylase phosphatase, there are several steps where insulin could act. First, it could, for instance, activate F_A and cause an increase in the rate of phosphorylation of the inactive complex. A drawback to this scenario is that F_A would also phosphorylate and _inactivate_ glycogen synthase, an effect opposite to what would be expected from insulin. Second, it could freeze the enzyme in one of its active structures. Lastly, and perhaps most attractive, insulin (or perhaps a second messenger of the hormone if one did exist) could allow expression of the potential activity hidden in the inactive form III by causing this complex to relax or dissociate, or by acting on another factor that would allow it to dissociate -- just like cAMP causes the dissociation and activation of cAMP-dependent protein kinase. We know that at low concentrations of F_A, 90% of the enzyme can exist in this form. As of yet, no direct effect of insulin and its receptor on the purified phosphatase have been demonstrated. An insulin-dependent tyrosine phosphorylation of the regulatory subunit does occur, but this reaction does not significantly affect enzyme activity. Similar results were obtained with purified EGF receptors. Experiments are being carried out under more physiological conditions, for instance with intact 3T3 cells in culture, where one sees a nice activation of glycogen synthase following stimulation with either insulin or EGF, and some concommitant decrease in phosphorylase activity, but the mechanism of this type of regulation is still unclear. A further understanding of the structural and regulatory properties of phosphorylase phosphatase should help clarify how these hormones operate within the cell.

IV. Type 2 Protein Phosphatases

(1) Protein Phosphatase 2A

Protein phosphatase 2A is a cytosolic enzyme that can be resolved into 3 forms termed $2A_0$, $2A_1$ and $2A_2$ by ion exchange chromatography on DE 52. They have been purified to homogeniety from a number of tissue sources and the properties of the skeletal muscle enzymes are presented in Table 2.

They all contain the same 36 kDa catalytic subunit bound to various other components. This has been suggested to be distinct from the catalytic subunit of protein phosphatase 1 by peptide mapping studies, although it is likely that some structural homology exists. The A subunits are also identical in each form of the enzyme, whereas the B subunit ($2A_1$) and the B' subunit ($2A_0$) are distinct. $2A_2$ is believed to be generated

Table 2. Protein Posphatase-2A

	$2A_0$	$2A_1$	$2A_2$
Subunit Composition	$AB'C_2$	ABC_2	AC
Subunit M_r	A 60 kDa B' 54 kDa C 36 kDa	A 60 kDa B 55 kDa C 36 kDa	A 60 kDa C 36 kDa
High M_r Form	Inactive	Active	Active
Substrate Specificity	Broad	Broad	Broad
Specificity for Phosphorylase Kinase	α	α	α
Inhibition by I-1 & I-2	No	No	No

from $2A_1$ and/or $2A_0$ by dissociation and/or proteolytic removal of the B or B' subunits respectively, and thus $2A_0$ and $2A_1$ are thought to be the forms of the enzyme present in vivo.

An important characteristic of these enzymes is their ability to be activated by polycationic compounds/basic proteins (e.g. polylysine) and polyamines (e.g. spermine). In fact, in its high molecular weight form, $2A_0$ is largely inactive unless assayed in the presence of such basic compounds or after dissociation of the catalytic subunit by ethanol treatment. One attractive hypothesis is that polyamines may be a second messenger involved in insulin action and that some of the effects of insulin may be mediated by this activation of protein phosphatase 2A. However, this remains to be substantiated.

The function of 2A is unclear but it may play a role in the control of the key enzymes regulated by phosphorylation in the pathways of glycolysis, glycogen metabolism, gluconeogenesis and aromatic amino acid breakdown.

(2) Protein Phosphatase 2B

Protein phosphatase 2B is a Ca^{2+}/calmodulin dependent phosphatase of narrow substrate specificity. It has been purified to homogeneity from a number of sources. One such procedure, involving the use of a thiophosphorylated myosin P light chain substrate column, purified the phosphatase from sheep brain and confirmed its identity with a major soluble calmodulin binding protein of brain extracts, calcineurin. It comprises 2 subunits, A and B, present in a 1:1 molar ratio. The B subunit (Mr 19,200) has been sequenced and shown to possess 35% identity with calmodulin (CaM), the regions of homology being most prevalent in the putative Ca^{2+} binding domains. Like CaM, it binds 4 mols of Ca^{2+} per mol with µmolar affinity. The A subunit (Mr 61,000) is the CaM binding subunit and binds this protein with a 1:1 molar ratio. It also contains the phosphatase catalytic site.

In the absence of Ca^{2+} and CaM, the enzyme is inactive. The binding of Ca^{2+} to the B subunit is essential for enzyme-substrate interaction and alone stimulates phosphatase activity to approximately 5% of maximum. Further binding of Ca^{2+}/CaM to the A subunit stimulates activity approximately 10 fold by increasing the Vmax while leaving the Km for substrate unaffected. Maximal activity appears to be achieved in the presence of an additional second metal ion, e.g. Ni^{2+} or Mn^{2+}, which interacts directly with the A subunit. In vivo, it is possible that Mg^{2+} serves as this second metal. Like many CaM regulated enzymes, limited trypsinization converts calcineurin to a form which is spontaneously active, even in the absence of CaM. These changes in the properties of the enzyme are associated with cleavage of the A subunit from a molecular weight of 61,000 to approximately 45,000. Calcineurin is specifically inhibited by trifluoroperazine (a calmodulin antagonist) and this has proved a useful tool when assaying the enzyme in crude extracts.

The function of this enzyme also remains to be elucidated. It is most prevalent in brain with particularly high levels in the neostriatum. It has been reported that the major period of synthesis of the enzyme corresponds to the period of synapse formation, which coupled with its immunohistochemical localisation at postsynaptic densities and dendritic microtubules suggests a role in synaptic function. It has also been proposed that it may function as a mechanism by which signals which act through Ca^{2+} may attenuate the effects of hormones which act through cAMP, via the dephosphorylation of inhibitor 1 in muscle or its brain homologue DARPP. However, experiments to confirm this idea have yet to be performed.

(3)   Protein Phosphatase 2C

This enzyme has also been purified to homogeneity from several sources. Its gel filtration behavior on Sephadex G-100, and migration on SDS PAGE, suggest it is a monomeric enzyme of apparent molecular mass of 45,000. However, recent evidence indicates its behavior on Sephadex G-100 may be anomalous and that is actually a dimer. It is completely dependent upon Mg^{2+} (Ka approximately 1.0 mM), although this can be replaced partly by Mn^{2+}. At present, mechanisms for regulating protein phosphatase 2C have not been identified.

V.   Tyrosine Protein Phosphatases - An Overview

It was not until 1979 that the phosphorylation of proteins on tyrosyl residues was first observed - in immunoprecipitates containing the transforming proteins of certain oncogenic retroviruses. Since then, transforming proteins from a dozen retroviruses have been identified as tyrosine-specific protein kinases (TPK), the most famous example being pp60^{v-src}. A second category of TPK was discovered in association with various growth factor receptors, namely, those for epidermal growth factor (EGF), platelet derived growth factor (PDGF), insulin and insulin-like growth factor (IGF-1) and colony stimulating factor (CSF). Although the last five years has seen a burgeoning of knowledge of tyrosine phosphorylation, an exact understanding of its physiological role has yet to be achieved. However, the association of this activity with growth factor and viral transforming proteins suggests a major role in the control of cell proliferation.

The dephosphorylation of tyrosine residues in proteins was first detected in A431 cell membranes. Treatment of these cells with EGF stimulated the phosphorylation of the EGF receptor and was then followed by slow dephosphorylation. Similarly, data were obtained with cells transformed with temperature-sensitive mutants of Rous Sarcoma Virus. To date, some progress has been made in characterizing these phosphatases.

A major limitation in their study has been the selection and preparation of suitable substrates. Their physiological substrates have yet to be clearly identified; furthermore, they are present in such limited amounts that their use in routine assays would be prohibitive. Therefore, artificial substrates have been used. These include the synthetic acidic copolymer poly Glu-Tyr (4:1), bovine serum albumin, casein, histone, tubulin, IgG and proteins that have been chemically modified to expose their tyrosine side chains. A major drawback has been the low levels of phosphorylation and the fact that phosphorylation can occur at many sites along the polypeptide chain, resulting in non-linear kinetics of dephosphorylation. The choice of an appropriate artificial substrate is crucial since it has been reported that when acidic proteins are used, the major cytosolic TPP activity is measured, whereas with histone, only alkaline phosphatase is detected. Furthermore, although synthetic peptides (e.g. RR-SRC) are readily phosphorylated by the kinases, they appear to be unreactive with the phosphatase.

1. Non-Specific Phosphatases which Dephosphorylate Tyrosine Residues In Proteins

Very weak activities were detected in types 2A, 2B (calcineurin) and 2C. In general, they required metal ions (Mn^{2+} or Mg^{2+}), with optimum pH ranging from 7 to 9. However, subsequently highly purified preparations of protein phosphatase 2B have been shown to dephosphorylate myosin P-light chain that has been phosphorylated on tyrosine with very similar kinetic constants to its dephosphorylation of the same protein phosphorylated on serine.

Alkaline phosphatases from calf intestine, beef liver and E. Coli contain an inherent TPP activity, dephosphorylating PTyr-histones at 5-10 times the rate of PSer-histones, as well as A431 cell membrane proteins labelled on tyrosine residues. Questions of substrate selectivity and high pH optimum will have to be addressed further before their significance as TPPs in vivo can be established.

The same can be said of two categories of acid phosphatases (tartrate sensitive and insensitive). Both have been shown to possess TPP activity.

The tartrate-sensitive human prostatic enzyme specifically dephosphorylates several tyrosine-labelled proteins, showing little activity toward phosphoserine proteins.

Tartrate-insensitive phosphatase from human astrocytoma cell membranes and bovine bone also dephosphorylate free phosphotyrosine and phosphotyrosyl proteins with little activity toward other phosphorylated amino acids or PSer histones. A pH optimum around 7 toward phosphotyrosyl proteins suggests some possible physiological role, but further characterization is required.

2. Distinct Novel Tyrosine Protein Phosphatase

Multiple and distinct TPPs have been detected in a number of tissues and cell lines, variously distributed between the soluble and particulate fractions; they have been separated from the major serine and threonine enzymes.

In rabbit, using modified BSA as substrate, the following prevalence was found: kidney > liver >> brain > lung = heart > skeletal muscle. The concentration in kidney (420 u/g protein; 1 unit = 1 nmol Pi released per minute) is of the same order of magnitude as that of type 1 and 2A phosphorylase phosphatase (approx. 2000 and 1300 u/g, respectively) found in rabbit skeletal muscle.

Sixty per cent of the kidney cytosolic TPP could be adsorbed onto Zn^{2+}-IDA agarose; subsequent chromatography resolved two forms of the activity. Form I (M_r 34k) displayed a neutral pH optimum and an absolute requirement for sulfhydryl compound, whereas form II (37k) exhibited an acidic pH optimum and was active without SH compounds. After a purification of approx. 1,000-fold, the final material had a specific activity of 200-300 units/mg protein. These are the most highly purified preparations of TPP reported to date.

3. Properties of Tyrosine Protein Phosphatases

The specific TPPs have neutral pH optima. Like the acid phosphatases, they are maximally active in the presence of EDTA whereas the alkaline phosphatases, which are Mg^{2+}-dependent, are inhibited by EDTA. They are unaffected by fluoride and tetramisole (an alkaline phosphatase inhibitor) but inhibited by vanadate and molybdate and Zn^{2+} in the uM range; Zn^{2+}-IDA-agarose affinity columns have been used for their purification. Vanadate has been shown to mimic the effects of insulin and stimulate the phosphorylation on a Tyr residue of the 95k subunit of the insulin receptor [in rat adipocytes]; it also increased the level of PTyr and reduced transformation of NRK-1 cells. Such observations further stress the importance of obtaining an understanding of the structure and regulation of these enzymes.

4. Future Prospectives

Since the partial purification of the two TPPs from rabbit kidney by Shreiner and Brautigan, little further progress has been reported in the study of this class of enzymes. The literature is replete with reports of multiple forms of TPPs which have been separated by various means from a vast array of tissues and cell lines - a situation reminiscent of that which faced investigators working on serine and threonine specific phosphatases some twenty years ago. So far, no TPPs have been reported to be obtained in a homogeneous state. Thus, the purification and characterization of these enzymes must be the first goal. One may then address the question of possible structural relationships between the various forms of TPP and the Ser/Thr specific enzymes. Are there differences in the properties that would allow one to distinguish them from one another, thus allowing an investigation of their potential regulation in response to hormones and growth factors? Such studies should further our understanding of the role of tyrosine phosphorylation in cell proliferation.

REFERENCES

Fischer, E.H., Brautigan, D.L., (1982) TIBS, Vol 7, No. 1, p. 3-4

Fischer, E.H., (1983) Bulletin de L'Institut Pasteur 81, p. 7-31.

Krebs, E.G., (1981) Curr. Topics in Cell. Reg., 18, 401-419.

Cohen, P. (1982) Nature 296, 613-620.

Ballou, L.M., Fischer, E.H., (1986) The Enzymes, Vol. 17, 311-361.

HIGH MOLECULAR WEIGHT PHOSPHORYLASE PHOSPHATASE FROM MUSCLE

GLYCOGEN PARTICLES

Emma Villa-Moruzzi

Istituto di Patologia Generale, Scuola Medica
Via Roma 55, 56100 Pisa, Italy

INTRODUCTION

Type-1 Protein Phosphatase (Phosphorylase Phosphatase) (1) is a key enzyme of glycogen metabolism: it activates Glycogen Synthase and inactivates Phosphorylase, thus stimulating glycogen synthesis. One would expect Phosphatase to be the target of both insulin and adrenaline, the major hormonal regulators of this metabolism. And in fact it is known that adrenaline inhibits phosphatase through the activation of phosphatase inhibitor-1 (2) and possibly also by removing the enzyme from its substrates (3). On the other hand, stimulation of Phosphatase by insulin has never been proven. In the recent years type-1 Phosphatase has been purified to homogeneity and there were great advances in the understanding of the Phosphatase system and of its activation by the kinase F_A (1,4). It has also been realized recently that the same Phosphatase is present at different cell sites (cytosol, glycogen particles and microsomal membranes) where the catalytic subunit is complexed to different regulatory subunits (5). Since the studies on the purified enzyme did not clarify yet some aspects of its regulation, and specifically those related to insulin stimulation, it may be useful to investigate Phosphatase in more crude fractions. It is known in fact that Phosphatase in a crude system, such as muscle extract, has different properties from the purified enzyme and displays a much higher molecular weight (250,000 by gel filtration). High molecular weight Phosphatase complexes are present not only in the cytosol (6), but also in the Phosphatase obtained from glycogen particles (7) and from microsomal membranes (E. Villa-Moruzzi and L.M.G. Heilmeyer Jr., in preparation). It is possible that such complexes are the primary target of the hormone-related signals directed toward Phosphatase. Here it is reported the isolation of high molecular weight Phosphatase from skeletal muscle glycogen particles and its disruption in smaller complexes by affinity chromatography on poly-lysine.

RESULTS AND DISCUSSION

Glycogen particle Phosphorylase Phosphatase was obtained following digestion of glycogen by α-amylase, removal of the microsomal membranes by ultracentrifugation, and batch chromatography on DEAE-Sepharose in order to remove the other enzymes of glycogen metabolism present in the fraction (8). The Phsphatase thus obtained is all of type-1 and displays Mr of ≈300,000 when subjected to gel filtration chromatography on FPLC using a Superose 6 column (Fig.1, a). This Mr is higher than the 250,000 reported by Khatra and Soderling (7). The

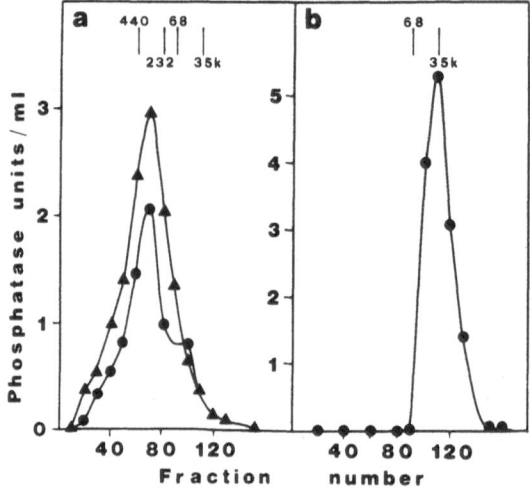

Fig. 1. Gel filtration on FPLC (Pharmacia) of Phosphorylase Phosphatase from the glycogen particles (after digestion of glycogen and removal of the other enzymes of glycogen metabolism, see text and ref. 8) before (a) and after (b) treatment with trypsin (20 µg/ml for 10 min at 30°C). A Superose™ 6 HR 10/30, Pharmacia (25 ml) column equilibrated in 20 mM imidazole, pH 7.5, 5% glycerol, 0.01% Brij-35, 15 mM 2-mercapto-ethanol was used at a flow rate of 24 ml/h. The volume of the sample was 200 µl, the fraction size was 64 µl and fraction collecting was started after 12 ml. Assays:●, spontaneous activity; ▲, activity after trypsin (4,8). The marker proteins were: ferritin (440k), catalase (232k), bovine serum albumin (68k) and phosphatase catalytic subunit (35k by gel filtration).

discrepancy may be due either to the different gel filtration conditions or it may indeed reflect a real difference between complexes (e.g. partial proteolysis of some component, see below). The enzyme obtained from the particles is mostly active, displaying ≈20% further activation by trypsin treatment. The activation by trypsin alone, and not by trypsin and Mn^{2+} may indicate the presence of some regulatory component other than inhibitor-2 (discussed in 4,8,10). Moreover, several sources reported that inhibitor-2 is not present in the particles (7,9,8). Although the peak fractions display Mr 300,000, the whole profile of the chromatogram may indicate the presence of complexes of different sizes, possibly resulting from multiple aggregation of some basic components. Whatever components are present other than catalytic subunit, they are sensitive to proteolysis. In fact trypsin treatment of the Mr 300,000 complex yields an isolated catalytic subunit of Mr 35,000 by gel filtration (Fig. 1,b).

Alternatively to tryptic proteolysis, a milder way of disrupting the Mr 300,000 complex is by chromatography on poly-lysine-Affi-Gel 10 (Fig. 2, a). This type of chromatography allows to separate one peak of almost inactive phosphatase (A) from a second peak containing enzyme which is mostly active (B). When subjected to gel filtration, the enzyme of peak A displays Mr of 70,000 (Fig. 2,b) and requires Mn^{2+} in addition to trypsin for activation. Both these characteristics are in common with the inactive phosphatase complex (made out of catalytic subunit and inhibitor-2) which is isolated from the cytosol following acetone treatment (4,10). The finding of this complex in the glycogen particles is unexpected and may indicate either some contamination from the cytosol or that some inhibitor-2 has been absorbed to the particles and bound to free catalytic subunit. The enzyme of peak B (Fig. 2,a) is mostly active and some additional activation can be achieved by using trypsin either alone or together with Mn^{2+}. This type of activation would indicate the presence of regulatory components other than inhibitor-2 (see above). Analysis of peak B by gel filtration shows that it is indeed an aggregation of different components of Mr ≈130,000, 52,000 and 35,000. While the latter clearly represents isolated catalytic subunit, the 130,000 peak may well be a complex between catalytic subunit and the G-subunit recently described by Cohen and coll. (9). The G-subunit is very sensitive to proteolysis (9) either during purification, which may account for the 52,000 component of Fig. 2,c, or by trypsin, releasing free catalytic subunit (as shown in Fig. 1,b). In turn the free catalytic subunit shown in Fig. 2,c may be either present in the particles from the beginning or be produced during preparation, following endogenous proteolysis of the G-subunit. The smaller complexes (Mr 130,000 or lower) can be fairly explained by combinations of catalytic subunit and G-subunit (or its degradation products). On the other hand it is not known whether the Mr ≈300,000 complex is due to aggregation of two Mr 130,000 complexes or if some other unknown component is present. The fact that also in other cell fractions (cytosol and microsomal membranes) where the G-subunit is not expected to be present Phosphorylase Phosphatase displays Mr of 250,000-300,000 may be more in

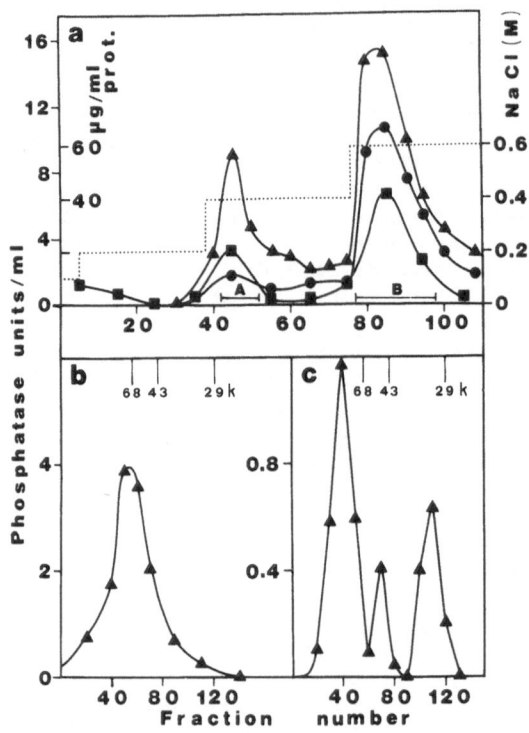

Fig. 2. a. Poly-lysine-Affi-Gel 10 chromatography of the same
phosphatase fraction used in Fig. 1,a. The
chromatography medium was prepared as described in (8)
and the 2.5 ml column was equilibrated in the
buffer described in the legend of Fig.1, but
containing also 0.1 mM EDTA, 0.1 mM benzamidine,
0.002% PMSF (phenylmethanesulfonyl fluoride).
The concentrations of NaCl used are shown on
the chromatogram. Flow rate was 5 ml/h and fraction
size 0.5 ml. Assays: ● , spontaneous activity; ▲ ,
activity after trypsin and Mn^{2+} (4,8); ■ , protein
(assayed according to Bradford, 11, using bovine serum
albumin as standard).
b and c. Gel filtration of phosphatase pool A (b) and
pool B (c) of Fig. 2 a, on a 0.95 x 45 cm column of
Bio-Gel A-0.5 m equilibrated in the same buffer
described in the legend of Fig. 1. The volume
of the sample applied was 180 µl, the flow rate was
3.6 ml/h and the fraction size was 100 µl. The marker
proteins used were: bovine serum albumin (68k),
ovalbumin (43k) and carbonic anhydrase (29k).

favour of the presence of some additional unknown subunit.

Further studies on these different complexes may help to clarify (i) the role of regulatory subunits in Phosphatase activation-inactivation and (ii) the relationship between the Phosphatase from the glycogen particles and Phosphorylase Phosphatase from cytosol or microsomal membranes.

ACKNOWLEDGMENTS

Supported by Muscular Dystrophy Association and M. P. I. Rome

REFERENCES

1. Ballou, L. M. & Fischer, E. H. Phosphoprotein Phosphatases, in: "The Enzymes", vol. XVII, Part A, P. D. Boyer and E. G. Krebs, eds., Academic Press, New York, (1986), pp. 312-361.
2. Nimmo, G. A. & Cohen, P. Regulation of glycogen metabolism. Phosphorylation of inhibitor-1 from rabbit skeletal muscle, and its interaction with protein phosphatase-III or -I. Eur. J. Biochem. 87: 353 (1978).
3. Villa-Moruzzi, E. Effects of streptozotocin-diabetes, fasting and adrenaline on phosphorylase phosphatase activities of rat skeletal muscle. Mol. Cell. Endocr. 47:43 (1986).
4. Villa-Moruzzi, E., Ballou, L. M. & Fischer, E. H. Phosphorylase phosphatase. Interconversion of active and inactive forms. J. Biol. Chem. 259: 5857 (1984).
5. Villa-Moruzzi, E. & Heilmeyer, L. M. G. Jr. Distribution of phosphorylase phosphatase catalytic subunit among cytosol, glycogen and membranes. A model for phosphatase regulation in skeletal muscle. Adv. Prot. Phosphatases 3: 225 (1986).
6. Lee, E. Y. C., Silberman, S. R., Ganapathi, M. K., Petrovic, S. & Paris, H. The phosphoprotein phosphatases: properties of the enzymes involved in the regulation of glycogen metabolism. Adv. Cycl. Nucl. Res. 13: 95 (1980).
7. Khatra, B. S. & Soderling, T. R. Rabbit muscle glycogen-bound phosphoprotein phosphatases:substrate specificities and effects of inhibitor-1. Arch. Biochem. Biophys. 227: 39 (1983).
8. Villa-Moruzzi, E. Purification and inactivation-reactivation of phosphorylase phosphatase from the protein-glycogen complex. Arch. Biochem. Biophys. 247: 155 (1986).
9. Strålfors, P., Hiraga, A. & Cohen, P. The protein phosphatases involved in cellular regulation. Purification and characterization of the glycogen-bound form of protein phosphatase-1 from rabbit skeletal muscle. Eur. J. Biochem. 149: 295 (1985).
10. Ballou, L. M., Villa-Moruzzi, E. & Fischer, E. H. Subunit structure and regulation of phosphorylase phosphatase. Curr. Topics Cell. Regul. 27: 183 (1985).
11. Bradford, M. M. A rapid and sensitive method for quantitation of microquantities of protein utilizing the principle of protein-dye binding. Anal. Biochem. 72:248(1976).

SUBCELLULAR DISTRIBUTION AND CHARACTERISTICS OF HEPATIC PROTEIN

PHOSPHATASES ACTING ON GLYCOGEN PHOSPHORYLASE AND ON GLYCOGEN SYNTHASE

Mathieu Bollen, Jackie Vandenheede, Jozef Goris and Willy Stalmans

Afdeling Biochemie
Faculteit Geneeskunde
Katholieke Universiteit Leuven
Belgium

INTRODUCTION

An important part of the hepatic phosphorylase phosphatase activity is associated in a largely latent form with the nuclear fraction.[1] In contrast, the phosphatase activity that is responsible for the activation of glycogen synthase is located in the post-mitochondrial supernatant, which comprises the glycogen fraction, the microsomes, and the post-microsomal supernatant. In our laboratory the synthase-phosphatase activity is recovered partly in the supernatant and partly as an enzyme-glycogen complex.[2] These two enzyme fractions have distinct regulatory properties.[3] No significant synthase-phosphatase activity could be detected in the microsomal fraction.[3] In contrast, Cohen's group[4] recently reported the presence of synthase-phosphatase activity in the microsomal as well as in the glycogen fraction; they adduced evidence in favour of a single protein phosphatase (type-1) that also displays phosphorylase phosphatase activity. In view of these discrepant data we have reassessed the characteristics of the protein phosphatases present in the glycogen fraction, the microsomal fraction, and the cytosol. The present report summarizes the results and offers some explanations.

EXPERIMENTAL PROCEDURES

A post-mitochondrial supernatant (8000 g for 10 min) was prepared from glycogen-depleted livers (fasted, glucagon-treated rats) in the presence of protease inhibitors. Microsomes were sedimented at high speed. Purified particulate liver glycogen was added to the high-speed supernatant, and a subsequent centrifugation separated the glycogen pellet from the cytosol.[2] The sedimented fractions were washed once. Unless otherwise specified, all cell fractions were assayed for protein phosphatase activities at a final concentration equivalent to a 10% liver homogenate. After treatment with trypsin, the fractions were diluted 5-fold before the assay of phosphorylase phosphatase. The latter enzyme was assayed by its ability to dephosphorylate ^{32}P-labeled phosphorylase,[1] and glycogen-synthase phosphatase by its ability to activate glycogen synthase b isolated from dog liver.[2] Synthase phosphatase was partially purified from the glycogen fraction.[3] Homogeneous preparations of modulator[5] and of deinhibitor[6] were obtained as described.

RESULTS AND DISCUSSION

Table 1. Distribution of protein phosphatase activities in the post-
         mitochondrial supernatant

Cell fraction	Phosphorylase phosphatase		Synthase phosphatase
	% of total recovered activity[a]		
Cytosol	39 ± 1	(13 ± 2)[b]	18 ± 4
Glycogen	38 ± 2	(7 ± 1)[b]	73 ± 5
Microsomes	23 ± 2	(18 ± 2)[b]	9 ± 2

[a] Mean values ± S.E.M. (4 preparations).
[b] -fold increase of activity by trypsin treatment.

     Table 1 illustrates the distribution of the protein phosphatase activi-
ties in the three cell fractions that make up the post-mitochondrial supernatant. Of the total phosphorylase phosphatase activity, the major part was
about equally distributed among the cytosol and the glycogen fraction, a
lesser part being associated with the microsomes. In contrast, the vast majority of the synthase phosphatase activity was bound to the glycogen fraction; much less was soluble, and the activity associated with the microsomes
was barely above the detection level. Table 1 also shows that the phosphorylase phosphatase activity could be enhanced manyfold in every fraction by
incubation with trypsin. However, this treatment destroyed all synthase
phosphatase activity (not shown). These results indicate that, besides a
catalytic subunit, protein phosphatases require a (trypsin-sensitive)
"specifying subunit" in order to recognize glycogen synthase _b_ as substrate.

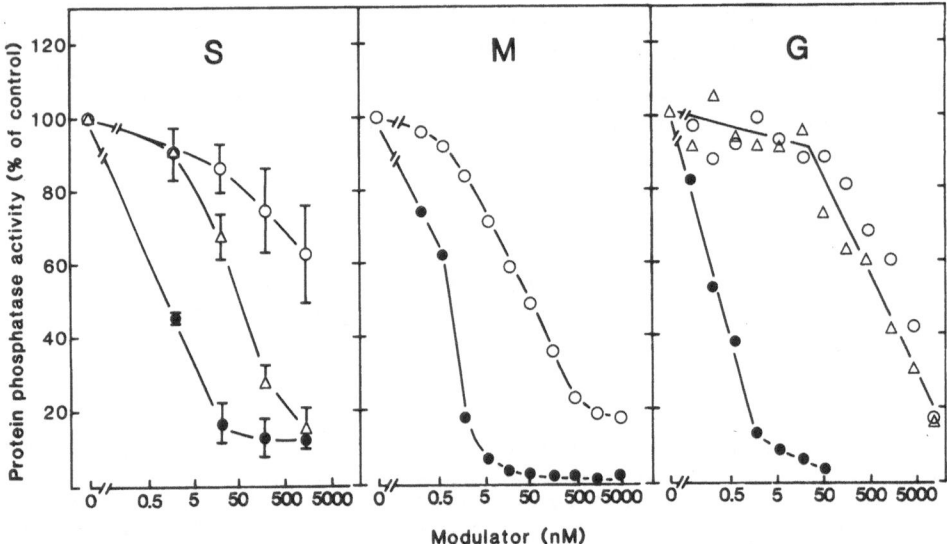

Fig. 1. Effect of modulator concentration on the protein phosphatase activi-
        ties in subcellular fractions. The post-microsomal supernatant (S),
        the microsomal fraction (M) and the glycogen fraction (G) were pre-
        incubated (20 min at 25°C) at a concentration of 14% with the indi-
        cated concentrations of modulator, and then assayed for synthase
        phosphatase ($\triangle$) and phosphorylase phosphatase (O). The latter ac-
        tivity was also assayed in cell fractions that had been treated with
        trypsin prior to the addition of modulator (●).

In contrast, this subunit appears to restrain considerably the phosphorylase phosphatase activity. The high synthase phosphatase activity of the microsomal enzyme, as reported by Alemany and Cohen,[4] could then be due to their use of minimally phosphorylated synthase.

Complete inhibition by modulator protein (inhibitor-2) is a common feature of the class of ATP,Mg-dependent[7] or type-1[8] protein phosphatases. Over 80% of the synthase phosphatase activities (soluble as well as glycogen-bound) belong to the latter class (Fig. 1). In the microsomal fraction the low activity precluded an accurate characterization. The phosphorylase phosphatase activities associated with glycogen and with the microsomes were again largely (over 80%) inhibited by modulator. However, a variable part (20-80%) of the soluble phosphorylase phosphatase activity was clearly insensitive to modulator (Fig. 1). The latter enzyme has a distinct chromatographic behaviour, and displays negligible synthase phosphatase activity (not illustrated).

Trypsin treatment increased the modulator-sensitivity of phosphorylase phosphatase to a common value in each fraction (Fig. 1). However, the native enzymes differed markedly in their resistance to the modulator. Indeed, trypsin treatment increased the sensitivity of the microsomal and soluble enzymes about 50-fold, whereas the glycogen-bound phosphatase became about 2000 times more sensitive. The major difference with the report by Alemany and Cohen[4] is that in our conditions the native enzymes are circa 50-fold less sensitive to the modulator. It is shown in Fig. 2 that this quantitative difference is presumably due to the dilution at which the enzyme is used. Dilution over a 25-fold range increased the modulator-sensitivity of the glycogen-bound phosphorylase phosphatase over a 1000-fold range (Fig. 2A). At the highest dilution the native enzyme became as sensitive as the trypsin-treated phosphatase; the modulator-sensitivity of the latter enzyme was not affected by dilution over the same range (Fig. 2B). It is also shown in Fig. 3 that the progressive dilution of the glycogen-bound enzyme over a

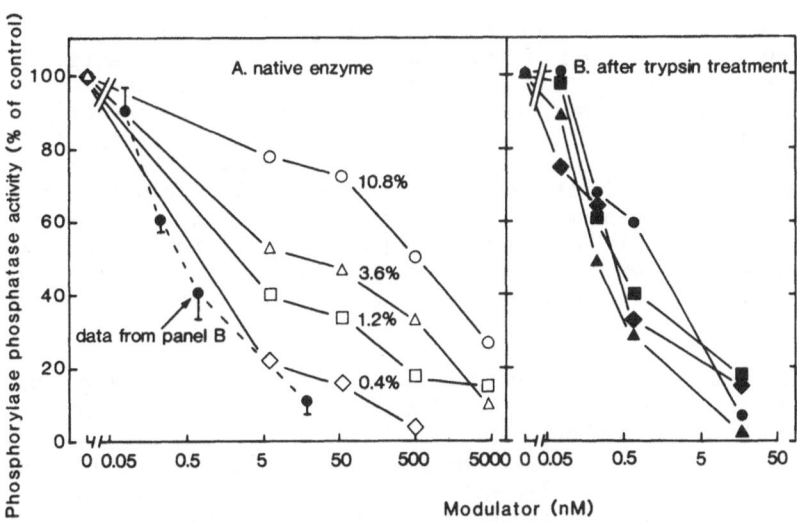

Fig. 2. Effect of the concentration of the glycogen-bound phosphorylase phosphatase on its sensitivity to inhibition by the modulator. The native (A) and trypsin-treated preparation (B) were diluted over a 25-fold range prior to incubation with modulator as in Fig. 1. The figures in the graph represent the final concentrations in the phosphatase assay.

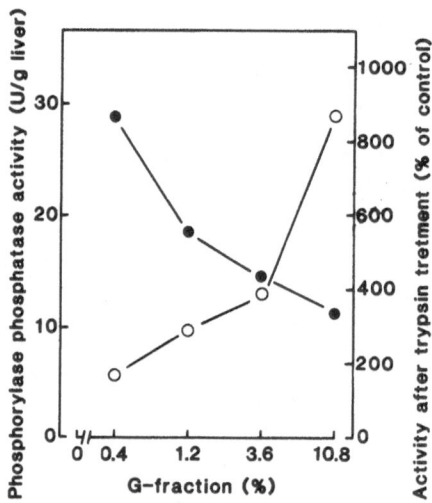

Fig. 3. Effect of dilution on the
specific activity of the
glycogen-bound phosphorylase
phosphatase (●) and on its
activation by trypsin (O).
Data obtained from the experiment shown in Fig. 2.

25-fold range caused a 3-fold increase of its specific activity and a 5-fold
decrease in the extent of activation by trypsin.

We[4] reported that the addition of a deinhibitor protein, which antagonizes the inhibitory action of the modulator, increased the activity of
phosphorylase phosphatase 5-fold in a post-mitochondrial supernatant, but
did not affect synthase phosphatase activity. We have now investigated the
behaviour of the enzymes in the individual cell fractions (Fig. 4). The
deinhibitor increased the activity of phosphorylase phosphatase 7-fold in
the cytosol and 5-fold in the microsomal fraction, but only 1.5-fold in the
glycogen fraction. Synthase phosphatase activity was not affected by the deinhibitor. Interestingly, after further purification from the enzyme-glycogen complex, the phosphorylase phosphatase activity was markedly stimulated
by the deinhibitor, but the synthase phosphatase activity remained insensitive (Fig. 4).

We conclude that the structure of the glycogen-bound enzyme differs
considerably from that of the cytosolic and microsomal modulator-sensitive
protein phosphatases, although they all share apparently the (type 1) catalytic subunit of the ATP,Mg-dependent phosphatase. The present data support a
multi-subunit model of the glycogen-bound protein phosphatase, where the
catalytic subunit is associated with one or several other polypeptides that
can dissociate upon dilution; these subunits ensure recognition of liver
glycogen synthase b, inhibition of phosphorylase phosphatase activity,
resistance to inhibition by modulator protein, high-affinity binding to
glycogen, and allosteric inhibition by phosphorylase a.[3,4]

ACKNOWLEDGEMENTS

Supported by the Belgian Fund for Scientific Medical Research (Grant
3.0051.82) and by the *Onderzoeksfonds KULeuven.*

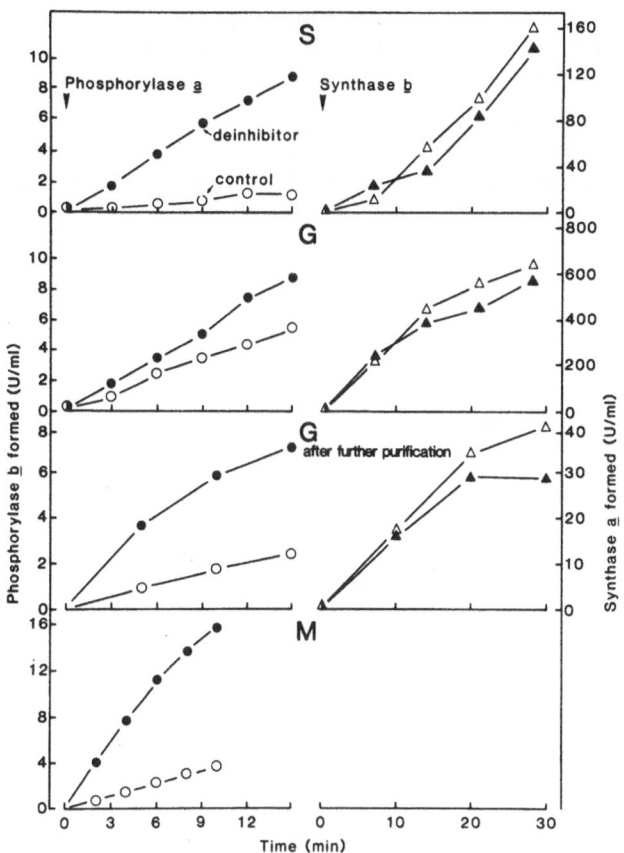

Fig. 4. Effect of the deinhibitor protein on the
activity of protein phosphatases in subcellular fractions. Time course of the dephosphorylation of phosphorylase (left) and the
activation of glycogen synthase (right) by
phosphatases present in the post-microsomal
supernatant (S), the microsomal fraction
(M), the glycogen fraction as such (G), and
partially purified enzyme thereof. Open symbols: no addition; filled symbols: presence
of a saturating amount of deinhibitor.

REFERENCES

1. F. Doperé and W. Stalmans, Release and activation of phosphorylase phosphatase upon rupture of organelles from rat liver, Biochem. Biophys.
   Res. Commun. 104:443 (1982).
2. F. Doperé, F. Vanstapel, and W. Stalmans, Glycogen-synthase phosphatase
   activity in rat liver, Eur. J. Biochem. 104:137 (1980).
3. W. Stalmans, L. Mvumbi, and M. Bollen, Properties and regulation of glycogen synthase phosphatase in the liver, Adv. Prot. Phosphatases 2:333
   (1985).
4. S. Alemany, S. Pelech, C.H. Brierley, and P. Cohen, The protein phosphatases involved in cellular regulation, Eur. J. Biochem. 156:101 (1986).
5. S.D. Yang, J.R. Vandenheede, and W. Merlevede, A simplified procedure for
   the purification of the protein phosphatase modulator (inhibitor-2)
   from rabbit skeletal muscle, FEBS Lett. 132:293 (1981).

6 .M. Bollen, F. Doperé, J. Goris, W. Merlevede, and W. Stalmans, The nature of the decreased activity of glycogen synthase phosphatase in the liver of the adrenalectomized starved rat, <u>Eur. J. Biochem.</u> 144:57 (1984)

7. W. Merlevede, Protein phosphates and the protein phosphatases. Landmarks in an eventful century, <u>Adv. Prot. Phosphatases</u> 1: 1 (1985).

8. T.S. Ingebritsen and P. Cohen, Protein phosphatases: Properties and role in cellular regulation, <u>Science</u> 221:331 (1983).

ACUTE HORMONAL REGULATION OF LIPOLYSIS AND STEROIDOGENESIS

S.J. Yeaman, S.R. Cordle, R.J. Colbran, A.J. Garton and
R.C. Honnor

Department of Biochemistry, University of Newcastle
Newcastle upon Tyne, NE1 7RU, U.K.

Adipose tissue triacylglycerol is the major energy store within the body, with approximately 90% of energy reserves being stored in this form. Triacylglycerol accumulates in highly-specialised adipocytes where it occupies the bulk of the intracellular space. Not surprisingly the synthesis and mobilisation of triacylglycerol is under acute hormonal and neural control. Adipose tissue is subject to sympathetic innervation, with catecholamine, (primarily noradrenaline), released from the nerve endings acting as a potent lipolytic agent. Circulating hormones which have lipolytic actions include adrenaline, ACTH and glucagon, whilst the major antilipolytic hormone is insulin, with other agents such as adenosine and nicotinic acid also having antilipolytic effects.

The enzyme responsible for lipolysis is hormone-sensitive lipase (HSL) which is capable of cleaving all three ester bonds in triacylglycerol[1]. However adipose tissue also contains relatively large amounts of a distinct monoacylglycerol lipase[2] and it is thought that, in vivo, this lipase is mainly responsible for cleavage of the 2-monoacylglycerol intermediate. Hence the two lipases interact to cause complete hydrolysis of the triacylglycerol stores[3], but it seems clear that the control of lipolysis is exerted via regulation of HSL. HSL has higher activity against diacylglycerol than against triacylglycerol and so cleavage of the primary bond of triacylglycerol is truly the rate-limiting step in the hydrolysis.

To date, the lipolytic actions of the various hormones can be fully accounted for by their ability to stimulate adenylate cyclase and thereby raise the intracellular concentration of cyclic AMP. This in turn activates cyclic AMP-dependent protein kinase which then phosphorylates and activates HSL. Although this scheme was first postulated as early as 1970[4] it is only recently that details of the molecular events involved in regulation of HSL have become understood.

Using the activation state of cyclic AMP-dependent protein kinase as an indicator of the intracellular concentration of cyclic AMP it has been demonstrated that there is excellent correlation between the level of cyclic AMP and the rate of lipolysis, at activation ratios up to 0.3-0.4[5]. At these values lipolysis is maximal and further increases in the activity ratio are without effect on the rate of lipolysis. This relationship between concentrations of cyclic AMP and lipolysis is true, independent of the agonists and inhibitors used to vary the levels of cyclic AMP. Insulin appears to have two antilipolytic actions[6]. When the activity ratio of cyclic AMP-dependent protein kinase is below 0.2 the action of insulin is primarily to decrease this ratio and hence reduce phosphorylation of HSL. At higher activity ratios however insulin appears to exert an additional antilipolytic action which is independent of observed changes in the levels of cyclic AMP. The mechanism of this effect remains unknown but one possibility is that insulin causes activation of one of the cytosolic protein phosphatases.

HSL was first identified and partially purified from rat adipose tissue in 1977[7] and subsequently has been purified to near-homogeneity[1]. It consists of a single subunit type of apparent molecular weight 84,000. The purified enzyme is capable of hydrolyzing tri-, di and monoacylglycerol and also cholesterol esters. Phosphorylation of the enzyme at a single site by cyclic AMP-dependent protein kinase causes several-fold activation[8] and subsequent dephosphorylation by one or more cytosolic adipose tissue protein phosphatases reverses this activation[9].

In common with other lipases HSL has a high turnover number and the enzyme protein is present in the cell at low concentration. This, coupled with the fact that the enzyme is closely associated with phospholipids, has made rat HSL resistant to purification in sufficient quantities to allow studies of the physicochemical properties of the enzyme. This led us to seek an alternative source of HSL and we have subsequently purified HSL from bovine perirenal adipose tissue[10]. The characteristics of the bovine lipase are essentially identical to those reported for the enzyme from rat[1] and swine[11] but utilization of bovine tissue allows milligram quantities of purified enzyme to be obtained relatively quickly and easily. This has allowed production of polyclonal antibodies against HSL and also commencement of protein chemical studies on the enzyme. In this regard we have isolated a thermolytic phosphopeptide containing the serine residue phosphorylated by cyclic AMP-dependent protein kinase. This phosphopeptide, containing approximately 5 residues has been purified by gel filtration and reverse-phase HPLC. Preliminary sequence analysis indicates that it possesses two consecutive arginine residues on the amino terminal side of the phosphorylated serine (A.J. Garton, S.J. Yeaman, D.G. Campbell and P. Cohen, unpublished). The presence of two adjacent basic residues is known to be a recognition site for the cyclic AMP-dependent protein kinase[12].

Phosphorylation of HSL in response to catecholamines takes place at a single serine residue recovered on a small phosphopeptide with chromatographic properties identical to the phosphopeptide isolated following phosphorylation of HSL in vitro by cyclic AMP-dependent protein kinase[13]. A second "basal" site also becomes labelled in adipocytes in the absence of any added hormone. The role of this "basal" site has yet to be established but it is apparently not involved directly in controlling the activity of HSL as the extent of phosphorylation of that site does not change when lipolysis is stimulated by catecholamines. This "basal" site can be phosphorylated in vitro by cyclic GMP-dependent protein kinase[14] and by a cyclic nucleotide-independent protein kinase termed glycogen synthase kinase 4 (P. Strålfors, unpublished) but it is not known which, if any, of these kinases is responsible for phosphorylation of that site in the adipocyte.

An exciting recent development in the study of lipolysis is the finding that the transport of free fatty acids out of the adipocyte is also under acute hormonal control. Transport across the membrane is stimulated by adrenaline[15], a process apparently mediated by phosphorylation of a transport protein by cyclic AMP-dependent protein kinase[16]. As with the regulation of HSL the effect of the lipolytic hormones is antagonised by insulin[17], but it is not yet clear whether this is a direct result of a reduction in the intracellular concentration of cyclic AMP.

Steroidogenesis occurs in specialised cell types in tissues such as adrenal cortex, ovaries and testes and is under the control of a variety of hormones, with the particular hormone being determined by the receptor type present on the cell surface. Steroidogenesis is stimulated both by hormones acting through cyclic AMP (e.g. ACTH, LH) and by hormones acting through IP_3 and Ca^{2+} ions (e.g. angiotensin II).

One of the rate-limiting steps in steroidogenesis is the supply of free cholesterol to the intramitochondrial side-chain cleavage system. Several steps are involved in delivery of the free cholesterol. One such step involves the activity of a neutral cytosolic cholesterol ester hydrolase which releases cholesterol from the stores of cholesterol esters present in lipid droplets in the cytoplasm. It was first shown in the 1970s that this enzyme could be activated by steroidogenic hormones in vivo[18] and that its activity in a partially-purified preparation could be elevated by incubation in the presence of cyclic AMP-dependent protein kinase and Mg-ATP[19,20]. The cholesterol ester hydrolase protein was first identified by this laboratory[21] using $[^3H]$ DFP, an inhibitor which covalently modifies the protein. This approach led to identification of the protein as a polypeptide of M_r 84,000, the same as that of adipose tissue HSL. In collaboration with the group of Per Belfrage in Lund we then compared the properties of the adipose tissue lipase with those of the adrenal cholesterol ester hydrolase and reached the conclusion that the two enzymes were very similar if not identical[22]. Again the enzyme has been shown directly to act as a substrate for cyclic AMP-dependent protein kinase with phosphorylation causing activation[23]. The properties of the enzyme in corpus luteum are also identical to those of the enzyme from adrenal and adipose tissue[24]. Hence HSL may have an additional physiological role in steroid-producing tissues in controlling the supply of cholesterol for steroidogenesis.

In addition to the stimulation of HSL and the resulting breakdown of cholesterol esters, several other steps are involved in increasing the rate of delivery of cholesterol to the mitochondrial side-chain cleavage system. In particular there is evidence for a sterol carrier protein involved in transport of the cholesterol to the inner membrane of the mitochondria[25,26]. The action of this protein may explain some observations concerning the role of a postulated "short half-life protein" which, when its synthesis is blocked by the protein synthesis inhibitor cycloheximide, results in loss of the ability of steroidogenesis to be stimulated by ACTH. Rats treated with cycloheximide accumulate cholesterol in their outer mitochondrial membranes but addition of SCP_2 to these isolated mitochondria causes transfer of cholesterol from the outer to the inner membranes[26]. Another related finding is that ACTH induces rapid synthesis of a protein which is co-translationally phosphorylated[27,28]. Induction of this protein is blocked by cycloheximide. However the molecular weight (M_r 28,000) of this protein is different from that of SCP$_2$ (M_r 13,500) and the physiological role of this induced protein in steroidogenesis has yet to be elucidated.

One final point of some interest is that a hormone-sensitive cholesterol ester hydrolase has also been identified in arterial smooth muscle cells and in macrophages [29,30]. In both cases the activity of the cholesterol ester hydrolase is increased following incubation with Mg-ATP and cyclic AMP-dependent protein kinase. Both these cell types have been implicated in the formation of lipid-rich foam cells which are present in atherosclerotic lesions and which contain large amounts of cholesterol esters and triacylglycerol. Furthermore herpesvirus infection of arterial smooth muscle cells leads to an inability of the cholesterol ester hydrolase to become activated and also to accumulation of cholesterol esters[31]. The possible relationship of this activatable cholesterol ester hydrolase to HSL has yet to be investigated but it is an exciting possibility that a defect in the regulation of HSL may be involved in pathological states such as atherosclerosis.

Acknowledgements   Work in the authors' laboratory was supported by grants from the Medical Research Council, U.K. and the British Diabetic Association. Dr. S.J. Yeaman is a Lister Institute Research Fellow.

References

1. G. Fredrikson, P. Strålfors, N.Ö. Nilsson and P. Belfrage, "Hormone-sensitive lipase of rat adipose tissue. Purification and some properties", J. Biol. Chem. 256: 6311 (1981).
2. H. Tornqvist and P. Belfrage, "Purification and some properties of a monoacylglycerol-hydrolyzing enzyme of rat adipose tissue", J. Biol. Chem. 251: 813 (1976).
3. G. Fredrikson, H. Tornqvist and P. Belfrage, "Hormone-sensitive lipase and monoacylglycerol lipase are both required for complete degradation of adipocyte triacylglycerol", Biochim. Biophys. Acta 876: 288 (1986).
4. J.D. Corbin, E.M. Reimann, D.A. Walsh and E.G. Krebs, "Activation of adipose tissue lipase by skeletal muscle cyclic adenosine 3',5'-monophosphate-stimulated protein kinase", J. Biol. Chem. 245: 4849 (1970).
5. R.C. Honnor, G.S. Dhillon and C. Londos, "cAMP-dependent protein kinase and lipolysis in rat adipocytes: II. Definition of steady-state relationship with lipolytic and antilipolytic modulators", J. Biol. Chem. 260: 15130 (1985).

6. C. Londos, R.C. Honnor and G.S. Dhillon, "cAMP-dependent protein kinase and lipolysis in rat adipocytes: III. Multiple modes of insulin regulation of lipolysis and regulation of insulin responses by adenylate cyclase regulators", J. Biol. Chem. 260: 15139 (1985).

7. P. Belfrage, B. Jergil, P. Strålfors and H. Tornqvist, "Hormone-sensitive lipase of rat adipose tissue: Identification and some properties of the enzyme protein", FEBS Lett 75: 259 (1977).

8. P. Strålfors and P. Belfrage, "Phosphorylation of hormone-sensitive lipase by cyclic AMP-dependent protein kinase", J. Biol. Chem. 258: 15146 (1983).

9. H. Olsson, P. Strålfors and P. Belfrage, "Direct evidence for protein phosphatase-catalyzed dephosphorylation/deactivation of hormone-sensitive lipase from adipose tissue", Biochim. Biophys. Acta 794: 488 (1984).

10. S.R. Cordle, R.J. Colbran and S.J. Yeaman, "Hormone-sensitive lipase from bovine adipose tissue", Biochim. Biophys. Acta 887: 51 (1986).

11. F.-T. Lee, S.J. Yeaman, G. Fredrikson, P. Strålfors and P. Belfrage, "Hormone-sensitive lipase from swine adipose tissue: Identification and some properties", Comp. Biochem. Physiol. 80B: 609 (1985).

12. P. Cohen, "The role of protein phosphorylation in the hormonal control of enzyme activity", Eur. J. Biochem. 151: 439 (1985).

13. P. Strålfors, P. Bjorgell and P. Belfrage, "Hormonal regulation of hormone-sensitive lipase in intact adipocytes: Identification of phosphorylated sites and effects on the phosphorylation by lipolytic hormones and insulin", Proc. Natl. Acad. Sci. USA 81: 3317 (1984).

14. P. Strålfors and P. Belfrage, "Phosphorylation of hormone-sensitive lipase by cyclic GMP-dependent protein kinase", FEBS Lett. 180: 280 (1985).

15. N.A. Abumrad, R.R. Perry and R.R. Whitesell, "Stimulation by epinephrine of the membrane transport of long chain fatty acid in the adipocyte", J. Biol. Chem. 260: 9969 (1985).

16. N.A. Abumrad, C.R. Park and R.R. Whitesell, "Catecholamine activation of the membrane transport of long chain fatty acids in adipocytes is mediated by cyclic AMP and protein kinase", J. Biol. Chem. 261: 13082 (1986).

17. N.A. Abumrad, R.R. Perry and R.R. Whitesell, "Insulin antagonizes epinephrine activation of the membrane transport of fatty acids. Potential site for hormonal suppression of lipid mobilization", J. Biol. Chem. 261: 2999 (1986).

18. W.H. Trzeciak and G.S. Boyd, "The effect of stress induced by ether anaesthesia on cholesterol content and cholesteryl-esterase activity in rat adrenal cortex", Eur. J. Biochem. 37: 327 (1973).

19. G.J. Beckett and G.S. Boyd, "Purification and control of bovine adrenal cortical cholesterol ester hydrolase and evidence for the activation of the enzyme by a phosphorylation", Eur. J. Biochem. 72: 223 (1977).

20. S. Naghshineh, C.R. Treadwell, L.L. Gallo and G.V. Vahouny, "Protein kinase-mediated phosphorylation of a purified sterol ester hydrolase from bovine adrenal cortex", J. Lipid Res. 19: 561 (1978).

21. K.G. Cook, F.-T. Lee and S.J. Yeaman, "Hormone-sensitive cholesterol ester hydrolase of bovine adrenal cortex. Identification of the enzyme protein", FEBS Lett. 132: 10 (1981).

22. K.G. Cook, S.J. Yeaman, P. Strålfors, G. Fredrikson and P. Belfrage, "Direct evidence that cholesterol ester hydrolase from adrenal cortex is the same enzyme as hormone-sensitive lipase from adipose tissue", Eur. J. Biochem. 125: 245 (1982).

23. R.J. Colbran, A.J. Garton, S.R. Cordle and S.J. Yeaman, "Regulation of cholesterol ester hydrolase by cyclic AMP-dependent protein kinase", FEBS Lett. 201: 257 (1986).
24. K.G. Cook, R.J. Colbran, J. Snee and S.J. Yeaman, "Cytosolic cholesterol ester hydrolase from bovine corpus luteum. Its purification, identification, and relationship to hormone-sensitive lipase", Biochim. Biophys. Acta 752: 46 (1983).
25. R. Chanderbhan, B.J. Noland, T.J. Scallen and G.V. Vahouny, "Sterol carrier protein$_2$. Delivery of cholesterol from adrenal lipid droplets to mitochondria for pregnenolone synthesis", J. Biol. Chem. 257: 8928 (1982).
26. G.V. Vahouny, P. Dennis, R. Chanderbhan, G. Fiskum, B.J. Noland and T.J. Scallen, "Sterol carrier protein (SCP$_2$)-mediated transfer of cholesterol to mitochondrial inner membranes", Biochem. Biophys. Res. Commun. 122: 509 (1984).
27. R.J. Krueger and N.R. Orme-Johnson, "Acute adrenocorticotropic hormone stimulation of adrenal corticosteroidogenesis. Discovery of a rapidly induced protein", J. Biol. Chem. 258: 10159 (1983).
28. L.A. Pon, J.A. Hartigan and N.R. Orme-Johnson, "Acute ACTH regulation of adrenal corticosteroid biosynthesis. Rapid accumulation of a phosphoprotein", J. Biol. Chem. 261: 13309 (1986).
29. D.P. Hajjar, C.R. Minick and S. Fowler, "Arterial neutral cholesteryl esterase. A hormone-sensitive enzyme distinct from lysosomal cholesteryl esterase", J. Biol. Chem. 258: 192 (1983).
30. J.C. Khoo, E.M. Mahoney and D. Steinberg, "Neutral cholesterol esterase activity in macrophage and its enhancement by cAMP-dependent protein kinase", J. Biol. Chem. 256: 12659 (1981).
31. D.P. Hajjar, "Herpesvirus infection prevents activation of cytoplasmic cholesteryl esterase in arterial smooth muscle cells", J. Biol. Chem. 261: 7611 (1986).

THE ROLE OF PHOSPHORYLATION IN THE REGULATION OF ACETYL-CoA

CARBOXYLASE ACTIVITY BY INSULIN AND OTHER HORMONES

Andrew C. Borthwick

Department of Biochemistry, University of Bristol

University Walk, Bristol BS8 1TD, U.K.

Insulin stimulates fatty acid synthesis in adipose and other tissues by increasing acetyl-CoA carboxylase activity. [1,2,3,4,5] In rat epididymal fat cells and liver cells this activation is associated with increased phosphorylation of the enzyme at specific sites, particularly within a peptide designated the I-peptide. [6,7,8,9] Inhibition of fatty acid synthesis by hormones such as adrenaline and glucagon involves decreases in acetyl-CoA carboxylase activity. Inhibition is also associated with increased phosphorylation of the enzyme however at sites distinct from those found after insulin treatment. There is good evidence that the adrenaline and glucagon stimulated phosphorylation of the enzyme is carried out by cyclic AMP-dependent protein kinase [10,8,11,12] whereas the insulin stimulated phosphorylation is carried out by an as yet uncharacterised cyclic AMP-independent protein kinase.

Acetyl-CoA carboxylase activity can be altered, at least in vitro, by both reversible phosphorylation and allosteric regulators. [13,14] In the case of the latter, the enzyme is activated by citrate and inhibited by fatty acyl-CoA esters. These effectors appear to modulate enzyme activity by altering the ratio of the inactive dimeric form of the enzyme to the active polymeric form which can be composed of up to 20 monomers. [14] The effect of hormones on the polymerisation state of acetyl-CoA carboxylase has been examined using density gradient centrifugation to separate the dimeric and polymeric forms of the enzyme. Both adrenaline[15] and glucagon[16] were shown to diminish the level of polymerisation, whereas insulin would appear to increase the proportion of the enzyme in a polymeric form[2].

In the present work a new approach has been used to examine the relationship between activity, phosphorylation and polymerisation of acetyl-CoA carboxylase in adipose tissue exposed to insulin or other hormones. This method involves separating the polymeric and dimeric forms of the enzyme rapidly on a Superose 6 gel filtration column fitted to the Pharmacia Fast Protein Liquid Chromatography system (FPLC). Full details are given in reference 17.

Initial studies on the behaviour of acetyl-CoA carboxylase on Superose 6 were carried out with enzyme purified on Sepharose-avidin either from rat mammary or adipose tissue. [18] Enzyme from both sources eluted as a polymer with apparent M_r greater than 4×10^6 if the enzyme was first preincubated with potassium citrate (20mM) and the same concentration of citrate was present in the column buffer. However in the absence of citrate the enzyme eluted largely in the dimeric form with an M_r close to 0.5×10^6. Phosphorylation with cyclic AMP-dependent protein kinase greatly diminished the extent of polymerisation that could be induced by citrate (20mM).

The effects of insulin on the polymerisation of acetyl-CoA carboxylase in rat epididymal adipose tissue were examined as follows. Tissue pieces were incubated with or without insulin and cell free extracts were prepared as described by Brownsey et al. [18] The extracts were taken to 40% saturated ammonium sulphate at $0°C$, the precipitated protein was resuspended and dialysed for 30 min against the basic column buffer, which was Mops (20mM) pH 7.2 containing EDTA (2mM), $MgCl_2$ (10mM), dithiothreitol (1mM), 2% (w/v) glycerol, 0.02% (w/v) sodium azide plus protease inhibitors (Pepstatin A, antipain and leupeptin each at 1µg/ml). These concentrated extracts contained virtually all the acetylCoA carboxylase, pyruvate carboxylase, fatty acid synthetase and most of the ATP-citrate lyase and only a small proportion of the albumin which is present in the extaction buffer. Samples of these concentrated extracts were incubated with or without potassium citrate (20mM) before application to a Superose 6 column. Columns were developed at a flow rate of 0.4ml/min at $4°C$ and elution of protein was monitored at A_{280} (Fig. 1). Fractions (1ml) were also collected and acetyl-CoA carboxylase activity was determined as described by Brownsey et al. [10].

Results for extracts chromatographed in the absence of citrate are shown in Figure 1. The A_{280} profile for the insulin extract (Fig. 1b) shows a substantial increase in protein with a M_r greater than 4×10^6 (peak 1). Activity measurements (Fig. 1) together with SDS/polyacrylamide gel electrophoresis and specific immunoprecipitation indicated that these changes in A_{280} absorption were largely accounted for by changes in the amount of acetyl-CoA carboxylase eluting with an M_r greater than 4×10^6. After treatment with citrate the A_{280} and activity profiles for both control and insulin extracts were essentially the same as Figure 1b thus indicating that citrate treatment abolishes the effect of insulin on these extracts. Peak 2 corresponds to the elution of fatty acid synthetase, pyruvate carboxylase and ATP-citrate lyase (approximate M_r of 0.5×10^6) and peak 3 is largely due to albumin.

The relationship between phosphorylation and polymerisation of acetyl-CoA carboxylase was studied using the same approach except that prior to hormone treatment the tissue pieces were incubated in medium containing [^{32}P] Pi (0.1 mCi/ml) for 2 hr to ensure a steady state of incorporation of ^{32}P into all intracellular proteins. As before concentrated extracts were prepared and subjected to gel filtration on Superose 6. The ^{32}P-labelled proteins present in each fraction were recovered by TCA precipitation and separated on SDS/polyacrylamide gels. Subsequent radioautography of these gels indicated that insulin treatment results in a marked increase in [^{32}P]-acetyl-CoA carboxylase eluting with a M_r greater than 4×10^6 whereas treatment with the β-adrenalinergic agonist isoprenaline increased incorporation of ^{32}P into the enzyme eluting with an M_r of 0.5×10^6. Prior treatment of

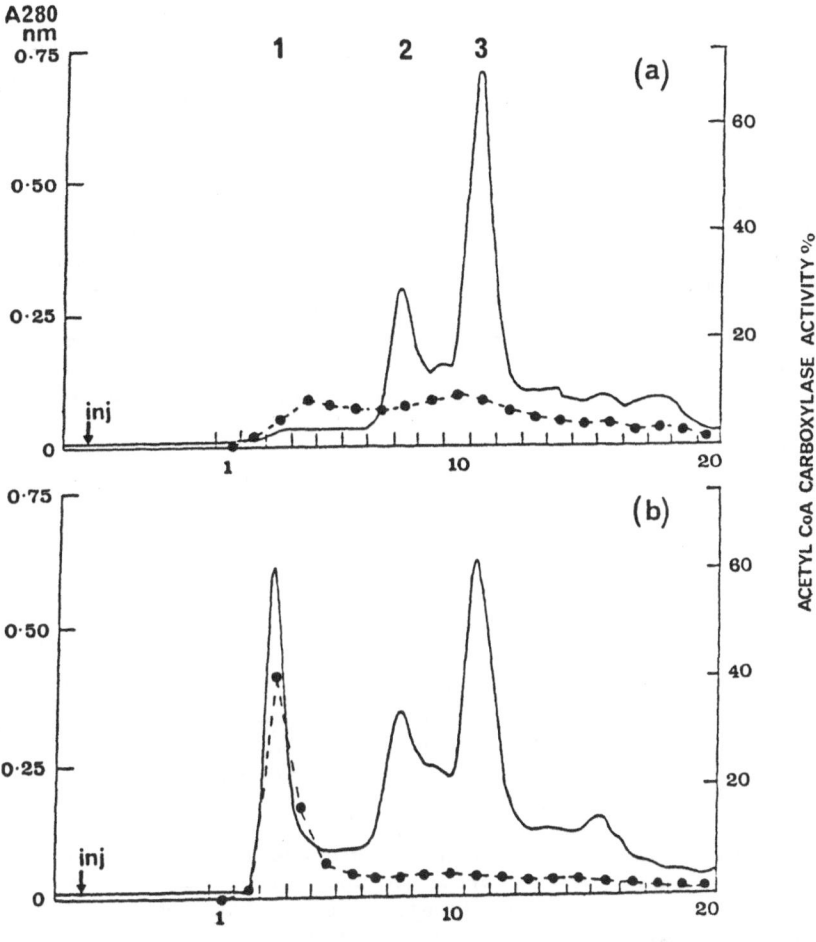

Fig. 1. Gel filtration on Superose 6 of concentrated extracts
prepared from control and insulin-treated epididymal
adipose tissue. Paired groups of pads were incubated
for 15 min with (b) or without (a) insulin (10 munits/
ml) and concentrated extracts were prepared as described in the text. Samples (200µl) (equivalent to 0.5g
original tissue) were applied to a Superose 6 column equilibrated in basic column buffer. In each case the eluate
was monitored at 280mn (——) and acetyl-CoA carboxylase
activity was assayed in each fraction (●). (Reprinted by
persission from Biochemical Journal Vol. 241, pp. 773
copyright © 1987, The Biochemical Society, London).

these extracts with citrate results in virtually all the $[^{32}P]$-acetyl-
CoA carboxylase eluting as a polymer in samples from control and insulin
treated extracts. However, the behaviour of acetyl-CoA carboxylase
following isoprenaline treatment of the tissue was similar to the
behaviour of purified acetyl-CoA carboxylase phosphorylated by cyclic
AMP-dependent protein kinase and a marked diminution in polymerisation
in the presence of 20mM-citrate was evident.

Finally [^{32}P]-acetyl - CoA carboxylase was specifically recovered by immunoprecipitation from the fractions separated by Superose 6 chromotography, digested with trypsin and the resulting ^{32}P-peptides resolved by two-dimensional analysis on thin later cellulose plates[7]. A major difference in the amount of ^{32}P-labelled I-peptide[9] was observed following insulin treatment.

This increased phosphorylation of the polymeric form from insulin treated tissue on a specific peptide (I-peptide) supports the idea that increased phosphorylation of a site/sites within this peptide may be the mechanism through which insulin exerts its control of acetyl-CoA carboxylase activity. However, further studies are necessary in order to obtain definitive proof. In particular these must include the isolation of the kinase and also demonstration in a purified system that phosphorylation of the appropriate site/sites in the enzyme cause activation and polymerisation.

ACKNOWLEDGEMENTS

These studies were supported by grants from Medical Research Council and British Diabetic Association to Dr R.M. Denton.

REFERENCES

1.   A.P. Halestrap and R.M. Denton, Hormonal regulation of adipose tissue acetyl-CoA carboxylase by changes in the polymeric state of the enzyme, Biochem, J. 142:365 (1974).
2.   A.L. Witters, D. Moriarty, and D.B. Martin, Regulation of hepatic acetyl-CoA carboxylase by insulin and glucagon, J. Biol. Chem. 254:6644, (1979).
3.   J.G. McCormack and R.M. Denton, Evidence that fatty acid synthesis in the intrascapular brown adipose tissue of cold adapted rats is increased in vivo by insulin, Biochem. J. 166:627, (1977).
4.   K.F. Beuchler, A.C. Beynen and M.J.H. Geelan, Studies on the assay, activity and sedimentation behaviour of acetyl-CoA carboxylase from isolated hepatocytes incubated with insulin and glucagon, Biochem. J. 221:869, (1984).
5.   M.R. Munday and D.H. Williamson, Effects of starvation, insulin or prolactin deficiency on the activity of acetyl-CoA carboxy- lase in mammary gland and liver of lactating rats, FEBS Lett. 138:285 (1982).
6.   R.W. Brownsey, W.A. Hughes, R.M. Denton and R.J. Mayer, Demon- stration of the phosphorylation of acetyl-CoA carboxylase within intact rat epididymal fat cells, Biochem. J. 168:441, (1977).
7.   R.W. Brownsey and R.M. Denton, Evidence that insulin activates acetyl-CoA carboxylase by increased phosphorylation of a specific site, Biochem. J. 202:77, (1982)
8.   L.A. Witters, J.P. Tipper and G.W. Bacon, Stimulation of site- specific phosphorylation and acetyl-CoA carboxylase by insulin and epinephrine, J. Biol. Chem. 258:5643, (1983).
9.   R. Holland and D.G. Hardie, Both insulin and epidermal growth factor stimulates fatty acid synthesis and increase phosphory- lation of acetyl-CoA carboxylase and ATP-citrate lyase in isolated hepatocytes, FEBS Lett. 181:308, (1985).
10. R.W. Brownsey, W.H. Hughes, and R.M. Denton, Adrenaline and the regulation of acetyl-CoA carboxylase in rat epididymal adipose tissue. Inactivation of the enzyme is associated with phos- phorylation and can be reversed by dephosphorylation, Biochem. J. 184:22 (1979).

242

GLYCOGEN METABOLISM IN SMOOTH MUSCLE

Theodore G. Sotiroudis, Stathis Nikolaropoulos
and Athanasios E. Evangelopoulos

The National Hellenic Research Foundation
Institute of Biological Research
48 Vassileos Constantinou Avenue, Athens 116 35, Greece

INTRODUCTION

In the last few years there has been a surge of interest in smooth muscle biochemistry. There are two major reasons why this tissue is of particular interest. First, although the total high energy phosphagen pool (phosphocreatine + ATP) is substantially lower than that of skeletal muscle, however energy metabolism in smooth muscle is finely tuned to meet contractile energy requirements[1]. Second, it has recently become evident that reversible phosphorylation of myosin, via an enzyme cascade similar to that involved in the activation of glycogen phosphorylase, plays a major role in regulating the interaction of actin and myosin in smooth muscle[2,3].

Glycogen is the most conspicuous source of endogenous carbohydrate metabolism observed in the smooth muscle cells of different organs. Although its function is not fully understood, glycogen has been considered as one of the main available energy stores which takes part in maintaining the vascular tone under stimulated conditions in the absence of glucose[4]. A large number of proteins are known to be involved in glycogen metabolism and its regulation[5]. Glycogenolysis is catalyzed by phosphorylase and debranching enzyme and glycogen synthesis from UDPG by glycogen synthase and branching enzyme[6]. Studies of the control of glycogen metabolism in mammalian skeletal muscle have revealed that a regulatory network of a number of protein kinases and phosphatases and several thermostable regulatory proteins determine the activation state of glycogen phosphorylase and glycogen synthase, the rate-limiting enzymes of this metabolic process[5]. Nevertheless, little is known on the catalytic and structural characteristics of the enzymes involved in the cascade of reactions associated with glycogen synthesis or degradation in smooth muscle and the overall picture of the control of glycogen metabolism in this type of muscle is not well defined.

It is the purpose of this review article to briefly summarize what is known about glycogen metabolism and its regulation in smooth muscle, with particular emphasis on the chemical and regulatory properties of the key enzymes of glycogen metabolism, which include phosphorylase and phosphorylase kinase, glycogen synthase and synthase kinases, and protein phosphatases. Moreover current knowledge of the relationships between smooth muscle contractile activity and glycogen metabolism will be also considered, although the functional (pharmacological and physiological) as well as structural heterogeneity of smooth muscles does not permit facile generalizations between different smooth muscles in the same species or between different species[7]. Finally, recent findings concerning functional compartmentation of glycogen metabolism within smooth muscle cells will be addressed.

GLYCOGEN PHOSPHORYLASE AND PHOSPHORYLASE KINASE

Available information about the properties of glycogen phosphorylases of different tissues suggests that more than two phosphorylase isozymes exist in mammalian tissues. Electrophoretic, immunological and other physicochemical and catalytic studies have shown that five phosphorylase isozymes can be separated and were designated as phosphorylase L, LI, I, II and III[8]. The L and III enzymes were the only forms found in liver and skeletal muscle respectively, while the I enzyme was dominant in brain, lung and smooth muscles. The II and LI enzymes were found to be the hybrid molecules between I and III enzymes and between I and L enzymes respectively[8].

Preliminary immunological and/or kinetic studies on glycogen phosphorylase from bovine tracheal and intestinal smooth muscles have shown that the smooth muscle enzymes differ from their skeletal counterpart, although controversial results regarding their affinity for AMP were reported[9,10]. Victorova and Ramensky[11,12] were first able to isolate an homogeneous smooth muscle phosphorylase preparation from cow myometrium using hydrophobic and affinity chromatographic procedures. Their results have clearly shown that uterine isozyme I presented significant pecularities which distinguish smooth muscle phosphorylase from skeletal muscle and liver isozymes (Table 1).

Table 1. Comparison of Properties of Uterine, Skeletal Muscle and Liver Phosphorylase

	Phosphorylase		
	Uterine	Skeletal Muscle	Liver
Tetramerization[a]	$-$[12]	$+$[17]	$-$[13]
Ka for AMP(b-form)	7.3 μM[12]	50 μM[15]	0.36 mM[16,b]
Na$_2$SO$_4$-Activation (in absence of AMP)	$-$[12]	$+$[14]	$+$[13]
AMP-Activation	$+$[12]	$+$[15]	$-$[13]
Km for Glucose-1-P (a-form)	43 mM[12]	1.6 mM[17]	0.7 mM[18,c]

[a]After phosphorylation with phosphorylase kinase.

[b]($A_{0.5}$) at which v=Vmax/2, determined from Hill plots; in presence of 0.15 M acetate.

[c]In presence of 5 mM AMP.

Uterine phosphorylase b is a dimer with a Mr of ~200,000, but the conversion of phosphorylase b to a was not accompanied by tetramerization, as in the case of liver isozyme[13]. Smooth muscle inactive phosphorylase differs also from liver and skeletal muscle enzymes in that it undergoes additional activation by Na2SO4 only in the presence of AMP[13,14]. Moreover, the Km of phosphorylase a of the myometrium for the substrate glucose-1-P is 27-fold and 61-fold higher than the Km for the skeletal and liver enzymes respectively, while its Ka for the activator AMP is significantly lower (Table 1). It is suggested that the forms cited in the literature as brain phosphorylase, heart isozyme 1, fetal isozyme and uterine smooth muscle phosphorylase should be considered as the same protein species.

Phosphorylase kinase is one of the regulatory enzymes involved in the cascade of reactions associated with glycogenolysis. The enzymes from rabbit[19,20], dogfish[21] and chicken[22,23] skeletal muscle, bovine heart[24] and rat liver[25] have been purified to apparent homogeneity and are known to be all hexadecamers composed of four types of subunits $(\alpha\beta\gamma\delta)_4$, although there appear to exist some distinct differences among the above characterized isozymes.

Since smooth muscle phosphorylase kinase has not been purified or studied to any extent, we had undertaken the purification and characterization of the enzyme from chicken gizzard to facilitate understanding of the hormonal control of glycogen metabolism in smooth muscle. Phosphorylase kinase was purified to near homogeneity by a procedure involving glycerol density gradient ultracentrifugation, 5´AMP-Sepharose 4B and calmodulin-Sepharose 4B affinity chromatography steps[26]. It was interesting as well as puzzling that gizzard phosphorylase kinase showed a completely different subunit pattern to that of rabbit skeletal muscle enzyme (one main protein band of 61 kDa and no protein staining at the position of γ subunit) upon dodecyl sulfate acrylamide gel electrophoresis, although the Mr of chicken gizzard and rabbit skeletal muscle kinases were found to be similar. Gizzard phosphorylase kinase showed a high pH 6.8/8.2 activity ratio of 0.53, it was stimulated by Ca^{2+}, inhibited up to 80% by EGTA and it was activated about 1.9-fold by calmodulin. In a recent report Tsutou et al.[27] using a partially purified uterine smooth muscle phosphorylase kinase preparation have also reached the conclusion that their smooth muscle kinase was activated about 1.5-2.0-fold by exogenous calmodulin and that Ca^{2+}-activation of the enzyme was greatly inhibited by EGTA. Further characterization of gizzard phosphorylase kinase has shown that the Km value for ATP was 0.45 mM, while the $K_{0.5}$ for rabbit muscle phosphorylase b was extremely low, more than 100-fold lower than the Km of activated skeletal muscle phosphorylase kinase for its protein substrate (Table 2).In this respect, it is interesting to report that the gizzard kinase activity was drastically decreased at phosphorylase b concentrations greater than 0.5 mg/ml (at 10 mg/ml of the protein substrate a 5-fold lower activity was observed)[26]. This phenomenon may be due either to the existence of more than one type of binding sites for the protein substrate or that the effector molecules present in the assay mixture (ATP, Mg^{2+}) induce on phosphorylase b a different conformation (at high protein substrate concentrations),which negatively affects b to a conversion. The possibility that the biphasic effect of phosphorylase $\overline{b}$ on gizzard kinase activity may be due to the presence of an unknown inhibitor can rather be excluded, because phosphorylase b (four times crystallized) chromatographed on a DEAE Sephadex A-50 column (pH 8.2) and gel filtrated through a Sepharose 6B column, was still able to present the inhibitory effect (unpublished results).

A very interesting feature of chicken gizzard phosphorylase kinase is that the enzyme cannot be activated by phosphorylation with cAMP-dependent protein kinase or by autophosphorylation[26], a property also shared by dogfish[21] and chicken[23] skeletal muscle phosphorylase kinase. Preliminary experiments of Ozawa[28,29] with partially purified smooth muscle phosphorylase kinase have shown that the enzyme was not stimulated by cAMP, in accordance with our results, while in contrast, Mohme-Lundholm[10] using an undefined tracheal muscle phosphorylase kinase preparation reported that the kinase was activated by cAMP. Although Mohme-Lundholm's results are not enough experimentally supported, in fact, it is possible that mammalian smooth muscle phosphorylase kinase can be regulated by a phosphorylation-dephosphorylation process similar to that found in skeletal muscle kinase, and that the inability of chicken gizzard kinase to show such a regulatory pattern may be a common property of ancient vertebrate muscles[21,23].

Concerning other regulatory properties of chicken gizzard phosphorylase kinase, the enzyme activity at pH 6.8 was activated 2-fold by limited proteolysis[26], inhibited by ethanol and was not significantly affected by

Table 2. Comparison of Properties of Phosphorylase Kinases Isolated from Various Tissues

	Phosphorylase Kinase[a]					
	Chicken Gizzard	Rabbit Muscle	Dogfish Muscle	Chicken Muscle	Rat Liver	Bovine Heart
M_r	1.3×10^6 [26]	1.3×10^6 [20]	1.2×10^6 [32]	1.3×10^6 [22]	1.3×10^6 [25]	1.3×10^6 [24]
Subunit Pattern	61 kDa [26] (main protein band)	α,145 kDa;[20] β,130 kDa; γ, 45 kDa; δ, 17 kDa;	α,138 kDa;[32] β,118 kDa; γ, 45 kDa; δ, 17 kDa;	α,140 kDa;[22] β,129 kDa; γ, 44 kDa; δ, 17 kDa;	α,140 kDa;[25] β,116 kDa; γ, 45 kDa; δ, 17.5 kDa;	α,134 kDa;[24] β,125 kDa; γ, 48 kDa; δ, ?
Activity Ratio (6.8/8.2)	0.5[26]	0.07[20]	0.3-0.5[21]	0.1-0.4[23]	Optimum pH=[25] 6.8-7.2	0.04-0.2[24]
Km for ATP	0.45 mM[26]	0.4 mM[20]	2 mM[21]	0.2 mM[23]	0.1 mM[33]	0.2 mM[24]
Km for Muscle Phosphorylase	< 0.04 mg/ml[26]	4 mg/ml[19,b]	16.5 mg/ml[21]	4 mg/ml[23]	2.5 mg/ml[33]	5.5 mg/ml[35]
Activation by Phosphorylation	-[26]	+[19]	-[21]	-[23]	+[25]	+[24]
Activation by Calmodulin and/or Troponin	+[26]	+[20]	-[21]	+[23]	+[34]	-[35]

[a]Nonactivated phosphorylase kinase; [b]Activated phosphorylase kinase

heparin (unpublished results). In contrast, rabbit muscle phosphorylase kinase activity at pH 6.8 is known to be drastically stimulated by proteolysis[19], organic solvents[30] and heparin[31]. Table 2 presents a comparison of significant properties of phosphorylase kinases isolated from various tissues.

GLYCOGEN SYNTHASE AND GLYCOGEN SYNTHASE KINASES

Glycogen synthase is the rate-limiting step in mammalian glycogenesis and hence is central in the hormonal control of glycogen metabolism. It has been established that glycogen synthase is regulated by a phosphorylation-dephosphorylation mechanism, the phosphorylation being catalyzed by at least three types of protein kinases:(i) cyclic nucleotide-dependent protein kinases, (ii) Ca^{2+}-dependent protein kinases and (iii) Ca^{2+}- and cyclic nucleotide-independent protein kinases[36].

Although glycogen synthase has been purified extensively from a number of eucaryotic sources[36] and at least two different isozymes of glycogen synthase are present in a number of tissues examined by immunoblot analysis[37], most structural and functional knowledge stems from studies of the rabbit skeletal muscle enzyme[36,38], while very little is known on the enzymatic system involved in glycogen synthesis in smooth muscles. Preliminary studies on smooth muscle glycogen synthesis and its control have been performed in uterine and endometrial tissues[39,40], but to our knowledge, glycogen synthase has not been purified or studied to any extent from any type of pure smooth muscle. However, Huang and Robinson[41] have reported the extensive purification of glycogen synthase from human placenta, a tissue which is known to contain large amounts of vascular and extravascular smooth muscle[42]. The human placental glucose-6-P -dependent form of glycogen synthase (D-form) can be converted to the glucose-6-P -independent form (I-form) by incubating the synthase with a copurified synthase phosphatase which shows a stringent requirement for Mn^{2+} [41]. The D to I conversion can be reversed by the addition of cAMP-dependent protein kinase. It is noteworthy that the purified D-form enzyme, which is free of synthase phosphatase activity, shows a unique activation by metal sulfates in the absence of glucose-6-P[41].

A number of cAMP-dependent and independent glycogen synthase kinases has been shown to be present in several tissues[43]. Two isozymic forms of cAMP-dependent protein kinase (types I and II) have been identified in coronary arterial smooth muscle[44], while in a variety of smooth muscles,cAMP-dependent protein kinase has been postulated to play a pivotal role in the β-adrenergic control of contractile activity,and alterations in the extent of phosphorylation of the type II regulatory subunit could affect the response of this tissue to stimulation by β-adrenergic agents[45]. Recently, a multisubstrate Ca^{2+}- and cyclic nucleotide-independent kinase (Mr=47,000) was purified from bovine aortic smooth muscle and has been shown to phosphorylate a number of proteins,which include glycogen synthase, phosphorylase kinase and type II regulatory subunit of cAMP-dependent protein kinase[46]. Phosphorylation of skeletal muscle glycogen synthase by this enzyme was polycation modulable. Glycogen synthase converted to D-form following phosphorylation in either the presence (7 mol ^{32}P/mol synthase) or absence (4 mol ^{32}P/mol synthase) of polylysine. These results suggest that the enzyme may participate in regulating arterial glycogen metabolism. In addition, vascular smooth muscle has been shown to contain protein kinase F$_A$, the activation factor of ATPMg-dependent protein phosphatase, which has been identified as glycogen synthase kinase 3[47,5].

Smooth muscle phosphorylase kinase may also be considered as a potential Ca^{2+}-dependent glycogen synthase kinase, knowing that skeletal muscle phosphorylase kinase efficiently phosphorylates and inactivates glycogen synthase[36], although the substrate specificity of this smooth muscle kinase towards glycogen synthase has not been examined. For completeness, cGMP-dependent protein kinase and Ca^{2+}- and phospholipid-dependent protein kinase (protein kinase C) must also be cited as two other kinases which are present in smooth muscles[48] and are able to phosphorylate glycogen synthase[36]. Of course, assesment of the physiological significance of glycogen synthase phosphorylation by protein kinase C and cGMP-dependent protein kinase must be made critically, all the more so since it is now evident that glycogen synthase is an extremely promiscuous protein kinase substrate in vitro. Nonetheless, the fact that the smooth muscle of vas deferens together with spleen and brain tissues contain the highest levels of protein kinase C, from a large number of tissues examined[48], strongly suggests a direct action of protein kinase C to phosphorylate and inactivate glycogen synthase in several types of smooth muscle.

PHOSPHOPROTEIN PHOSPHATASES

Recent studies indicate that four enzymes termed protein phosphatases 1, 2A, 2B and 2C account for virtually all the protein phosphatase activity in cellular extracts, acting on phosphorylated proteins involved in the control of glycogen metabolism and other metabolic pathways[49]. Four phosphatases designated SMPI-IV have been identified in the smooth muscle of turkey gizzard[50,51]. Three of these have been purified to apparent homogeneity :SMP-I consists of three subunits (60 kDa, 55 kDa, 38 kDa) in a molar ratio 1:1:1 and it is a protein phosphatase $2A_1$. SMP-II is a Mg^{2+}-dependent monomeric enzyme (Mr=43,000) and could be designated as protein phosphatase 2C. SMP-IV is composed of two subunits (58 kDa and 40 kDa), has a Mr of 150,000 and cannot be classified as either a type 1 or type 2 phosphatase, because although it dephosphorylates the β-subunit of phosphorylase kinase, is not inhibited by the heat stable inhibitor-2 and has low activity toward phosphorylase a . The ability of SMP-I and SMP-IV to dephosphorylate phosphorylase a and the observation that SMP-IV dephosphorylates preferentially the β subunit of phosphorylase kinase may suggest a role of these enzymes in glycogen metabolism[50,51]. In addition, it has been suggested that SMP-III and IV are involved in the process of relaxation in vivo[50].

Multiple forms of protein phosphatases have also been identified in mammalian vascular smooth muscle. Werth et al.[52] described an aortic phosphatase consisting of two subunits (67 kDa and 38 kDa) in a molar ratio 1:1, which was active against native myosin and exhibited relatively low activity against phosphorylase a. The properties and subunit structure of this enzyme indicate that it is a protein phosphatase $2A_2$. DiSalvo et al.[53] have also identified several protein phosphatases in aortic smooth muscle. One of these was the multisubstrate ATPMg-dependent protein phosphatase, which is apparently the inactive form of the major physiologically relevant phosphatase involved in coordinating glycogen synthesis and breakdown[54] and which did not dephosphorylate myosin light chains (MLC) in either the presence or absence of modulator protein[55]. A second, apparently "latent" phosphorylase phosphatase, which migrated as a protein of Mr of 130,000 during sucrose density gradient centrifugation, has been shown to exist in vascular smooth muscle and its activity against phosphorylase a was markedly stimulated by histone-H_1 and polylysine. This "latent" phosphorylase phosphatase activity has been also identified as a major inhibitor-1 phosphatase in vascular smooth muscle[56], while its substrate protein phosphatase inhibitor-1 was also shown to be present in the same tissue[56]. Since the physical properties of this enzyme are not yet known, comparison with other smooth muscle phosphatases is not possible at this stage. Recently, an apparent different form of aortic polycation modulated phosphatase(s) was also identified[57], which migrated as a protein of Mr of 63,000 and showed

Table 3. Characteristics of Smooth Muscle Phosphatases

Muscle	Subunit Composition[a]	Total Mr	Activators/Inhibitors	Substrates	Type[b]	References
Turkey Gizzard	I. 60 kDa; 55 kDa; 38 kDa (1:1:1)	165,000	-	MLC[c] (100)[d]; Phosphorylase Kinase(α Subunit)(19); Inhibitor 1 (21);Glycogen Synthase(Sites 3)(13); Phosphorylase a (4)	$2A_1$	51,59
	II. 43 kDa (one subunit)	43,000	Mg^{2+}-Dependent	MLC (100); Phosphorylase Kinase (α Subunit)(50); Glycogen Synthase:Sites 3 (4),Site 2(42),Site 1(12); Inhibitor 1 (11); Phosphorylase a (1)	2C	51,60
	III. Unknown	-	-	Myosin≫MLC	-	50,58
	IV. 58 kDa; 40 kDa	150,000	Ca^{2+}, Mg^{2+}	MLC (62);Myosin (100); Heavy Meromyosin (92); Phosphorylase a (2); Phosphorylase Kinase (β Subunit) (1)	Cannot be classified	50
Chicken Gizzard	67 kDa; 54 kDa; 34 kDa (1:1.8:0.6)	-	-	Only MLC tested	-	61
Bovine Aorta	Unknown	140,000	ATPMg-Dependent	Phosphorylase a	-	47,55
	Unknown	130,000	"Latent", Histone and Polylysine Activators	Phosphorylase a (100); Inhibitor 1 (9)	-	56,62
	72 kDa; 53 kDa; 35 kDa	63,000	Histone and Polylysine Activators	MLC ; Phosphorylase a	Type 2 for MLC Type 1 for Phosphorylase a	57
	67 kDa; 38 kDa (1:1)	-	-	MLC(50);Myosin(100) Phosphorylase a(9)	$2A_2$	52

[a]Molar ratio in parentheses; [b]Classification system proposed by Cohen's group[49]; [c]Abbreviation:MLC, myosin light chain(s); [d]Values in parentheses represent relative activities (%).

a subunit structure (72 kDa, 53 kDa, 35 kDa) similar to that found for SMP-I[51]. This aortic phosphatase exhibits relatively low basal activity against phosphorylase a, 5-9-fold lower than that expressed against MLC. However, low concentrations of polylysine (or histone H_1) stimulate phosphorylase phosphatase activity 6-15-fold, while MLC phosphatase activity is virtually abolished[57]. The close parallel of ion exchange chromatographic patterns of MLC phosphatase and phosphorylase phosphatase in aortic smooth muscle suggests that the enzyme(s) may define an important functional link in coordinating contractility and glycogen metabolism in vascular smooth muscle [58]. Some important characteristics of protein phosphatases from different smooth muscles are summarized in Table 3.

THE HORMONAL CONTROL OF SMOOTH MUSCLE GLYCOGEN METABOLISM: COORDINATION WITH CONTRACTILITY

The utilization of glycogen by skeletal and cardiac muscle during contractile activity is well documented, however relatively few studies have been directed towards an elucidation of its role as an energy source and of the mechanisms controlling its synthesis or degradation in smooth muscle. Whereas cardiac and skeletal muscle phosphorylase activation has been shown to be induced by β-adrenergic stimulants, this has not been a consistent observation with all smooth muscles. Epinephrine-induced phosphorylase activation was demonstrated in uterine, tracheal, intestinal but not in arterial smooth muscle[63]. An increase in phosphorylase a activity was also observed following β-adrenergic stimulation of rat portal vein[64] and rabbit stomach[65] smooth muscle. The absence of phosphorylase activation in response to β-adrenergic stimulation in arterial muscle may indicate that the phosphorylase activating system of this smooth muscle differs significantly from that of other muscle types[63]. It possibly involves a phosphorylase kinase isozyme similar to that we found in chicken gizzard smooth muscle[26], which is unable to be regulated by a protein phosphorylation-dephosphorylation mechanism, in as much as cAMP or dibutyryl cAMP in concentrations up to 100 μM were ineffective in changing the percentage of phosphorylase a activity of the rabbit aorta[63]. In any case, β-adrenergic agonists do not affect either resting tension or MLC phosphorylation[63,66].

It is now well supported that the inward current of the action potential in vertebrate smooth muscle is entirely different from that of nerve or skeletal muscle and is carried by Ca^{2+} [29]. It is released from intracellular storage sites or enters through channels gated by potential or by receptor stimulation[67]. The intracellular concentration of Ca^{2+} regulates the contraction-relaxation cycle in smooth muscle, mainly through the regulation of a kinase catalyzed phosphorylation of the 20 kDa phosphorylatable light chain of myosin, allowing actin-myosin interactions to occur[2,3]. In parallel, it is known that phosphorylase kinase can be activated at the same range of Ca^{2+} concentration, which activated the contractile system[29]. Silver and Stull[66] studying the relationship between MLC phosphorylation, phosphorylase a formation and isometric development during cholinergic (carbachol) stimulation of intact tracheal smooth muscle, they showed that contraction of the muscle is accompanied by increases in phosphorylation of MLC and phosphorylase. Maximal increase in phosphorylation of both substrates preceded the maximal development of isometric tension, while long-term maintenance of isometric tension was accompanied by an immediate decline in MLC phosphate content and a slower decrease of phosphorylase a formation.

Recent studies with coronary arterial smooth muscle[68] suggest that the decline in phosphorylase a formation, after long periods of stimulation, may also occur in vascular smooth muscle, when a distinct contractile agent (KCl) was used. Nevertheless, Silver and Stull[69] have reported for this particular type of stimulation (KCl depolarization) in tracheal smooth muscle, that the extent of phosphorylase a formation was maintained at a maximal value. In addition, it has been demonstrated that the activation of phosphorylase during K^+-induced contraction of vascular smooth muscle is independent of the cAMP system[68], a result obtained also during the spontaneous contraction of uterine smooth muscle[70]. Thus, the tight temporal coupling found between activation of phosphorylase and development of isometric force in smooth muscles suggests that these events may be functionally coordinated. Such coordination is unlikely to involve activation of cAMP-dependent protein kinases and may be linked to the fact that stimulation of phosphorylase kinase and initiation of contraction are both Ca^{2+}-dependent processes[26,3]. The critical role of Ca^{2+} as a key regulator of phosphorylase activity in smooth muscle is also indicated by the experiments of Pettersson[71], who showed that phosphorylase activation by anoxia or 2,4 dinitrophenol, in bovine mesenteric artery and rabbit colon, is not dependent on cAMP, AMP, ATP or ADP and that mitochondrial Ca^{2+}-release is one of the regulatory factors of the anoxic induced glycogenolysis. In addition, nitroprusside and 8-Br-cGMP, two compounds supposed to be involved in the removal of Ca^{2+} from the cytoplasm of smooth muscle cell, through until now unknown cGMP-dependent mechanisms, have been shown to inhibit contraction and phosphorylase activation induced by α-adrenergic stimulation of KCl[72]. As far as the nature of Ca^{2+}-mediated process of phosphorylase activation in smooth muscle is concerned, Silver and Stull[73] observed that the calmodulin antagonist fluphenazine did not inhibit KCl-mediated phosphorylase a formation in intact tracheal smooth muscle. This finding suggests that calmodulin may be bound to tracheal phosphorylase kinase in the absence of Ca^{2+}, as in the case of rabbit muscle enzyme[20]. Nevertheless, such a functional property might not be a feature of other smooth muscle phosphorylase kinases[26].

ORGANIZATION OF GLYCOGEN METABOLIZING ENZYMES IN SMOOTH MUSCLE

It now seems generally accepted that structural proteins of skeletal muscles, inner surfaces of cytoplasmic membrane and the surface of the membranes of subcellular structures may serve as support providing reversible binding of cytoplasmic enzymes[74]. In skeletal muscle, a significant amount of glycogen metabolizing enzymes is associated with glycogen and represent the so called "glycogen particles", which as demonstrated by electron microscopy, are either free or membrane bound[75]. In addition, it is known that the glycolytic enzymes of skeletal muscle form a multienzyme complex, which leads to the compartmentation of glycolytic process[74].

Recently, Lynch and Paul[76,77,1] have shown that exogenous glucose (and not glycogen) is the sole precursor of aerobic lactate production both in unstimulated and KCl-activated porcine carotid arteries, in spite of the substantial glycogenolysis observed in the last case. Moreover, glycogen utilization is coordinated with the increase in the rate of oxidative metabolism, which is associated with the stimulation of mechanic activity[77]. When ouabain was used to inhibit lactate production, the activity of phosphorylase a was elevated. The paradox of activation of phosphorylase in the presence of an inhibition of glycolysis, lead the authors[77] to the hypothesis that glycolysis and glycogenolysis operate in separate compartments in vascular smooth muscle.

Given the lack of basic informations concerning the molecular properties of key enzymes of glycogen metabolism in smooth muscles, their subcellular localization and their mode of interaction with glycogen, it is impossible at the present time to define the nature of the compartment in which glycogenolysis predominates in vascular smooth muscle or to extend the proposed compartmentation hypothesis to other types of smooth muscle.

CONCLUDING REMARKS

A striking feature of recent investigations into the regulation of contractile tone in smooth muscles has been the discovery that phosphorylation-dephosphorylation mechanisms play a major role in the control of actin-myosin interaction. This is a principal reason why this tissue is of particular interest especially for the study of glycogen metabolism, a metabolic process known to be synchronized with contractile energy requirements in various muscle types.

Although relatively insignificant progress has been made in the study of molecular properties of glycogen metabolizing enzymes in smooth muscle, however interesting pecularities and intricate questions concerning the regulation of this metabolic pathway have emerged during the last few years research work. Is smooth muscle phosphorylase kinase structurally and functionally dissimilar to the skeletal muscle isozyme and if so, how many isoforms of phosphorylase kinase exist in the various smooth muscle tissues ? Our data[26] indicate that if hormonal control of glycogenolysis exists in chicken gizzard smooth muscle, this control cannot be exercised by a direct interaction between the cAMP-dependent protein kinase and phosphorylase kinase, as has been shown in mammalian skeletal muscle and liver tissues. Is this a unique feature of primitive vertebrate tissues or it is a property shared also by a number of mammalian smooth muscle cells? Is the hormonal control of glycogen synthesis in smooth muscle mediated through a phosphorylation-dephosphorylation mechanism similar to that found in skeletal muscle and if so which is the exact role of the various glycogen synthase kinases and phosphatases?.

Finally, the recent finding[76] that carbohydrate metabolism is compartmentated in vascular smooth muscle rise the question whether this compartmentation includes the formation of functional "glycogen particles" associated with membrane and/or contractile components of the smooth muscle cell. Additional work with purified enzymes and regulatory proteins of smooth muscle(s) glycogen metabolism cascade are needed to resolve the details of this metabolic system and to further understand the coordination of metabolic and contractile activity in smooth muscle.

REFERENCES

1. R.M. Lynch and R.J. Paul, Energy metabolism and transduction in smooth muscle, Experientia 41:970 (1985).
2. R.S. Adelstein, M.D. Pato and M.A. Conti, The role of phosphorylation in regulating contractile proteins, Adv.Cycl.Nucl.Res. 14:361 (1981).
3. K.E. Kamm and J.T. Stull, The function of myosin and myosin light chain kinase phosphorylation in smooth muscle, Ann.Rev.Pharmacol.Toxicol. 25:593 (1985).
4. R.J. Paul, Chemical energetics of vascular smooth muscle, in:"Handbook of Physiology.The Cardiovascular System," Am.Physiol.Soc., Bethesda, MD (1980).
5. P.Cohen, Protein phosphorylation and the control of glycogen metabolism in skeletal muscle, Phil.Trans.R.Soc.Lond. B302:13 (1983).
6. F. Huijing, Glycogen metabolism and glycogen-storage diseases, Physiol. Rev. 55:609 (1975).
7. R.E. Garfield and A.P. Somlyo, Structure of smooth muscle, in:"Calcium and Contractility," A.K. Grover and E.E. Daniel, eds., The Humana Press, Inc., Clifton, New Jersey (1985).
8. S. Yonezawa and S.H. Hori, Electrophoretic studies on the phosphorylase isozymes, J.Histochem.Cytochem. 23:745 (1975).
9. E. Bueding, N. Kent and J. Fischer, Tissue specificity of glycogen phosphorylase b of intestinal smooth muscle, J.Biol.Chem. 239:2099 (1964).

10. E. Mohme-Lundholm, Smooth muscle phosphorylase and enzymes affecting its activity, Acta Physiol.Scand. 59:74 (1963).

11. L.N. Viktorova and E.V. Ramensky, Glycogen phosphorylase of smooth muscles, isolation and certain properties, Dokl.Acad.Nauk.SSSR 222: 1463 (1975).

12. L.N. Viktorova and E.V. Ramensky, Further molecular and catalytic characterization of uterine phosphorylase b, FEBS Lett. 115:239 (1980).

13. M.M. Appleman, E.G. Krebs and E.H. Fischer, Purification and properties of inactive liver phosphorylase, Biochemistry 5:2101 (1966).

14. H.D. Engers and N.B. Madsen, The effect of anions on the activity of phosphorylase b, Biochem.Biophys.Res.Commun. 33:49 (1968).

15. T.G. Sotiroudis, C.T. Cazianis, N.G. Oikonomakos and A.E. Evangelopoulos, Effect of sodium cholate on the catalytic and structural properties of phosphorylase b, Eur.J.Biochem. 131:625 (1983).

16. W. Stalmans and G. Gevers, The catalytic activity of phosphorylase b in the liver, Biochem.J. 200:327 (1981).

17. A.E. Melpidou and N.G. Oikonomakos, Effect of glucose-6-P on the catalytic and structural properties of glycogen phosphorylase a, FEBS Lett. 154:105 (1983).

18. B. Lederer and W. Stalmans, Human liver glycogen phosphorylase. Kinetic properties and assay in biopsy specimens, Biochem. J. 159:689 (1976).

19. G.M. Carlson, P.J. Bechtel and D.J. Graves, Chemical and regulatory properties of phosphorylase kinase and cyclic AMP-dependent protein kinase, Adv.Enzymol. 50:41 (1979).

20. P. Cohen, Phosphorylase kinase from rabbit skeletal muscle, Meth. Enzymol. 99:243 (1983).

21. S.Pocinwong, H. Blum, D. Malencik and E.H. Fischer, Phosphorylase kinase from dogfish skeletal muscle. Purification and properties, Biochemistry 20:7219 (1981).

22. I.E. Andreeva, G.V. Silonova, N.B. Livanova, T.B. Eronina, V.E. Morozov and B.F. Poglazov, Purification, quaternary structure and certain immunological properties of phosphorylase kinase from chicken skeletal muscles, Biokhimiya 50:1504 (1985).

23. I.E. Andreva, N.B. Livanova, T.B. Eronina and B.F. Poglazov, Regulatory properties of phosphorylase kinase from chicken skeletal muscles, Biokhimiya 50:1646 (1985).

24. R.H. Cooper, H.S. Sul, E. McCullough and D.A. Walsh, Purification and properties of the cardiac isoenzyme of phosphorylase kinase, J.Biol. Chem. 255:11794 (1980).

25. T.D. Chrisman, J.E. Jordan and J.H. Exton, Purification of rat liver phosphorylase kinase, J.Biol.Chem. 257:10798 (1982).

26. S. Nikolaropoulos and T.G. Sotiroudis, Phosphorylase kinase from chicken gizzard. Partial purification and characterization, Eur.J.Biochem. 151:467 (1985).

27. A. Tsutou, S. Nakamura, A. Negami, K. Mizuta, E. Hashimoto and H. Yamamura, Calcium- and calmodulin-dependent phosphorylase kinase activity in porcine uterine smooth muscle, Biochem.Biophys.Res. Commun. 126:544 (1985).

28. E. Ozawa, Energetics of smooth muscle, J.Jap.Med.Ass. 72:1322 (1974).

29. S. Ebashi, Ca^{2+} in biological systems, Experientia 41:978 (1985).

30. T.J. Singh and J.H. Wang, Stimulation of glycogen phosphorylase kinase from rabbit skeletal muscle by organic solvents, J.Biol.Chem. 254: 8466 (1979).

31. Z. Hessová, M. Varsányi and L.M.G. Heilmeyer, Jr, Dual function of calmodulin (δ) in phosphorylase kinase, Eur.J.Biochem. 146:107 (1985).

32. D.A. Malencik and E.H. Fischer, Structure, function and regulation of phosphorylase kinase, in:"Calcium and Cell Function," W.Y. Cheung, ed., Academic Press, New York (1982).

33. J.R. Vandenheede, H. DeWulf and W. Merlevede, Liver phosphorylase b kinase. Cyclic AMP-mediated activation and properties of the partially purified rat liver enzyme, Eur.J.Biochem. 101:51 (1979).

34. S. Nakamura, A. Tsutou, K. Mizuta, A. Negami, T. Nakaza, E. Hashimoto and H. Yamamura, Calcium-calmodulin-dependent activation of porcine liver phosphorylase kinase, FEBS Lett. 159:47 (1983).

35. S.D. Killilea and N.M. Ky, Purification and partial characterization of bovine heart phosphorylase kinase, Arch.Biochem.Biophys. 221: 333 (1983).

36. P.J. Roach, Glycogen synthase and glycogen synthase kinases, Curr.Top. Cell.Regul. 20:45 (1982).

37. H.R. Kaslow, D.D. Lesikar, D. Antwi and A.W.H. Tan, L-type glycogen synthase. Tissue distribution and electrophoretic mobility, J.Biol. Chem. 260:9953 (1985).

38. P. Cohen, The role of protein phosphorylation in neural and hormonal control of cellular activity, Nature 296:613 (1982).

39. A. Rubulis, R.D. Jacobs and E.C. Hughes, Glycogen synthesis in mammalian uterus, Biochim.Biophys.Acta 99:584 (1965).

40. A. Milwidsky and A. Gutman, Glycogen metabolism of normal human myometrium and leiomyoma. Possible hormonal control, Gynecol.Obstet. Invest. 15:147 (1983).

41. K.-P. Huang and J.C. Robinson, Purification and properties of the glucose-6-phosphate-dependent form of human placental glycogen synthase, Arch.Biochem.Biophys. 175:583 (1976).

42. R.J. Babcoc, Smooth muscle in the human placenta, Am.J.Obst.Gynec. 105:612 (1969).

43. K.K. Schlender and E.M. Reimann, Glycogen synthase kinases distribution in mammalian tissues of forms that are independent of cyclic AMP, J.Biol.Chem. 252:2384 (1977).

44. P.J. Silver, C. Schmidt-Silver and J. DiSalvo, β-adrenergic regulation and cAMP kinase activation in coronary arterial smooth muscle, Am.J.Physiol. 242:H177 (1982).

45. C.W. Scott and M.C. Mumby, Phosphorylation of typeII regulatory subunit of cAMP-dependent protein kinase in intact smooth muscle, J.Biol. Chem. 260:2274 (1985).

46. J. DiSalvo, D. Gifford and A. Kokkinakis, A multisubstrate Ca^{2+} and cyclic nucleotide independent kinase from vascular smooth muscle. Modulation of activity by polycations, Biochem.Biophys.Res.Commun. 136:789 (1986).

47. J. DiSalvo, J.M. Jiang, J.R. Vandenheede and W.Merlevede, The ATPMg-dependent phosphatase is present in mammalian vascular smooth muscle, Biochem.Biophys.Res.Commun. 108:534 (1982).

48. J.F. Kuo, R.G.G. Andersson, B.C. Wise, L. Mackerlova, I. Salomonsson, N.L. Brackett, N. Katoh, M. Shoji and R.W. Wrenn, Calcium-dependent protein kinase:Widespread occurrence in various tissues and phyla of the animal kingdom and comparison of effect of phospholipid, calmodulin and trifluoperazine, Proc.Natl.Acad.Sci.U S A 77:7039 (1980).

49. T.S. Ingebritsen and P. Cohen, Protein phosphatases:Properties and role in cellular regulation, Science 221:331 (1983).

50. M.D. Pato and E. Kerc, Purification and characterization of a smooth muscle myosin phosphatase from turkey gizzards, J.Biol.Chem. 260: 12359 (1985).

51. M.D. Pato, R.S. Adelstein, D. Crouch, B. Safer, T.S. Ingebritses and P. Cohen, The protein phosphatases involved in cellular regulation. 4. Classification of two homogeneous myosin light chain phosphatases from smooth muscle as protein phosphatase-2A₁ and 2C and a homogeneous protein phosphatase from reticulocytes active on protein synthesis initiation factor eIF-2 as protein phosphatase-2A₂, Eur.J.Biochem. 132:283 (1983).

52. D.K. Werth, J.R. Haeberle and D.R. Hathaway, Purification of a myosin phosphatase from bovine aortic smooth muscle, J.Biol.Chem. 257: 7306 (1982).

53. J. DiSalvo, D.Gifford and A. Kokkinakis, Modulation of aortic protein phosphatase activity by polylysin, Proc.Soc.Exp.Biol.Med. 177:24 (1984).
54. W. Merlevede, J.R. Vandenheede, J. Goris and S.-D. Yang, Regulation of ATPMg-dependent protein phosphatase, Cur.Top.Cell.Regul. 23:177 (1984).
55. J. DiSalvo, D. Gifford, J.R. Vandenheede and W. Merlevede, Spontaneously active and ATPMg-dependent protein phosphatase activity in vascular smooth muscle, Biochem.Biophys.Res.Commun. 111:912 (1983).
56. E. Waelkens, J. Goris, J. DiSalvo and W. Merlevede, Inhibitor-1 phosphatase activity in vascular smooth muscle, Biochem.Biophys.Res. Commun. 120:397 (1984).
57. J. DiSalvo, D.Gifford and A. Kokkinakis, Properties and function of a bovine aortic polycation modulated protein phosphatase, in:"Advances in Protein Phosphatases," W.Merlevede and J.DiSalvo,eds., Leuven University Press, Leuven (1985).
58. J. DiSalvo, D.Gifford and M.J. Jiang, Properties and function of phosphatases from vascular smooth muscle, Fed.Proc. 42:67 (1983).
59. M.D. Pato and R.S. Adelstein, Purification and characterization of a multisubunit phosphatase from turkey gizzard smooth muscle. The effect of calmodulin binding to myosin light chain kinase on dephosphorylation, J.Biol.Chem. 258:7047 (1983).
60. M.D. Pato and R.S. Adelstein, Characterization of a Mg^{2+}-dependent phosphatase from turkey gizzard smooth muscle, J.Biol.Chem. 258: 7055 (1983).
61. H. Onishi, J. Umeda, H. Uchiva and S. Watanabe, Purification of gizzard myosin light chain phosphatase, and reversible changes in the ATPase and superprecipitation activities of actomyosin in the presence of purified preparations of myosin light chain phosphatase and kinase, J.Biochem. 91:265 (1982).
62. J. DiSalvo, E. Waelkens, D. Gifford, J. Goris and W. Merlevede, Modulation of latent protein phosphatase activity from vascular smooth muscle by Histone-H$_1$ and polylysine, Biochem.Biophys.Res. Commun. 117:493 (1983).
63. D.H. Namm, The activation of glycogen phosphorylase in arterial smooth muscle, J.Pharmacol.Exper.Ther. 178:299 (1971).
64. R.J. Paul and P.Hellstrand, Dissociation of phosphorylase a activation and contractile activity in rat portal vein, Acta Physiol.Scand. 121:23 (1984).
65. J. Debowy, Adrenergic regulation of phosphorolysis and hydrolysis of glycogen in smooth muscle of rabbit stomach in situ, Arch.Immunol. Ther.Exper. 25:863 (1977).
66. P.J. Silver and J.T. Stull, Regulation of myosin light chain and phosphorylase phosphorylation in tracheal smooth muscle, J.Biol.Chem. 257:6145 (1982).
67. T.B. Bolton, Calcium exchange in smooth muscle, in:"Control and manipulation of calcium movement," J.R. Parratt, ed., Raven Press, New York (1985).
68. P.E. Galvas, C. Kuettner, R.J. Paul and J. DiSalvo, Temporal relationships between isometric force, phosphorylase and protein kinase activities in vascular smooth muscle, Proc.Soc.Exp.Biol.Med. 178:254 (1985).
69. P.J. Silver and J.T. Stull, Phosphorylation of myosin light chain and phosphorylase in tracheal smooth muscle in response to KCl and carbachol, Mol.Pharmacol. 25:267 (1984).
70. J. Diamond, Phosphorylase, calcium and cyclic AMP in smooth muscle contraction, Am.J.Physiol. 225:930 (1973).
71. G. Pettersson, Effects of dinitrophenol on phosphorylase a activity, adenine nucleotide levels and tension in rabbit colon smooth muscle, Acta Pharmacol.Toxicol. 56:302 (1985).
72. T.M. Lincoln and R.M. Johnson, Possible role of cyclic GMP-dependent protein kinase in vascular smooth muscle function, Adv.Cyclic

Nucleotide Res. 17:285 (1984).

73. P.J. Silver and J.T. Stull, Effect of the calmodulin antagonist, fluphenazine, on phosphorylation of myosin and phosphorylase in intact smooth muscle, Mol.Pharmacol. 23:665 (1983).

74. B.I. Kurganov, N.P. Sugrobova and L.S. Mil'man, Supramolecular organization of glycolytic enzymes, J.Theor.Biol. 116:509 (1985).

75. S.J.W. Busby and G.K. Radda, Regulation of the glycogen phosphorylase system. From physical measurements to biological speculations, Cur.Top.Cell.Regul. 10:89 (1976).

76. R.M. Lynch and R.J. Paul, Compartmentation of glycolytic and glycogenolytic metabolism in vascular smooth muscle, Science 222:1344 (1983).

77. R.J. Paul and R.M. Lynch, Integration of metabolism and contractility in vascular smooth muscle:Role of phosphorylation-dephosphorylation mechanisms in a functionally compartmented system, in:"Advances in Protein Phosphatases," W. Merlevede and J. DiSalvo, eds., Leuven University Press, Leuven (1985).

NOTE

When the preparation of this paper has been completed, Andreeva et al (Eur.J.Biochem., 158:99 (1986)), reexamining their previous results[23], presented evidence that phosphorylase kinase from chicken skeletal muscle can be phosphorylated by the catalytic subunit of cAMP-dependent protein kinase and by itself. Comparing these results with ours, concerning chicken gizzard phosphorylase kinase[26], we can suggest that the inability of gizzard kinase to be regulated by phosphorylation-dephosphorylation may be a characteristic property of smooth muscle of this primitive vertebrate.

THE SITES OF INTERACTION OF CALMODULIN WITH PHOSPHOFRUCTOKINASE

Bärbel Buschmeier, Helmut E. Meyer, Hans-Hermann Kiltz,
Ludwig M. G. Heilmeyer, jr. and Georg W. Mayr

Inst. für Physiologische Chemie, Abt. für Biochemie
Supramolekularer Systeme

Ruhr-Universität, Universitätsstraße 150, 4630 Bochum, FRG

INTRODUCTION

Calmodulin is a multifunctional calcium-binding protein found in all eukaryotic cells (for review see 1). It consists of four Ca^{++}-binding helix-loop-helix domains which are homologous to each other and are often referred to as the EF-hand structures (2). Upon binding of calcium ions, calmodulin undergoes conformational changes (1,3-5) allowing it to interact with several target proteins.

Recently phosphofructokinase from rabbit skeletal muscle has been shown to interact with calmodulin (6). This enzyme plays a key role in regulation of glycolysis and is subject to allosteric activation and inhibition by various metabolites (7,8). In earlier studies (9,10) we have observed a high affinity binding of calmodulin to the inactive dimeric form of phosphofructokinase, whereas in tetrameric enzyme only sites of low affinity are available. Binding of calmodulin to the dimeric enzyme stabilizes this association state, shifting the dimer-tetramer-equilibrium towards this enzymatically inactive form (9). To clarify structural details of the interaction between calmodulin and phosphofructokinase, interacting fragments of both proteins were isolated. Using large tryptic fragments of calmodulin, the action of calmodulin domains on phosphofructokinase was characterized. From phosphofructokinase two CNBr-fragments were isolated, which apparently represent two different calmodulin-binding domains.

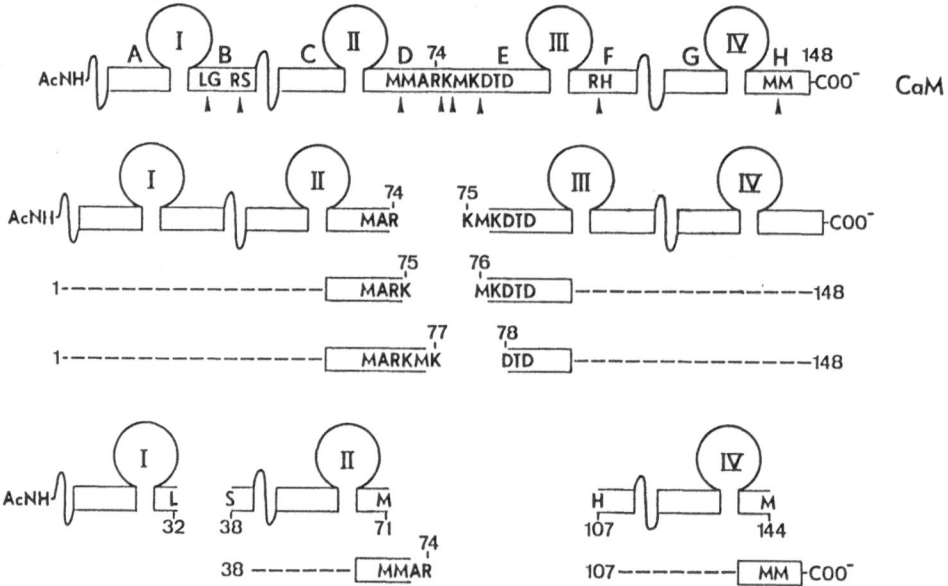

Fig. 1. Isolated tryptic fragments of Ca++-calmodulin, containing one or
two EF-hand domains.
Calmodulin was digested with trypsin at pH 7.8 in presence of
CaCl₂. Fragments were separated using gel filtration, anion
exchange chromatography and fluphenazine affinity chromatography.
The fragments were identified by amino acid analysis and partial
sequencing.

INTERACTION OF CALMODULIN AND CALMODULIN-FRAGMENTS WITH THE HIGH AFFINITY
SITES OF PHOSPHOFRUCTOKINASE

Limited tryptic digestion of calmodulin in the presence of CaCl₂
yielded a number of large fragments consisting either of N- or C-terminal
calmodulin-halfs or of single EF-hand domains (Fig.1). The binding of
calmodulin and of the isolated fragments was measured with a phosphofructokinase incubation-activity assay, carried out essentially as described
in (9). Phosphofructokinase at a subunit concentration of 1 μM was incubated in presence of increasing concentrations of calmodulin or calmodulin-
fragments in 0.2 M KCl, 20 mM Hepes/KOH, pH 7.5, 1 mM MgCl₂, 0.5 mM ATP,
100 μM CaCl₂, 1 mM K-phosphate and 5 mM dithioerythritol at 23°C. The
residual activity was repetetively determined by an enzyme coupled
assay (9).

Both N- and C-terminal calmodulin-halves bound to phosphofructokinase dimers with an apparent affinity only one magnitude of order
below that for intact calmodulin. Isolated helix-loop-helix domains were
200-300 fold less active than calmodulin. Recent x-ray diffraction studies

on calmodulin crystals have revealed a dumb-bell shape of the molecule with two globular lobes connected by a long central α-helix (11). The ability of phosphofructokinase to form complexes with both calmodulin-halves but with reduced affinity is in agreement with recent results about interaction of calmodulin with melittin and mastoparan. From NMR (5,12) and fluorimetric (13,14) studies it has been proposed, that calmodulin contains more than one binding site for these peptides and that these occur both in the N- and C-terminal halves of the molecule. The central α-helix of calmodulin may play a key role in calmodulin-target protein interactions, as 1) it is part of both calmodulin-halves; 2) it contains an anionic sequence at position 78-84 which may interact with the basic regions found in all high affinity calmodulin-binding peptides (15); 3) a recent investigation of the reactivities of the lysine residues in calmodulin towards acetic anhydride suggests that lysine-75 represents a critical residue of the calmodulin binding site for myosin light chain kinase (16).

An importance of parts of this central helix in binding to target sites is strengthened by our results. Among the N-terminal fragments of calmodulin (1-74, 1-75, 1-77) and the C-terminal ones (75-148, 76-148, 78-148) the shorter fragments always interact with phosphofructokinase with a higher affinity than the longer ones containing two to three basic residues at their ends. The excess of basic residues in the longer fragments may "neutralize" part of neighbouring anionic residues which contribute free energy of binding to the cationic target sites.

CALMODULIN-BINDING DOMAINS OF PHOSPHOFRUCTOKINASE

Carboxymethylated phosphofructokinase was digested with CNBr and the fragments soluble in 4 M urea were applied to a column of calmodulin-Sepharose 4B which had previously been equilibrated with a buffer containing 0.1 M ammonium acetate, 1 mM $CaCl_2$, 2 M urea, pH 6.7. After washing with this buffer, the bound material was eluted by the same buffer containing 2 mM EGTA instead of $CaCl_2$. The peptides eluting were further fractionated on a column of carboxymethyl cellulose (methods will be described in detail elsewhere).

Two fragments, M11 and M22, were isolated in this way, which were identified with the help of amino acid analysis and automatic Edman degradation according to the partial sequence published by Poorman et al. (17). Fragment M11 (Mr 3,080) connects the N- and C-terminal half of the polypeptide (Fig. 2) and is located in the region which has been proposed

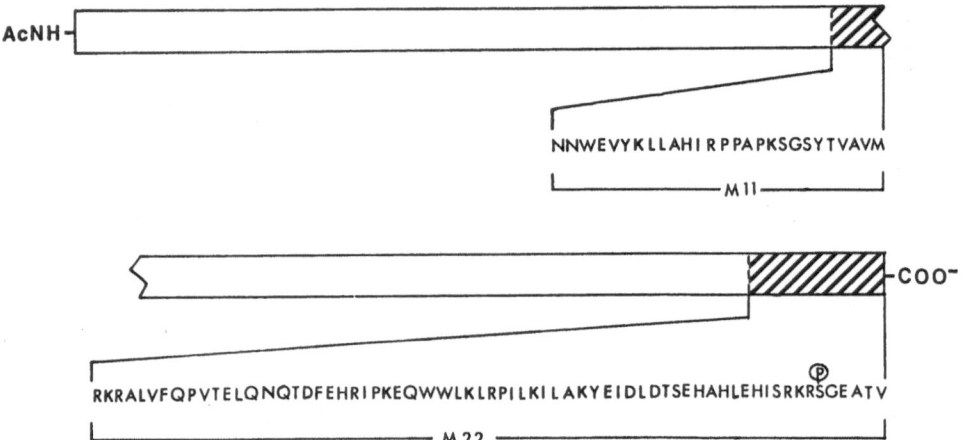

Fig. 2. Localization of the calmodulin-binding fragments of phosphofructo-kinase in the polypeptide

to form the contact zone between two dimers. Measuring the changes of the intrinsic tryptophan fluorescence of the isolated fragments upon addition of calmodulin, we could show that M11 binds calmodulin with a higher affinity than M22. This result together with the observation that calmodulin binds with high affinity only to the dimeric enzyme supports the assignment of fragment M11 to the high affinity calmodulin-binding domain of phosphofructokinase. M22 (Mr 8,060) corresponds to the C-terminus of the phosphofructokinase-subunit (Fig. 2) and contains the site phosphorylated by cAMP-dependent protein kinase (18). The isolated fragments, which are slightly homologous to each other, have a feature in common with other known calmodulin-binding peptides, namely an excess of hydrophobic and basic residues. (Results of these investigations will be published in detail elsewhere).

REFERENCES

1. C. B. Klee & T. C. Vanaman, Calmodulin, Adv. Protein Chem. 35, 213-321 (1982)
2. R. H. Kretsinger & C. E. Nockolds, Carp muscle calcium-binding protein, J. Biol. Chem. 248, 3313-3326 (1973)
3. D. W. Kupke & T. E. Dorrier, Volume changes upon addition of Ca^{2+} to calmodulin: Ca^{2+}-calmodulin conformational states, Biochem. Biophys. Res. Commun. 138 199-204 (1986)
4. S. R. Martin & P. M. Bayley, The effects of Ca^{2+} and Cd^{2+} on the secondary and tertiary structure of bovine testis calmodulin, Biochem. J. 238, 485-490 (1986)

5. R. E. Klevit, D. C. Dalgarno, B. A. Levine & R$_2$ J. P. Williams, H-NMR studies of calmodulin. The nature of the Ca^{2+}-dependent conformational change, Eur. J. Bioch. 139, 109-114 (1984)
6. G. W. Mayr & L. M. G. Heilmeyer, jr., Phosphofructokinase is a calmodulin binding protein, FEBS Lett. 159, 51-57 (1983)
7. K. Uyeda, Phosphofructokinase, Adv. Enzymol. 48, 193-244 (1979)
8. D. P. Bloxham & H. A. Lardy, Phosphofructokinase, in: "The Enzyme", P. D. Boyer, ed., vol.8, 239-278, Academic Press, New York (1973)
9. G. W. Mayr, Interaction of calmodulin with muscle phosphofructokinase. Changes of the aggregation state, conformation and catalytic activity of the enzyme, Eur. J. Biochem. 143, 513-520 (1984)
10. G. W. Mayr, Interaction of calmodulin with muscle phosphofructokinase. Interplay with metabolic effectors of the enzyme under physiological conditions, Eur. J. Biochem. 143, 521-529 (1984)
11. Y. S. Babu, J. S. Sack, T. J. Greenhough, C. E. Bugg, A. R. Means & W. J. Cook, Three-dimensional structure of calmodulin, Nature 315, 37-40 (1985)
12. S. Linse, T. Drakenberg & S. Forsén, Mastoparan binding induces a structural change affecting both the N-terminal and C-terminal domains of calmodulin, FEBS Lett. 199, 28-32 (1986)
13. C. G. Caday & R. F. Steiner, The interaction of calmodulin with melittin, Biochem. Biophys. Res. Commun. 135, 419-425 (1986)
14. R. F. Steiner, L. Marshall & D. Needleman, The interaction of melittin with calmodulin and its tryptic fragments, Arch. Biochem. Biophys. 246, 286-300 (1986)
15. D. A. Malencik & S. R. Anderson, Peptide binding by calmodulin and its proteolytic fragments and by troponin C, Biochemistry 23, 2420-2428 (1984)
16. A. E. Jackson, K. L. Carraway III., D. Puett & K. Brew, Effects of the binding of myosin light chain kinase on the reactivities of calmodulin lysines, J. Biol. Chem. 261, 12226-12232 (1986)
17. R. A. Poorman, A. Randolph, R. G. Kemp & R. L. Heinrikson, Evolution of phosphofructokinase - gene duplication and creation of new effector sites, Nature 309, 467-469 (1984)
18. R. G. Kemp, G. F. Lawrence, S. P. Latshaw, R. A. Poorman & R. L. Heinrikson, Studies on the phosphorylation of muscle phosphofructokinase, J. Biol. Chem. 256, 7282-7286 (1981)

PHOSPHORYLATION SITES ON TYROSINE HYDROXYLASE

P. R. Vulliet and D. G. Hardie

Dept. of Physiology, Colorado State University
Fort Collins, CO 80523 and Department of Biochemistry
University of Dundee, Dundee, Scotland DD1 4HN

INTRODUCTION

Tyrosine Hydroxylase (TH), the rate limiting enzyme in the biosynthesis of the catecholamine neurotransmitters, is regulated by protein phosphorylation. Previously, it has been demonstrated that tyrosine hydroxylase is phosphorylated by cyclic AMP-dependent protein kinase (PK A)(1-4) and calmodulin-dependent protein kinase (Cam K) (5-6). Recently, calcium and phospholipid-dependent protein kinase (PK C) has been reported to phosphorylate tyrosine hydroxylase (7,8). More recently, Roskoski et al. (9) have demonstrated that tyrosine hydroxylase is also a substrate of cyclic GMP dependent protein kinase (PK G). This communication compares the sites in tyrosine hydroxylase that are phosphorylated by each of these kinases. An endogenous kinase has been found to be present in the preparation of the purified tyrosine hydroxylase. Using the sequence of tyrosine hydroxylase derived from the cDNA and the sequences of the phosphopeptides produced from tryptic digestion of the phosphorylated enzyme, the specific sites of phosphorylation can be identified.

METHODS

Tyrosine hydroxylase was purified by a modification of the method of Vulliet et al. (6) utilizing ammonium sulfate fractionation, DEAE and heparin sepharose affinity chromatography. The protein kinases were purified as previously described in (6).

Tyrosine hydroxylase was phosphorylated by the various protein kinases, precipitated with 25 % (w/v) trichloroacetic acid and the pellet washed two times with 25 % TCA and two times with distilled water. The pellet was resuspended in 0.1 ml of 0.1 M ammonium bicarbonate, pH 8.0, and TPCK treated trypsin added at a ratio of 1:10. The digestion was performed for 2 to 12 hours at 37°C and the reaction was terminated by freezing until HPLC analysis. Tryptic phosphopeptides were analysed using a Gilson HPLC equipped with a 5 micron C_{18} reverse phase column in 0.1% trifluoroacetic acid. The.

peptides were eluted with a linear 0 to 40% acetonitrile gradient and
monitored by flow through counting of the radioactivity.

RESULTS

Purified tyrosine hydroxylase was incubated in the presence of
various protein kinases and [γ-^{32}P] ATP and then analysed by SDS gel
electrophoresis and autoradiography. Lane 1, 2, and 3 represent TH that
was phosphorylated by PK A for 5 min, 10 min and 30 min. Lane 4
represents the autoradiographic pattern obtained with PK A alone. Lane
5, 6, and 7 illustrate tyrosine hydroxylase that had been phosphorylated
by Cam K for 5, 10 and 30 min. Lane 8, 9 and 10 represent Cam K in the
absence of tyrosine hydroxylase. Lane 11 and 12 represent tyrosine
hydroxylase plus casein kinase and casein kinase alone. Lane 13
represents tyrosine hydroxylase plus calcium plus calmodulin.

Tyrosine hydroxylase was phosphorylated to 0.16 mol/ subunit with
the endogenous kinase; 0.49 mol/subunit with cyclic AMP dependent
protein kinase; and 0.9 mol/subunit with calmodulin dependent
multiprotein kinase and then digested with trypsin as described above.
Phosphopeptides were analysed using a Gilson HPLC equipped with a 5
micron C$_{18}$ reverse phase column equilibrated in 0.1% trifluoroacetic
acid. Peptides were eluted with a linear 0 to 40% acetonitrile gradient
and monitored with a radioactivity detector and representative elution
profiles are illustrated in fig 2.

Analysis of tryptic phosphopeptides from tyrosine hydroxylase

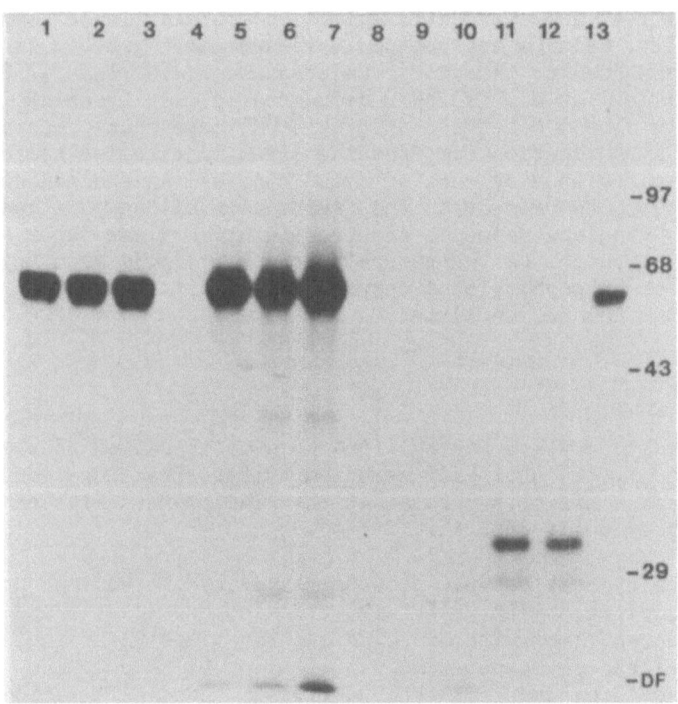

Fig 1. Polyacrylamide Gel Electrophoretic and Autoradiographic
        Analysis of Tyrosine Hydroxylase Phosphorylated by Various
        Protein Kinases.

phosphorylated in the absence of added kinase demonstrates the presence of a single peak of radioactivity labelled E. Phosphorylation of TH by PK A produces two additional phosphopeptides, termed site A1 and A2, that elute from the HPLC at a concentration of 22-24% acetonitrile. Phosphorylation of TH by Cam K and digestion with trypsin reveals the presence of two additional distinct phosphopeptides, labelled C1 and C2, eluting at 12% to 14% acetonitrile. Each of these peaks were purified by HPLC techniques and the sequenced as described in (10). It was found that A1/A2 and C1/C2 doublet represent the same phosphorylation site that had been cleaved at an alternate site as shown in figure 3. The solid line represents the major phosphopeptide produced by tryptic digestion and the dashed line represents an alternate cleavage product of the same phosphorylation site.

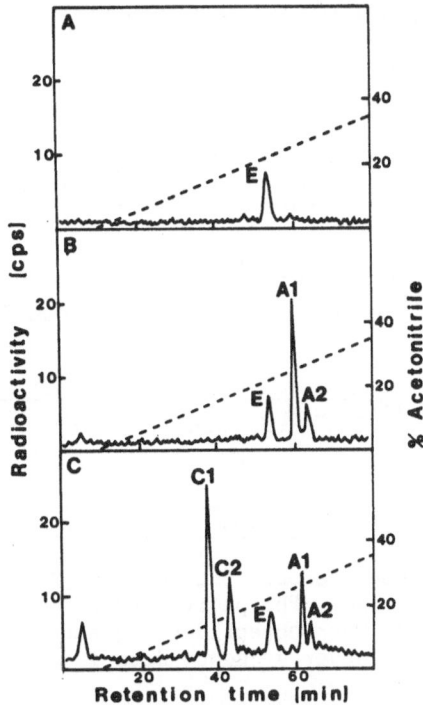

Fig 2. Reverse Phase HPLC Analysis of the Tryptic Phosphopeptides of Tyrosine Hydroxylase using [γ-^{32}P] ATP and: A, no added protein kinase; B, cyclic AMP dependent-protein kinase; and C, calmodulin-dependent multiprotein kinase.

DISCUSSION

Tyrosine hydroxylase, a protein that is present only in adrenergic tissues, is known to be a substrate for multiple protein kinases. Four of the kinases that phosphorylate the enzyme do so at serine 40. Calmodulin-dependent protein kinase preferentially phosphorylates tyrosine hydroxylase at serine 19 and secondarily at serine 40.

The physiological function of these phosphorylations is currently being investigated. It is clear from the data from a number of laboratories using a variety of tissue preparations that phosphorylation of TH is the mechanism responsible for its depolarization dependent activation of the enzyme. Treatment of PC12 cells with acetylcholine, the endogenous secretagogue, results in the phosphorylation of both serine 19 and serine 40, suggesting that Cam K as well as other kinases may play a role in mediating the response of chromaffin cells to this neurotransmitter.

Both SDS gel electrophoresis/autoradiography and HPLC techniques have identified endogenous kinase activity that is present in the preparation of purified tyrosine hydroxylase. Since the phosphorylation occurs in serine 8, a sequence that is not phosphorylated by other known kinases, this phosphorylation may represent a unique kinase activity. We are currently investigating the identity of this endogenous kinase. Using a synthetic peptide that is identical to the first 19 amino acids of tyrosine hydroxylase, it has been possibly to develop an assay for this kinase. Using this assay, the enodgenous activity and its role in the regulation of catecholamine biosynthesis will be examined.

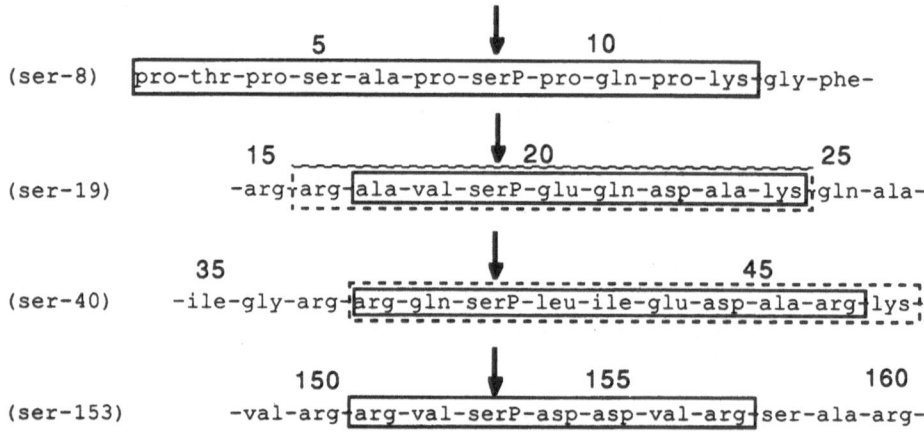

Fig 3. Amino acid sequence around the phosphorylation sites on tyrosine hydroxylase. Phosphorylated serine residues are shown by arrows. Phosphopeptides sequenced in the present study are boxed, other sequence data is from (11).

CONCLUSIONS

1. Tyrosine Hydroxylase is a good substrate for cyclic AMP-dependent, cyclic GMP-dependent protein kinase, calcium and phospholipid-dependent protein kinase and calmodulin-dependent protein kinase in vitro .

2. These four kinases phosphorylate at least two distinct sites, serine 19 (site C) and serine 40 (site A). Cyclic AMP and cyclic GMP-dependent protein kinases and calcium and phospholipid-dependent protein kinase phosphorylate tyrosine hydroxylase in site A. Calmodulin dependent protein kinase preferentially phosphorylates the enzyme at site C; however at higher concentrations of the kinase, it will also phosphorylate tyrosine hydroxylase in site A.

3. This preparation of TH also contains an endogenous kinase that phosphorylates TH in the serine 8 position.

4. The physiological role that "site C" phosphorylation plays in

regulating catecholamine biosynthesis is not clear; however it is known to be involved in mediating the effects of acetylcholine on adrenal chromaffin cells.

ACKNOWLEDGEMENTS

The research presented above involved the collaboration of the following investigators over the past few years: Dr. Philip Cohen, Dr. Robert Roskoski and Dr. James Woodgett. This research was supported by a NATO travel grant and grants from the National Science Foundation (BNS 81-18957) and the U. S. Air Force Office of Scientific Research (84-122).

REFERENCES

1. Joh, T., Park, D. and Reis, D. Proc. Natl. Acad. Sci. USA 75: 4744 (1978).
2. Yamuchi, T. and Fujisawa, H. J. Biol. Chem. 254: 503 (1979).
3. Vulliet, P., Langan, T. and Weiner, N. Proc. Natl. Acad. Sci. USA 77: 92 (1980).
4. Markey, K., Kando, S., Schenkman, L. and Goldstein, M. Mol. Pharmacol. 17: 79 (1980).
5. Yamauchi, T. and Fujisawa, H. Eur. J. Biochem. 132: 15 (1983).
6. Vulliet, P., Woodgett, J. and Cohen, P. J. Biol. Chem. 259: 13680 (1984).
7. Albert, K., Helmer-Matyjek, E., Nairn, A., Muller, T., Haycock, J., Greene, L., Golstein, M., and Greengard, P. Proc. Natl. Acad. Sci. USA. 81: 7713 (1984).
8. Vulliet, P., Woodgett, J., Ferrari, S. and Hardie, D. FEBS Letters 182: 335 (1985).
9. Roskoski, R., Vulliet, P. and Glass, D. J. Neurochem. (In press, 1987).
10. Campbell, D., Hardie, D. and Vulliet, P. R. J. Biol. Chem. 261: 10489 (1986).
11. Grima, B., Lamouroux, A., Blanot, F., Faucon-Biguet, N. and Mallet, J. Proc. Natl. Acad. Sci. USA 82: 617 (1985).

IV. CONTROL OF CELLULAR PROCESSES

REGULATION OF AMINO ACIDS

BIOSYNTHESIS IN PROKARYOTES

Georges N. Cohen

Unité de Biochimie Cellulaire
Département de Biochimie et Génétique
Moléculaire, Institut Pasteur
28 rue du Docteur Roux
75724 Paris Cedex 15, France

One day in 1959, François Jacob came down from the third floor to the first floor of the Institute where I was working and told me that it would be extremely useful if we could find constitutive mutants for amino acid biosynthesis, because of the possibility of extending the model of negative regulation established with the lac system to biosynthetic systems.

The results obtained from the study of the lactose system suggested that the repression of inducible systems by internal repressors might be basically similar to the repression effect observed in so-called constitutive biosynthetic systems. If this were so, a common feature of many protein-synthesizing systems would be inhibition by repressors formed under the control of specific genes. One could then expect to find a genetic control of the class of repression in which the end product of a biosynthetic pathway, such as an amino acid or a base, inhibited the synthesis of the specific enzymes of the pathway. This prediction has been confirmed by the isolation of "derepressed" mutants in several such repressible systems.

Derepressed mutants can be selected by use of certain analogs of the repressing metabolite. For instance, wild type E.coli does not grow in the presence of 5-methyltryptophan. A small fraction of the resistant mutants able to grow in the presence of this analog prove to be derepressed for all the enzymes involved in the biosynthetic pathway of tryptophan. Not only does the mutant synthesize as much of these enzymes whether grown in the presence or in the absence of tryptophan, but it produces more of these enzymes than does the wild type. In fact, the mutant synthesizes more of these enzymes in the presence of tryptophan than does the wild type in its absence. The loss of sensitivity to the repressing action of tryptophan can be attributed neither to a change in the concentration mechanism for tryptophan nor to the destruction of tryptophan by increased tryptophanase activity since the mutant and the wild type are similar in these respects. The properties of the mutant must therefore arise from an alteration to the mechanism of regulation (Cohen & Jacob, 1959).

Several independent mutants of the same type have been isolated and all these mutations (R_{try} - now renamed trpR) are located in a small segment of the chromosome, close to the threonine locus and, to our surprise (knowing the proximity of <u>lacI</u> and <u>lacZ</u>), far from the cluster of <u>trp</u> biosynthetic genes (Yanofsky & Lennox, 1959). It appeared, therefore, that the <u>trpR</u> gene determines the rate of synthesis for all of the enzymes in the sequence, whereas the cluster of <u>trp</u> genes determines the structure of the enzymes. Since <u>trpR</u> is located far from the cluster of structure-determining genes, it must operate through a cytoplasmic compound. The transient heterozygotes <u>trpR</u>⁺/<u>trpR</u>⁻ formed by conjugation, are sensitive to 5-methyltryptophan, which indicates that the repressible allele <u>trpR</u>⁺ is dominant over the <u>trpR</u>⁻ allele. These findings suggested that tryptophan itself is not the repressor, but that the repressor is synthesized under the control of the <u>trpR</u>⁺ and needs tryptophan to be operative (Cohen & Jacob, 1959). These results have been extended largely in the elegant work of Yanofsky (1981), which elucidated the structure and the mode of action of the <u>trp</u> aporepressor, which is coded by the <u>trpR</u> gene.

My main interest during the last 35 years has focused on the biosynthesis of amino acids which derive all or part of their carbon atoms from aspartic acid, and on the regulation of this biosynthesis. The relevant amino acids are lysine, threonine and methionine. Isoleucine derives from threonine and will not be considered here. The first reaction of the common pathway is catalyzed by three distinct aspartokinases, present in all Enterobacteriaceae so far examined (Cohen, Stanier & Le Bras, 1969), which differ in the way their synthesis and activity are regulated. Aspartokinase I activity is inhibited by threonine; its synthesis is repressed by threonine <u>plus</u> isoleucine. The synthesis of aspartokinase II is repressed by methionine. Lysine inhibits the activity and also represses the synthesis of aspartokinase III. The reduction of aspartate semialdehyde to homoserine is catalyzed by two distinct homoserine dehydrogenases. Synthesis of homoserine dehydrogenase I is repressed by threonine <u>plus</u> isoleucine; its activity is inhibited by threonine. Methionine represses the synthesis of homoserine dehydrogenase II. Expression of homoserine kinase and threonine synthase is repressed by threonine <u>plus</u> isoleucine.

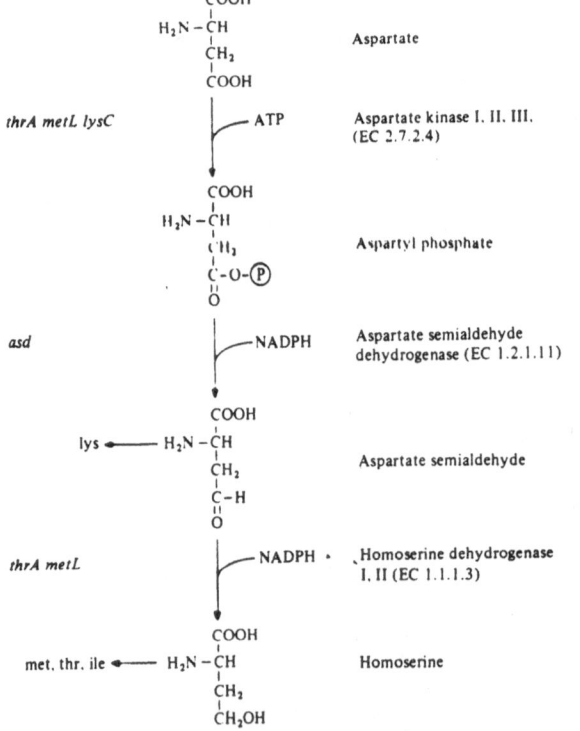

<u>Fig. 1</u>. The biosynthetic pathway common to lysine, methionine and threonine.

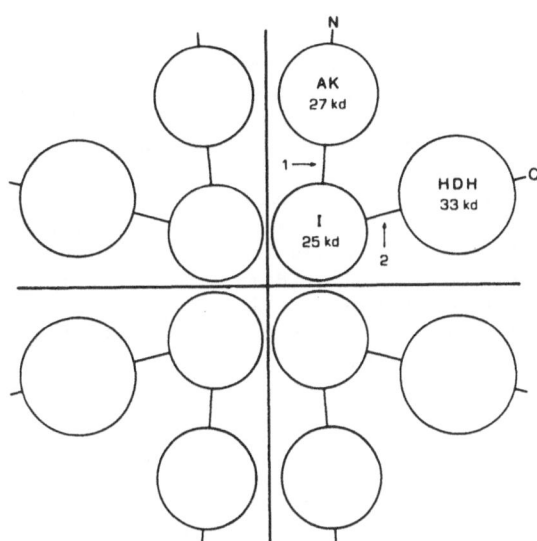

Fig. 2. Domain I is depicted as responsible for the sub-unit contacts which generate the tetrameric structure of the native enzyme. The actual polymeric fragments obtained by limited proteolysis or from extraction from an ochre mutan can be visualized as the sum of I and of either of the two catalytically active fragments HDH_D and AK. The arrows 1 and 2 indicate the sites of cleavages leading to the different fragments described in Fazel et al. (1983).

The three first proteins catalyze the synthesis of intermediates common to lysine, threonine and methionine or to threonine and methionine syntheses (Fig. 1). They are part of what has been called the common pathway (Cohen, 1983).

The three aspartokinases (namely aspartokinase I-homoserine dehydrogenase I, aspartokinase II-homoserine dehydrogenase II and aspartokinase III) have been cloned and sequenced (Cossart et al., 1979; Katinka et al., 1980; Zakin et al., 1983; Cassan et al., 1986). Extensive homologies between thrA and metL, coding for the bifunctional aspartokinase I-homoserine dehydrogenase I and aspartokinase II-homoserine dehydrogenase II respectively, point to a common ancestor for the two genes (Zakin et al., 1983), corroborating previous immunochemical observations (Zakin et al., 1978). Recent studies (Fazel et al., 1983; Belfaiza et al., 1984) show that each of the two bifunctional homoserine dehydrogenases is composed of three globular domains: the N-terminal domain possesses the kinase activity (AK); the C-terminal domain carries the dehydrogenase activity (HDH_D). These two parts of the polypeptide chains are separated by a central inactive domain (I) lacking enzyme activity but which may be essential for subunit interaction in these multimeric enzymes. Thus, the polypeptide chains of the two multifunctional proteins are homologous not only in their sequence but also in their triglobular domain structures. Figure 2 summarizes the structure of aspartokinase I-homoserine dehydrogenase I as derived from studies of limited proteolysis (Fazel et al., 1983).

The polypeptide chain of aspartokinase III is longer than the portion of the two aspartokinases-homoserine dehydrogenases corresponding to the aspartokinase domain. Examination of its sequence shows that its homology extends way beyond the aspartokinase domain, well into the inactive domain which is believed to play a role in subunit contacts. Therefore the possibility is excluded that the derivation of aspartokinase III from a common ancestor preceded the gene duplication and fusion leading to the two isofunctional enzymes. We prefer the model in which the aspartokinase III gene (lysC) was formed by deletion of one of the two bifunctional gene ancestors (Cassan et al., 1986).

The three structural genes coding for the threonine biosynthetic enzymes belong to the same operon (Thèze & Saint-Girons, 1974), located at 0 min on ghe genetic map of E.coli (Thèze et al., 1974). These genes, thrA, thrB and thrC, code for aspartate kinase I-homoserine dehydrogenase I, homoserine kinase and threonine synthase, respectively. The biosynthesis of these enzymes is subject to multivalent repression by threonine and isoleucine (Freundlich, 1963). Some complex regulatory mutants resistant to an isoleucine analog (Szentirmai, Szentirmai & Umbarger, 1968) and to the antibiotic borrelidin (Nass & Thomale, 1974) suggested the involvement of charged isoleucyl- and threonyl-ribonucleic acids, respectively, in the regulation of the threonine operon. The first was the product of three mutations at different loci. However, a single mutation in ilvS, coding for the isolceuyl-tRNA synthetase leads to the derepression of the threonine operon in Salmonella typhimurium (Blatt & Umbarger, 1972).

A mutant in thrS with a threonyl-tRNA synthetase having a 200-fold decreased apparent affinity for threonine, has regulatory properties which imply the involvement of charged threonyl-tRNA or of the related synthetase in the regulation of the threonine operon. A mutation in ilvA, coding for threonine deaminase, leads also to derepression of the thr operon, presumably as a resulf of isoleucine limitation (Johnson, Cohen & Saint-Girons, 1977).

The genetic fine structure of the threonine operon has been analyzed by deletion mapping (Saint-Girons & Margarita, 1978). Complementation analysis indicates that both thrB and thrC consist of a single cistron, whereas thrA is composed of two cistrons specifying a single polypeptide chain (Thèze & Saint-Girons, 1974), corroborating the bifunctional character of the thrA protein.

Regulatory mutations affecting the expression of the threonine operon were obtained by insertion of lambda at a secondary site between the threonine promoter and the first structural genes (Gardner et al., 1974; Gardner & Smith, 1975); other mutants were isolated on the basis of resistance to α-amino-β-hydroxyvaleric acid and identified by the excretion of threonine into the media. They are cis dominant, lie adjacent to thrA (Saint-Girons & Margarita, 1975) and are different from mutants resistant to the same analog affecting the thrA structural gene, previously described (Cohen & Patte, 1963; Cohen, Patte & Truffa-Bachi, 1965). They have been called thrO; the gene sequence is thrOABC, as suggested by earlier studies. This order was established by analysis of phage Mu insertion and nonsense mutations in the threonine operon (Thèze & Saint-Girons, 1974) and by analysis of a phage lambda insertion into the thr regulatory region (Gardner et al., 1974).

Regulatory mutants have also been isolated from thr-lac fusions (Saint-Girons, 1978). They are also localized upstream of thrA (Saint-Girons & Margarita, 1978).

All of these results identify a regulatory region situated upstream from the first structural gene of the operon, acting only in cis in merodiploids and causing pleiotropic effects on the expression of the threonine operon. Although the mutants were first thought to be classical operator mutants (Gardner & Smith, 1975; Saint-Girons & Margarita, 1975), it does not appear that the threonine operon is regulated by a repressor-operator mechanism; no aporepressor mutants have been isolated and all mutations analyzed to this date reside in the thr attenuator - as a consequence, the thrO locus should be renamed.

The sequence of the regulatory region of the <u>thr</u> operon has been determined (Gardner, 1979) by using DNA cloned from lambda transducing phages (Gardner & Reznikoff, 1978). The sequence was identified by the known amino-terminal sequence of aspartokinase I-homoserine dehydrogenase I (Véron & Cohen, 1974). The transcription initiation point has been determined as has been the promoter region (Gardner, 1979).

Downstream from the transcription initiation site, the threonine regulatory region contains extensive dyad symmetry and a stretch of A-T base pairs (bp) characteristic of rho-independent terminators (Rosenberg & Court, 1979). This region has been called the <u>thr</u> attenuator (Gardner, 1979) and is analogous to similar regions found in other biosynthetic operons (Yanofsky, 1981). It contains a sequence encoding a leader peptide, containing eight threonine and four isoleucine codons (threonine and isoleucine being the two amino acids which regulate expression of the operon). This region also contains the classical mutually exclusive secondary structures, which regulate RNA polymerase transcription of the structural genes according to the levels of the charged threonyl- and isoleucyl-tRNAs; the attenuation model thus accommodates the results showing that mutations in <u>ilvS</u> or <u>thrS</u> result in increased expression of the operon (Blatt & Umbarger, 1972; Johnson <u>et al.</u>, 1977). The attenuation region contains also an RNA polymerase pausing site, of unknown significance (Gardner, 1982). An internal promoter, situated within a 60 bp fragment at the 3' end of <u>thrA</u>, allows the expression of <u>thrB</u> in addition to the major promoter, but with an efficiency lower by at least one order of magnitude (Saint-Girons & Margarita, 1985).

All the constitutive mutations of the "operator" type analyzed to date are either point mutations (Gardner, 1979) or a deletion of the transcription termination signal of the attenuator (Parsot, Saint-Girons & Cossart, 1982). Among the α-amino-β-hydroxyvalerate resistant mutants (Cohen & Patte, 1963; Cohen <u>et al.</u>, 1965; Saint-Girons & Margarita, 1975) is a class which does not excrete threonine: homoserine dehydrogenase I activity is increased ten-fold and the operon is not regulated by threonine and isoleucine. The mutations are not located within the <u>thr</u> operon or in the <u>thrS</u> gene. These mutants (<u>thrX</u>) might affect one of the threonyl-tRNA genes (Comer, 1982) or some gene involved in tRNA modification or act in some unknown manner.

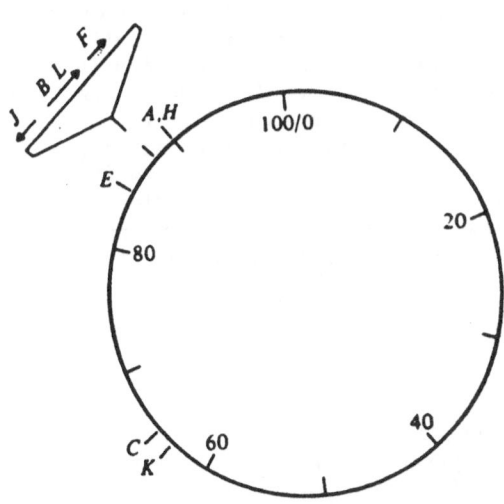

<u>Fig. 3</u>. Map of the <u>E.coli</u> methionine regulon. The scale has been expanded at 88 min to show the organization of the <u>metJBLF</u> gene cluster. <u>metA</u>: homoserine transuccinylase; <u>metB</u>: cystathionine--synthase; <u>metC</u>: -cystathionase; <u>metE</u>: B12-independent transmethylase; <u>metF</u>: 5,10-methylenetetrahydrofolate reductase; <u>metH</u>: B12-dependent transmethylase; <u>metJ</u>: <u>met</u> repressor; <u>metK</u>: S-adenosylmethionine synthetase; <u>metL</u>: aspartokinase II-homoserine dehydrogenase II.

In conclusion, regulation of threonine biosynthetic is at the level of enzyme activity and at the level of enzyme synthesis. Whereas aspartate kinase-homoserine dehydrogenase has been thoroughly studied and characterized, the other enzymes of the operon and aspartic semialdehyde dehydrogenase have been less studied (especially threonine synthase). DNA sequence analysis of the entire thr operon is complete including the promoter, attenuator and termination signals of the operon (Gardner, 1979; Parsot et al., 1983). Nuclease S_1 mapping experiments have allowed the precise localization of the termination of the operon mRNA transcript 60 nucleotides downstream the thrC stop codon (C. Parsot, personal communication).

The regulation of gene expression occurs at the level of termination of transcription via attenuation, and allows the cells to "sense" the levels of the two amino acids regulating the expression of the thr operon. Some problems remain however pending, such as the role of a repressor protein (Bogosian & Somerville, 1983; Johnson & Somerville, 1983, 1984).

REGULATION OF METHIONINE BIOSYNTHESIS

The various genes involved in methionine biosynthesis are scattered on the E.coli chromosome (Bachmann & Low, 1980; Fig. 3) and, although certain genes are clustered, most constitute independent transcription units. Addition of methionine to the growth medium causes repression of the synthesis of all enzymes specific of the methionine branch (Cohn, Cohen & Monod, 1953; Wijesundera & Woods, 1960; Rowbury & Woods, 1961, 1964), of S-adenosyl-methionine synthetase coded by the metK gene (Holloway, Greene & Su, 1970) and of aspartokinase II-homoserine dehydrogenase II, an enzyme of the common pathway coded by the metL gene (Patte, Le Bras & Cohen, 1967).

The study of the regulation of methionine synthesis has been clarified by the existence of E.coli and S.typhimurium mutants resistant to methionine analogs, such as norleucine (Cohen & Jacob, 1959), – methylmethionine and ethionine (Lawrence, Smith & Rowbury, 1968; for reviews, see Flavin, 1975 and Rowbury, 1983). Three classes of mutants have been isolated: (a) α-methylmethionine-resistant mutants localized in metA and affecting the allosteric control of homoserine succinyltransferase (Smith, 1961); (b) mutants where the synthesis of the methionine enzymes is constitutive (Cohen & Jacob, 1959; Patte et al., 1967; Lawrence et al., 1968) which have been localized in an E.coli locus called metJ, at 88 min (Su & Greene, 1971)* (c) mutants localized in metK (64 min on the E.coli chromosome). Different types of mutants have been obtained and, notwithstanding the complications introduced by the central role of S-adenosylmethionine synthetase (coded by metK), certain types of mutants exhibit a phenotype similar to that of the metJ mutants (Greene, Su & Holloway, 1970). An explanation could be that S-adenosylmethionine (or one of its derivatives) could be a co-repressor of the regulation system. This cannot be confirmed in vivo since E.coli is impermeable to S-adenosylmethionine, but studies in vitro (see below) strongly support this hypothesis.

---

* For both S.typhimurium and E.coli (Ahmed, 1973) metJ[+] is trans-dominant to metJ, indicating a diffusible product encoded by metJ[+]. This interpretation that the metJ[+] product is a protein was reinforced by the fact that certain metJ mutations were suppressed by nonsense suppressors.

It thus appears that regulation of methionine synthesis at the gene level is governed by the metJ gene coding for an aporepressor which controls in a non-coordinated fashion (Rowbury & Woods, 1966) the scattered genes of the methionine regulon. Methionyl-tRNA does not seem to play a regulatory role in the expression of the methionine genes since mutants of metG, which code for methionyl-tRNA synthetase, do not produce a defect in their level of expression (Gross & Rowbury, 1969; Ahmed, 1973).

The metJ gene has been mapped in E.coli (Su & Greene, 1971) and found to be close to the metBLF cluster (Thèze et al., 1974). Defective phages carrying the entire metJBLF cluster have been isolated (Press et al., 1971; Johnson et al., 1977) and the entire cluster has been cloned in plasmid vectors (Zakin et al., 1982). The physical maps have been studied and the four genes localized (Zakin et al., 1982) and sequenced (Zakin et al., 1983; Duchange et al., 1983; Saint-Girons et al., 1983; Tran et al., 1983; Saint-Girons et al., 1984).

The cluster is organized into three independent transcriptional units:

(1)   The metJ structural gene, transcribed counterclockwise, occupies 312 bp (Saint-Girons et al., 1984).

(2)   A complex 279 bp region is found between the metJ and metB structural genes; analysis of the 5' region of the metJ gene shows a regulatory region for two divergent transcriptional units (the metJ gene and the metBL operon); the promoter of the metBL operon has been identified as has been its transcription start (Duchange et al., 1983; Kirby, Hindenach & Greene, 1985). Our results showed the presence in this region of another promoter activity in an orientation opposite to that of the metBL operon but did not locate precisely the metJ promoter. This localization has now been established: several regulatory sites are present in this region; three promoter sequences were selected by their homology score and evidence has been obtained that transcripts in vivo begin near each of them. In cells with a single copy of metJ, promoter-1 is repressed, while that from the metJ promoter-2 still occurs. In plasmid-bearing cells, with many copies of the metJ gene, transcription from all three promoters is repressed, suggesting as the met repressor concentration increases other binding sites become occupied, thereby preventing transcription (Smith et al., 1985). This region also contains a putative operator region (see below), which might be active in the control of both metBL and metJ, which are transcribed on two opposite DNA strands.

(3)   The metB and metL genes form an operon; their structural genes occupy 1158 nucleotides for metB, and 2427 nucleotides for metL. The intracistronic region is 2 bp long. The metL gene is followed by a 351 bp segment within which a typical rho-independent terminator structure (Rosenberg & Court, 1979) preceded by a potential stem and loop structure very similar to the consensus structure described for several intercistronic and interoperonic regions (Gilson et al., 1984). This segment also contains the promoter and the transcriptional start of metF, which have been identified (Saint-Girons et al., 1983). Independent observations based on insertion mutagenesis by phage Mu or Tn5 show that metB and metL form an operon (Greene & Smith, 1984).

(4)   The metF structural gene comprises 888 nucleotides. No typical rho-independent terminator could be identified downstream of the structural gene; however, about 160 nucleotides beyond the end of the structural gene, some structural features of a rho-dependent

terminator (Rosenberg & Court, 1979) can be identified (Saint-Girons et al., 1983); the expression of metF is specifically inhibited in vitro by the metJ protein in the presence of S-adenosylmethionine. The inhibition is at the level of transcription (Shoeman et al., 1985).

All the methionine genes are subject to repression by methionine and the repression is mediated through the metJ gene product, a common aporepressor. It is therefore reasonable to assume that the repressor binding sites should be similar for all the met genes. Indeed, DNA sequences with a two-fold symmetry in the regions upstream from the metF and metBL transcriptional units were suggested as possible binding sites for the metJ gene product and were called "Met boxes" (Duchange et al., 1983; Saint-Girons et al., 1983). Michaeli, Mevarech & Ron (1984) compared then the 5' region of the metA gene to those of metF and metB and found an extensive homology although no common axis of symmetry was found. In the DNA upstream from the coding region of the metC gene, a "Met box" has also been detected (Belfaiza et al., 1985). Bearing in mind that the four "Met boxes" cited above are homologous, it is worthwhile to note that additional base matches are present between the "Met boxes" of metC and metB (Fig. 4a) on the one hand and between those of metA and metF (Michaeli et al., 1984) on the other hand. Moreover, the present comparison between the four "Met boxes" (for metC, B, A, F) revealed the existence of a repetitive unit (R) eight nucleotides long. As a consequence, multiple alignments could be drawn by sliding each sequence with an eight-nucleotide periodicity. In the alignment presented Fig. 4b, out of the 128 positions considered, 89 matches and 21 transitions are found when the repetitive units are compared to their consensus sequence

(a) metC : CGCGAATATTCATGCTAGTTTAGACATCCAGACGTATAAAAACAGGAATCCCG   -3
            ••• •• •••••••• •••••• •••• •
    metB : AATCTATACGCAAAGAAGTTTAGATGTCCAGATGTATTGACGTCCATTAACAC   ±56
            -50

(b) metC : ATATTCATGCTAGTTT A G A C A T C C   -3
                            A G A C G T A T AAAAACAGGAATCCCG
           -119
    metB : GGGATTTGCTCAATCT A T A C G C A A
                            A G A A G T T T
                            A G A T G T C C
                            A G A T G T A T   -48
                            T G A C G T C C ATTAACACAATGTTTA
           -97
    metF : CGCCCTTCGGCTTTTC C T T C A T C T
                            T T A C A T C T
                            G G A C G T C T
                            A A A C G G A T   -26
                            A G A T G T G C ACAACACAACATATAA
           -67
    metA : TTTTCTGGTTATCTTC A G C T A T C T
                            G G A T G T C T
                            A A A C G T A T   -4
                            A A G C G T A T GTAGTGAGGTAATCAG

    R :   A G A C G T C T

Bold characters indicate nucleotides matching the consensus sequence presented in line R. Numbers indicate positions relative to the A of the respective start codon taken as +1.

Fig. 4. Comparison of the 5' flanking regions of the met genes. (a) Comparison of the "Met boxes" of the metC and metB genes. Stars between the nucleotide sequences indicate identical residues. Numbers indicate positions relative to the A of the respective start codon taken as +1. (b) Comparison of the 5' flanking regions of the metC, B, F, A genes. The sequences 5' to the structural metC, metB, metF and metA genes (Belfaiza et al., 1985; Duchange et al., 1983; Saint-Girons et al., 1983; Michaeli et al., 1984) are presented discontinuously and have been aligned in order to focus on the presence of the underlying repetitive palindromic unit located within the "Met boxes".

R. This consensus sequence is a perfect palindrome AGACGTCT, which is present under an altered form, two to five times in the "Met boxes" (Fig. 4b). It is possible that the differences between the "Met boxes" are related to the different extents of repression elicited by the metJ gene product, as already mentioned by Michaeli et al. (1984); the ratios of derepressed versus fully repressed levels were found to be 6-12 for metC (Flavin, 1975; Belfaiza et al., 1985), 40 for metB, 100 for metF, and 300 for metA (Flavin, 1975).

We must recall that the metB and metJ genes are transcribed divergently and share the same "Met box" (Saint-Girons et al., 1984). In addition, after elucidation of the metK gene sequence coding for S-adenosylmethionine synthase, an enzyme that utilizes methionine as a substrate, Markham, De Parisis & Gatmaitan (1984) assigned a "Met box" to its 5' flanking region. However, the authors introduced a gap 13 nucleotides long in the 26 nucleotides long "Met boxes" of the metB and metF genes in order to maximize the overlap and to keep the same axis of symmetry. We have thus not considered the metK gene in our discussion. Further experiments are needed to localize precisely the regions of interactions and to determine the relative affinities of binding with the metJ protein.

Upstream of metA, two transcription starting points 74 bp apart, have been found. Only one is under negative control by methionine (Michaeli et al., 1984). In this case, as in that of the four transcriptional units metJ, metBL, metF and metC, no structure typical of attenuation control could be detected.

The metC gene of E.coli has been sequenced (Belfaiza et al., 1985). The metE gene of the same organism, coding for B12-independent transmethylase, has been isolated in a λdmetE transducing phage at the λ attachment site, cloned into a plasmid able to transform cells devoid of the enzyme (Chu et al., 1985). The metE gene of S.typhimurium has been recently cloned. Its expression in minicells yields a polypeptide of MW = 92,500 (Schulte, Stauffer & Stauffer, 1984).

The metJ gene product (104 amino acid residues; MW = 11,996) was identified in maxicell extracts of a strain carrying the proper plasmid (Smith & Greene, 1984). The gene has been completely sequenced (Saint-Girons et al., 1984). Its N-terminal half contains 67 of 100 of the basic residues which are often in tandem. Two trans dominant constitutive mutants (Liljestrand-Golden & Johnson, 1984) have been sequenced (Saint-Girons et al., 1984) and correspond respectively to a point mutation (Ala ⟶ Thr) and to a nonsense mutation (Trp ⟶ Amber). A metJ-lacZ fusion has been constructed, the study of which has provided evidence that metJ is autoregulated (Saint-Girons et al., 1984). The five first residues of the pure hybrid protein have been determined and correspond to the deduced protein sequence of the metJ product (Saint-Girons et al., 1984). The met repressor is a dimer (Smith et al., 1985).

The metJ gene has also been cloned and sequenced in S.typhimurium (Urbanowski & Stauffer, 1985a, b). The two deduced protein sequences differ very slightly: a Met residue in position 95 and an Asn residue in position 98 are replaced by a Leu and an Asp residue respectively in S.typhimurium. Fifteen identical amino acids are specified by different codons and the termination codons are different (TAA in E.coli, TGATAG in S.typhimurium).

The product of the metJ protein, the methionine aporepressor has been obtained in large quantities in the pure state by putting the gene under the control of a strong promoter. The parameters of its binding to its specific co-repressor, S-adenosylmethionine, and to a specific DNA target, the Met box upstream the metF structural gene, have been established (Saint-Girons et al., 1986). Protection experiments of the region between the metB and the metJ structural genes against DNase, intercalating agents or chemical cleavage have shown that the Met boxes are indeed the binding target of the aporepressor (Kirby et al., 1985 and unpublished). Finally, operator constitutive mutants of the metF gene have been obtained: all the mutations analyzed are located in the region defined above as the potential target of the methionine repressor by its similarity to other binding sites. The mutationally defined metF operator consists of a 40 bp long region with five 8 base pair imperfect palindromes spanning the metF transcription start. The altered operators do not recognize the purified repressor in an in vitro transcription-translation system, although the repressor binds efficiently to the metF wild type operator (Belfaiza et al., in the press).

The relationship between metJ, coding for the met repressor, metE and metH is complex. The expression of metE (coding for the B12-independent transmethylase) is repressed by the metJ regulatory system when cells are grown in the presence of methionine. The expression of metE is also repressed by the metH gene product (coding for the B12-dependent enzyme) in the presence of vitamin B12 (Greene et al., 1973). Evidence suggests that repression by vitamin B12 is due to the increased formation of the transmethylase holoenzyme rather than to increased methionine synthesis (Milner, Whitfield & Weissbach, 1969).

REFERENCES

Ahmed, A., 1973, Mechanism of repression of methionine biosynthesis in Escherichia coli. I. The role of methionine, S-adenosylmethionine and methionyl transfer ribonucleic acid in repression, Molecular and General Genetics, 123:299.

Bachmann,B.J. & Low,K.B., 1980, Linkage map of Escherichia coli K12, Edition 6, Microbiological Reviews, 44:1.

Belfaiza,J., Fazel,A., Müller,K. & Cohen,G.N., 1984, E.coli aspartokinase II-homoserine dehydrogenase II polypeptide chain has a triglobular structure, Biochemical Biophysical Research Communications, 123: 16.

Belfaiza,J., Parsot,C., Martel,A., Bouthier de la Tour,C., Margarita,D., Cohen,G.N. & Saint-Girons,I., 1986, Evolution in biosynthetic pathways: two enzymes catalyzing consecutive steps in methionine biosynthesis originate from a common ancestor and possess a similar regulatory region, Proceedings of the National Academy of Sciences USA, 83: 867.

Blatt,J.M. & Umbarger,H.E., 1972, On the role of isoleucyl-t-RNA synthetase in multivalent repression, Biochemical Genetics, 6:99.

Bogosian,G. & Somerville,R.L., 1983, Trp repressor protein is capable of introducing into other amino acid biosynthetic systems, Molecular and General Genetics, 191:51-58.

Cassan,M., Parsot,C., Cohen,G.N. & Patte,J.C., 1986, Nucleotide sequence of the lysC gene encoding the lysine sensitive aspartokinase III of E.coli K12: evolutionary pathway leading to three isofunctional enzymes, Journal of Biological Chemistry, 261:1052.

Chu,J., Shoeman,R., Hart,J., Coleman,T., Mazaitis,A., Kelker,N., Brot,N. & Weissbach,H., 1985, Cloning and expression of the metE gene in Escherichia coli, Archives of Biochemistry and Biophysics, 239:467.

Cohen,G.N., 1983, The common pathway to lysine, methionine and threonine, p. 147-171. In "Amino acids: Biosynthesis and Genetic Regulation", eds K.M. Herrmann & R.L. Somerville, London: Addison-Wesley Publishing.

Cohen,G.N. & Jacob,F., 1959, Sur la répression de la synthèse des enzymes intervenant dans la formation du tryptophane chez Escherichia coli, Comptes Rendus de l'Académie des Sciences, Paris, 248:3490.

Cohen,G.N. & Patte,J.C., 1963, Some aspects of the regulation of amino acid biosynthesis in a branched pathway, Cold Spring Harbor Symposia in Quantitative Biology, 28:13.

Cohen,G.N., Patte,J.C. & Truffa-Bachi,P., 1965, Parallel modifications caused by mutations in two enzymes concerned with the biosynthesis of threonine in Escherichia coli, Biochemical and Biophysical Research Communications, 19:546.

Cohen,G.N., Stanier,R.Y. & Le Bras,G., 1969, Regulation of the biosynthesis of the amino acids of the aspartate family in coliform bacteria and pseudomonads, Journal of Bacteriology, 99:791.

Cohn,M., Cohen,G.N. & Monod,J., 1953, L'effet inhibiteur spécifique de la méthionine dans la formation de la méthionine synthase chez E.coli, Comptes Rendus de l'Académie des Sciences, Paris, 236:746.

Comer,M.M., 1982, Threonine tRNAs and their genes in Escherichia coli, Molecular and General Genetics, 187:132.

Cossart,P., Katinka,M., Yaniv,M., Saint-Girons,I. & Cohen,G.N., 1979, Construction and expression of a hybrid plasmid containing the Escherichia coli thrA and thrB genes, Molecular and General Genetics, 175:39.

Duchange,N., Zakin,M.M., Ferrara,P., Saint-Girons,I., Park,I., Tran,S.V., Py,M.C. & Cohen,G.N., 1983, Structure of the metJBLF cluster in E.coli K12. Sequence of the metB structural gene and of the 5' and 3' flanking regions of the metBL operon, Journal of Biological Chemistry, 258:14868.

Fazel,A., Müller,K., Le Bras,G., Garel,J.R., Véron,M. & Cohen,G.N., 1983, A triglobular model for the polypeptide chain of aspartokinase I-homoserine dehydrogenase I of Escherichia coli K12, Biochemistry, 22:158.

Flavin,M., 1975, Methionine biosynthesis, p. 407-503. In "Metabolic Pathways, 3rd ed., vol. 7, Metabolism of Sulfur Compounds, ed. D.M. Greenberg, New York: Academic Press.

Freundlich,M., 1963, Multivalent repression in the biosynthesis of threonine in Salmonella typhimurium and Escherichia coli, Biochemical and Biophysical Research Communications, 10:277.

Gardner,J.F., 1979, Regulation of the threonine operon: tandem threonine and isoleucine codons in the control region and translation control of transcription termination, Proceedings of the National Academy of Sciences, USA, 76:1706.

Gardner,J.F., 1982, Initiation, pausing and termination of transcription in the threonine operon regulatory region of Escherichia coli, Journal of Biological Chemistry, 257:3896.

Gardner,J.F. & Reznikoff,W.S., 1978, Identification and restriction endonuclease mapping of the threonine operon regulatory region, Journal of Molecular Biology, 126:241.

Gardner,J.F. & Smith,O.H., 1975, Operator-promoter functions in the threonine operon of Escherichia coli, Journal of Bacteriology, 124:161.

Gardner,J.F., Smith,O.H., Fredricks,W.W. & McKinney,M.A., 1974, Secondary-site attachment of coliphage lambda near the thr operon, Journal of Molecular Biology, 90:613.

Gilson,E., Clément,J.M., Brutlag,D. & Hofnung,M., 1984, A family of dispersed repetitive extragenic palindromic DNA sequences in E.coli, EMBO Journal, 3:1417.

Greene,R.C. & Smith,A.A., 1984, Insertion mutagenesis of the metJBLF gene cluster of Escherichia coli: Evidence for an metBL operon, Journal of Bacteriology, 159:767.

Greene,R.C., Su,C.H. & Holloway,C.J., 1970, S-adenosylmethionine synthetase deficient mutants of Escherichia coli K12 with impaired control of methionine synthesis, Biochemical and Biophysical Research Communications, 38:1120.

Greene,R.C., Williams,R.D., Kung,H.F., Spears,C. & Weissbach,H., 1973, Effect of methionine and vitamin B12 on the activities of methionine biosynthetic enzymes in metJ mutants of Escherichia coli K12, Archives of Biochemistry and Biophysics, 158:249.

Gross,T.S. & Rowbury,R.J., 1969, Methionyl-transfer RNA synthetase mutants of Salmonella typhimurium which have normal control of the methionine biosynthetic enzymes, Biochimica et Biophysica Acta, 184:233.

Holloway,C.T., Greene,R.C. & Su,C.H., 1970, Regulation of S-adenosylmethionine synthetase in Escherichia coli, Journal of Bacteriology, 104:734.

Johnson,D.I. & Somerville,R.L., 1983, Evidence that repression mechanisms can exert control over the thr, leu, and ilv operons of Escherichia coli K12, Journal of Bacteriology, 155:49.

Johnson,D.L. & Somerville,R.L., 1984, New regulatory genes involved in the control of transcription initiation at the thr and ilv promoters of Escherichia coli K12, Molecular and General Genetics, 195:70.

Johnson,E.J., Cohen,G.N. & Saint-Girons,I., 1977, Threonyl-transfer ribonucleic acid synthetase and the regulation of the threonine operon in Escherichia coli, Journal of Bacteriology, 129:66.

Katinka,M., Cossart,P., Sibilli,L., Saint-Girons,I., Chalvignac,M.A., Le Bras,G., Cohen,G.N. & Yaniv,M., 1980, Nucleotide sequence of the thrA gene of Escherichia coli, Proceedings of the National Academy of Sciences, USA, 77:5730.

Kirby,T., Hindenach,B. & Greene,R., 1985, Location of promoters and the 5' ends of transcripts of the metB and metJ genes of E.coli K12, Federation Proceedings, 44:1416.

Lawrence,D.A., Smith,D.A. & Rowbury,R.J., 1968, Regulation of methionine biosynthesis in Salmonella typhimurium: mutants resistant to inhibition by analogues of methionine, Genetics, 58:473.

Liljestrand-Golden,C.A. & Johnson,J.R., 1984, Physical organization of the metJB component of the Escherichia coli K12 metJBLF gene cluster, Journal of Bacteriology, 157:413.

Markham,G.D., De Parisis,J. & Gatmaitan,J., 1984, The sequence of metK, the structural gene for S-adenosylmethionine in Escherichia coli, Journal of Biological Chemistry, 259:14505.

Michaeli,S., Mevarech,M. & Ron,E.Z., 1984, Regulatory region of the metA gene of Escherichia coli K12, Journal of Bacteriology, 160:1158.

Milner,L., Whitfield,C. & Weissbach,H., 1969, Effects of L-methionine and vitamin B12 on methionine biosynthesis in Escherichia coli, Archives of Biochemistry and Biophysics, 133:413.

Nass,G. & Thomale,J., 1974, Alteration of structure or level of threonyl-t-RNA synthetase in borrelidin resistant mutants of Escherichia coli, FEBS Letters, 39:182.

Parsot,C., Cossart,P., Saint-Girons,I. & Cohen,G.N., 1983, Nucleotide sequence of thrC and of the transcription termination region of the threonine operon in Escherichia coli K12, Nucleic Acids Research, 11: 7331.

Parsot,C., Saint-Girons,I. & Cossart,P., 1982, DNA sequence change of a deletion mutation abolishing attenuation control of the threonine operon of E.coli K12, Molecular and General Genetics, 188:455.

Patte,J.C., Le Bras,G. & Cohen,G.N., 1967, Regulation by methionine of the synthesis of a third aspartokinase and a second homoserine dehydrogenase in Escherichia coli K12, Biochimica et Biophysica Acta, 136:245.

Press,R., Glansdorff,N., Miner,M., de Vries,J., Kadner,R. & Maas,W.K., 1971, Isolation of transducing particles of φ80 bacteriophage that carry different regions of the Escherichia coli genome, Proceedings of the National Academy of Sciences, USA, 68:795.

Rosenberg,M. & Court,D., 1979, Regulatory sequences involved in the promotion and termination of RNA transcription, Annual Reviews of Genetics, 13:319.

Rowbury,R.J., 1983, Methionine biosynthesis and its regulation, p. 191-211. In "Amino Acids: Biosynthesis and Genetic Regulation", eds K.M. Herrmann & R.L. Somerville, Mass., USA: Addison-Wesley.

Rowbury,R.J. & Woods,D.D., 1961, Further studies in the repression of methionine synthesis in Escherichia coli, Journal of General Microbiology, 24:129.

Rowbury,R.J. & Woods,D.D., 1964, Repression by methionine of cystathionine formation in Escherichia coli, Journal of General Microbiology, 35:145.

Rowbury,R.J. & Woods,D.D., 1966, The regulation of cystathionine formation in Escherichia coli, Journal of General Microbiology, 42:155.

Saint-Girons,I., 1978, A new class of regulatory mutations affecting the expression of the threonine operon in Escherichia coli K12, Molecular and General Genetics, 162:95.

Saint-Girons,I., Belfaiza,J., Guillou,Y., Perrin,D., Guiso,N., Barzu,O. & Cohen,G.N., 1986, Interactions of the Escherichia coli methionine repressor with the metF operator and with its corepressor, S-adenosylmethionine, Journal of Biological Chemistry, 261:10936.

Saint-Girons,I., Duchange,N., Cohen,G.N. & Zakin,M.M., 1984, Structure and autoregulation of the metJ regulatory gene in E.coli, Journal of Biological Chemistry, 259:14282.

Saint-Girons,I., Duchange,N., Zakin,M.M., Park,I., Margarita,D., Ferrara,P. & Cohen,G.N., 1983, Nucleotide sequence of metF, the E.coli structural gene for 5-10 methylene tetrahydrofolate reductase and of its control region, Nucleic Acids Research, 11:6723.

Saint-Girons,I. & Margarita,D., 1975, Operator-constitutive mutants in the threonine operon of Escherichia coli K12, Journal of Bacteriology, 124:1137.

Saint-Girons,I. & Margarita,D., 1978, Fine structure analysis of the threonine operon in Escherichia coli K12, Molecular and General Genetics, 162:101.

Saint-Girons,I. & Margarita,D., 1985, Evidence for an internal promoter in the Escherichia coli threonine operon. Journal of Bacteriology, 161:461.

Schulte,L.L., Stauffer,T.L. & Stauffer,G.V., 1984, Cloning and characterization of the Salmonella typhimurium metE gene, Journal of Bacteriology, 158:928.

Shoeman,R., Redfield,B., Coleman,T., Greene,R.C., Brot,N. & Weissbach,H., 1985, Regulation of methionine synthesis in E.coli: effect of the metJ gene product and S-adenosylmethionine on the expression of the metF gene, Proceedings of the National Academy of Sciences, USA, 82:3601.

Smith,D.A., 1961, S-amino acid metabolism and its regulation in Escherichia coli and Salmonella typhimurium, Advances in Genetics, 16: 141.

Smith,A. & Greene,R.C., 1984, Cloning of the methionine regulatory gene, metJ, of E.coli K12 and identification of its product, Journal of Biological Chemistry, 259:14279.

Smith,A., Greene,R.C., Kirby,T.W. & Hindenach,B.R., 1985, Isolation and characterization of the product of the methionine regulatory gene, metJ, of E.coli K12, Proceedings of the National Academy of Sciences, USA, 82: 6104.

Su,C.H. & Greene,R.C., 1971, Regulation of methionine biosynthesis in Escherichia coli, mapping of the metJ locus and properties of a metJ$^+$-metJ diploid, Proceedings of the National Academy of Sciences, USA, 68: 367.

Szentirmai,A., Szentirmai,M. & Umbarger,H.E., 1968). Isoleucine and valine metabolism of Escherichia coli. Biochemical properties of mutants resistant to thiaisoleucine. Journal of Bacteriology, 95, 1672-1679.

Thèze,J., Margarita,D., Cohen,G.N., Borne,F. & Patte,J.C. (1974). Mapping of the structural genes of the three aspartokinases and of the two homoserine dehydrogenases of Escherichia coli K12. Journal of Bacteriology, 117, 133-144.

Thèze,J. & Saint-Girons,I. (1974). Threonine locus of Escherichia coli K12: genetic structure and evidence for an operon. Journal of Bacteriology, 118, 990-998.

Tran,V.S., Schaeffer,E., Bertrand,O., Mariuzza,R. & Ferrara,P. (1983). Purification, molecular weight and N-terminal sequence of cystathionine-γ-synthase of Escherichia coli. Journal of Biological Chemistry (Appendix), 258, 14872-14873.

Urbanowski,M.L. & Stauffer,G.V. (1985a). Cloning and initial characterization of the metB and metJ genes from Salmonella typhimurium LT2. Gene, 35, 187-197.

Urbanowski,M.L. & Stauffer,G.V. (1985b). Nucleotide sequence and biochemical characterization of the metJ gene from Salmonella typhimurium LT2. Nucleic Acids Research, 13, 673-685.

Véron,M. & Cohen,G.N. (1974). Intra- and interprotomeric interactions between the catalytic regions of aspartokinase I-homoserine dehydrogenase I from Escherichia coli K12. In Metabolic Interconversion of Enzymes, eds E.H. Fisher, E.G. Krebs, H. Neurath, and E.R. Stadtman, p. 335-347. Berlin: Springer Verlag.

Wijesundera,S. & Woods,D.D. (1960). Suppression of methionine synthesis in Escherichia coli by growth in the presence of this amino acid. Journal of General Microbiology, **22**, 229-241.

Yanofsky,C. (1981). Attenuation in the control of expression of bacterial operons. Nature, **289**, 751-758.

Yanofsky,C. & Lennox,E.S. (1959). Transduction and recombination study of linkage relationships among the genes controlling tryptophan synthesis in E.coli. Virology, **8**, 425-447.

Zakin,M.M., Duchange,N., Ferrara,P. & Cohen,G.N. (1983). Nucleotide sequence of the metL gene of Escherichia coli. Its product, the bifunctional aspartokinase II-homoserine dehydrogenase II and the bifunctional product of the thrA gene, aspartokinase I-homoserine dehydrogenase I derive from a common ancestor. Journal of Biological Chemistry, **258**, 3028-3031.

Zakin,M.M., Garel,J.R., Dautry-Varsat,A., Cohen,G.N. & Boulot,G. (1978). Detection of the homology among proteins by immunochemical cross-reactivity between denatured antigens: application to the threonine and methionine regulated aspartokinases-homoserine dehydrogenases from E.coli K12. Biochemistry, **17**, 4318-4323.

Zakin,M.M., Greene,R.C., Dautry-Varsat,A., Duchange,N., Ferrara,P., Py,M.C., Margarita,D. & Cohen,G.N. (1982). Construction and physical mapping of plasmids containing the metJBLF gene cluster of E.coli K12. Molecular and General Genetics, **187**, 101-106.

DESENSITIZATION OF GLUCAGON-STIMULATED ADENYLATE CYCLASE IS MEDIATED BY STIMULATION OF INOSITOL PHOSPHOLIPID METABOLISM

Gregory J. Murphy, Michael J.O. Wakelam and Miles D. Houslay

Molecular Pharmacology Group, Department of Biochemistry
University of Glasgow
Glasgow G12 8QQ, Scotland, U.K.

Desensitization is the process whereby responsiveness to a hormone or drug is reduced after repeated or prolonged exposure to that agonist. Although this phenomenon is widespread in biological regulation, the underlying molecular events have only recently begun to be defined. We have provided an initial characterisation of the molecular events leading to the rapid desensitization of glucagon-stimulated adenylate cyclase activity in rat hepatocytes. This, together with activation of specific cyclic AMP phosphodiesterases (1,2), accounts for the transient rise in intracellular cyclic AMP concentrations that occurs when hepatocytes are challenged with glucagon (Fig.1).

Thus, exposure of hepatocytes to glucagon causes a rise in cyclic AMP concentrations and, at the same time, a progressive desensitization of the glucagon-stimulated adenylate cyclase activity (1,2) observed in membranes that were isolated from such cells (Fig.1). Glucagon-stimulated adenylate cyclase activity is seen to slowly resensitize in bringing cyclic AMP concentrations back to their basal values (1).

Isolated, washed membranes from glucagon (10nM)-pretreated hepatocytes (desensitized for 5 min. at 10nM) displayed identical basal, NaF- and forskolin-stimulated adenylate cyclase activities and showed an identical capacity to bind [125I]-labelled glucagon specifically. This implied that the functioning of the catalytic unit (AC) was not impaired; coupling between the stimulatory guanine nucleotide regulatory protein G_s and AC was unimpaired; the ability of the glucagon receptor to bind glucagon was unimpaired; the total number of glucagon receptors was unaltered; and that no significant amounts of residual glucagon contaminated the washed membrane preparation (2). Indeed we and others have also been able to show that challenge of hepatocytes with glucagon did not lead to rapid internalisation of glucagon receptors (2). Desensitization, in hepatocytes, thus appears to be due to an inability of glucagon receptors to couple to G_s. This may involve a lesion occurring at either the coupling interface of G_s or at the glucagon receptor itself. This process is specific. Only the glucagon-stimulated response of adenylate cyclase was inhibited. This effect was exerted

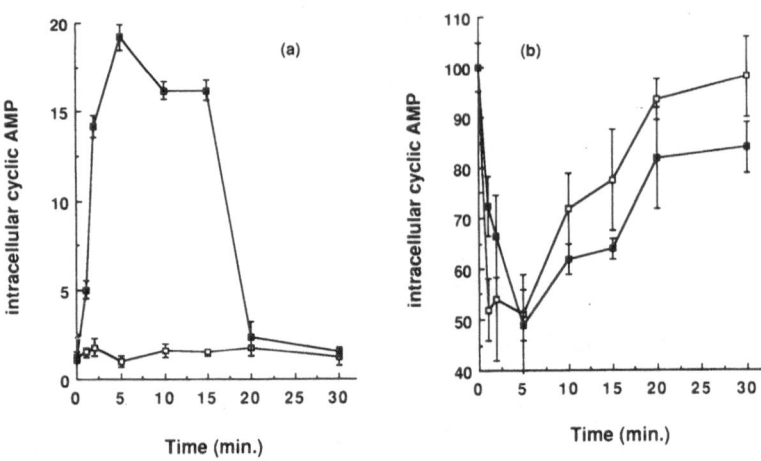

Figure 1 <u>Cyclic AMP concentrations and desensitization in intact hepatocytes</u>

Hepatocytes were challenged with 10nM glucagon (a) or 10nM TH-glucagon (b) and both the intracellular concentrations of cyclic AMP ( ) and the glucagon-stimulated adenylate cyclase activity in membranes isolated from these cells ( ) was measured. [Cyclic AMP] are given as pmol/mg dry wt of cells and glucagon (10nM)-stimulated adenylate cyclase activity is pmol/min/mg protein. Methods are as described in detail before (2,3).

predominately on the V_{max} of the activation process as the K_a for activation of adenylate cyclase, by glucagon, in membranes from desensitized cells was not very different from that found using membranes from control cells (2).

Glucagon caused the dose-dependent desensitization of adenylate cyclase with a K_a value of around 0.4nM. This value is very different from the K_a value of around 8nM observed for glucagon's activation of adenylate cyclase in both broken membranes and in intact hepatocytes. One interpretation of such differences is that two distinct receptor populations might mediate these two processes. We have obtained evidence for such a conclusion using the glucagon analogue, TH-glucagon (1-N-alpha-trinitrophenyl histidine, 12-homoarginine-glucagon). This analogue, whilst not causing any increase in intracellular concentrations of cAMP, was just as effective as glucagon in eliciting desensitization of adenylate cyclase (Fig.1). This implies strongly that the process of rapid desensitization is independent of intracellular concentrations of cAMP.

We have recently provided a molecular explanation for these observations by demonstrating that glucagon activates two signal-transduction systems in hepatocytes; one system being coupled to the stimulation of inositol phosphate production, the other to stimulation of adenylate cyclase (3). Figure 2 shows the dose-dependent stimulation of inositol phosphates, in hepatocytes, by glucagon and TH-glucagon. The K_a values for these processes were 0.2nM and 0.3nM for glucagon and TH-glucagon, respectively. These values correlate well with the K_a values derived for these ligands in eliciting the desensitization of adenylate cyclase (0.4nM and 0.7nM) for glucagon and TH-glucagon respectively.

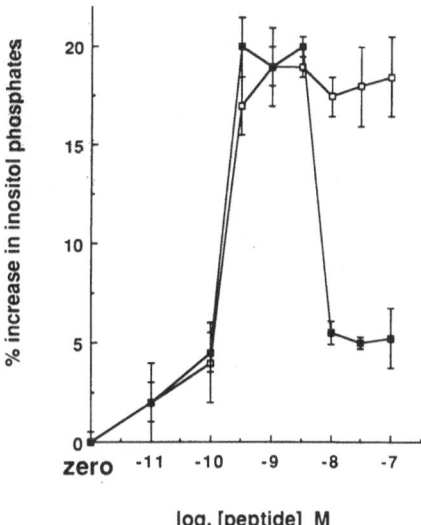

Figure 2   <u>Glucagon and TH-glucagon stimulate the production of inositol
phosphates in hepatocytes</u>

        Treatment of [^{3}H]-inositol labelled hepatocytes with glucagon
( ) or TH-glucagon ( ) caused an increase in the production of inositol
phosphates in Li$^+$-treated cells.   This is shown as the %-age
increase elicited as a function of log dose of peptide hormones (see 2).

        Interestingly, high concentrations of glucagon (but not
TH-glucagon) inhibited the production of inositol phosphates: the K_a
for this process (around 10nM) was similar to that observed for
glucagon's stimulation of adenylate cyclase.   However, although the
inositol phospholipid response was reduced somewhat, desensitization of
adenylate cyclase was still apparent (Fig.1 & 2).   Presumably, even
the small stimulation of inositol phospholipid metabolism occurring at
such concentrations, when coupled to a suitable amplification step was
sufficient to stimulate desensitization of adenylate cyclase.   This
probably involves C-kinase activation as treatment of hepatocytes with
phorbol esters mimicked the desensitization process (4).

        Glucagon thus appears to act on rat hepatocytes through two
functionally and, presumably, structurally distinct receptor
populations, which we have termed GR1 and GR2 receptors (Fig.3).
These can be discriminated by the glucagon analogue, TH-glucagon, as
well as other analogues (see 3); the different K_a values for the
activation of the different signalling systems (3); and by the complex
binding of glucagon to membranes (see 3).   The GR1 receptor is coupled
to the stimulation of inositol phosphate production, whilst the GR2
receptor is coupled to adenylate cyclase stimulation.   Activation of
the GR1 receptor (and others stimulating inositol phosphate production
e.g. angiotensin, vasopressin) leads to desensitization of GR2
receptor-stimulated adenylate cyclase, probably through the generation
of diacylglycerol and activation of C-kinase.   This may either modify
G_s or may activate G_i indirectly (Fig.3).   These signalling
pathways can attenuate each others function: GR1 as described (Fig.3),

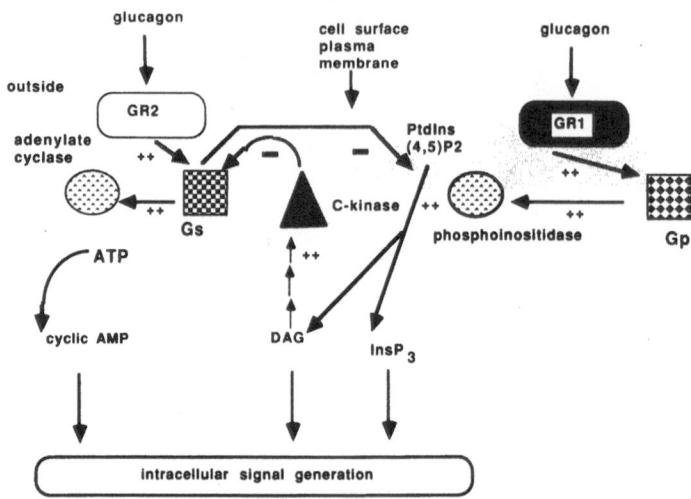

Figure 3 **The coupling of GR1 and GR2 receptors with distinct signalling mechanisms**

GR1 receptors are proposed to be coupled to stimulation of phospholipase C through the putative G-protein, G_p. GR2 receptors stimulate adenylate cyclase through the G-protein, G_s. It is suggested that C-kinase activation plays a key role in this process as phorbol esters inhibit glucagon-stimulated adenylate cyclase (4). Activation of G_s attenuates G_p functioning.

whilst GR2 activation leads to attenuation of the GR1 pathway probably through activation of G_s (3). Whilst desensitization plays a key role in the transience of the cAMP response, the other major factor which leads to the reduction in intracellular cyclic AMP is the activity and activation of specific cyclic AMP phosphodiesterases (1,5).

Acknowledgements

This work was supported by grants from the Medical Research Council, Scottish Home & Health Department and the California Metabolic Research Foundation. Both MDH and GJM thank the Wellcome Trust for Travel Grants given in support of this study.

References

1. Houslay, M.D. Biochem. Soc. Trans. 14, 183-193 (1986).
2. Heyworth, C.M. & Houslay, M.D. Biochem.J. 214, 93-98 (1983).
3. Wakelam, M.J.O., Murphy, G.J. & Houslay, M.D. Nature 323, 68-71 (1986).
4. Heyworth, C.M., Wilson, S.R., Gawler, D.J. & Houslay, M.D. FEBS Lett. 187, 196-200 (1985).
5. Heyworth, C.M., Wallace, A.V. & Houslay, M.D. Biochem. J. 214, 99-110 (1983).

PHOSPHOINOSITIDE-HYDROLYSIS IN A STIMULATED MURINE T CELL CLONE

Berthold Behl, Beate Schwinzer and Klaus Resch

Division of Molecular Pharmacology, Department of Pharmacology and Toxicology, Medical School Hannover, D-3000 Hannover 61, FRG

ABSTRACT

The time-dependent accumulation of inositolphosphates was measured in a Con A and IL 2 stimulated murine alloantigen specific, noncytolytic T cell clone. Con A provoked a sharp increase in the inositol triphosphate (IP_3) fraction up to 370 % of the control. There was also a response to Con A in the production of IP_1 (about 200 % of the control) and less in IP_2. IL 2 had no significant effect on the level of inositol phosphates, whereas under similar conditions cells showed a strong proliferative response to IL 2. Furthermore we investigated the period of exposure to IL 2, which is necessary for the induction of proliferation. Cells had to be incubated more than 4-6 hours to stimulate cell growth. These data suggest that IL 2 in opposite to Con A provokes a slowly increasing process, by which the IL 2 starved lymphocytes are stimulated to proliferation. This may be due to an induction of the expression of IL 2-receptors by IL 2 itself.

INTRODUCTION

Proliferation of T-lymphocytes is induced by stepwise activation by antigen and Interleukin 2 (1). The role of the antigen or antigen substituting lectin in this two step activation is probably the induction of IL 2 receptors[1]. The variable extent of IL 2 production induced by antigens or lectins may be responsible for proliferation which occurs after stimulation by antigen or lectin without addition of IL 2. The exact molecular mechanism of both the antigen- and the IL 2-activation is not completely elucidated. Con A was shown to activate the phosphoinositide turnover and to induce a subsequent increase in cytosolic Ca^{++} and pH[2,3,4].

Little is known about the primary events following the binding of IL 2 to its receptor. Several authors reported an early phosphorylation of cellular proteins after binding of IL 2 to lymphocytes[5,6,7]. Recently Farrar et al. proposed a mechanism

of IL 2 mediated lymphocyte stimulation, demonstrating the translocation of proteinkinase C from the cytosol to the plasma membrane [8].
It is not yet clear, whether this translocation is the consequence of
an increased phosphoinositide hydrolysis and a concomittant generation
of diacylglycerol. On the contrary there are some reports, that IL 2
did not provoke an enhanced PI-turnover [9]. It is however difficult to
assess the molecular mechanism of the interleukin 2 induced activation
as it could be shown that IL 2 receptors are not constitutively expressed in resting T-lymphocytes. They are induced by lectins, anti
T-cell-receptor antibodies and by IL 2 itself [10]. On the other hand
investigations of early effects like PI-breakdown and ion signals
require conditions, under which cells have high levels of unoccupied
receptors.

In the present paper we investigated changes in the phosphoinositide turnover of a murine alloantigen-specific noncytolytic T-
cell clone, because there are no examinations of Con A- and IL 2 -
effects on this signal transduction pathway in physiological and
homogenous T-cell population so far. In these cells IL 2 receptors are
constitutively expressed and they are therefore suitable for investigations of early signal transducing processes induced by IL 2.

MATERIAL AND METHODS

Culture of clone cells

The alloantigen-specific non-cytolytic murine T lymphocyte clone
II-8 was generated from a C57Bl/6 anti DBA/2 mixed lymphocyte culture
as described elsewere [11]. Cells were cultured in multiwell plates
(Falcon 3047) in medium 199 supplemented with 15 U/ml IL 2 and 10 %
fetal calf serum (FCS). They were maintained by weekly passage and
restimulation with 10^6 irradiated cells per well. Cells were used for
the investigations of lipid metabolism 6 to 7 days after the last
restimulation.

Proliferation assay

10^6 clone cells were incubated in 5-ml polystyrene tubes (Falcon
2058) in 1 ml RPMI 1640 supplemented with 16 U/ml IL 2 from a
concanavalin A stimulated rat spleen supernatant (Spiess and Rosenberg, 1981) for 2 - 20 h at 37 C. and 5 % CO_2. Then the cells were
washed twice with RPMI 1640, resuspended in 1 ml RPMI 1640
supplemented with 10 % FCS and transferred into microtiterwells (200
ul / Nunc 167008). Proliferation was measured after 20 h by incorporation of ^{3}H-TdR for the following 4 h (0.5 µCi / well).

Accumulation of inositolphosphates

Clone cells (IL 2 starved for 3 days) were washed 3 times with
inositol-free medium (BMED (modified)/Flow Laboratories supplemented
with 20 mM Tris, 24 mM sodiumbicarbonate, 200 mM glutamine). Cells
($2*10^6$/ml) were incubated with 50 µCi/ml ^{3}H-inositol (Amersham) for 18
h in inositol free medium plus 0.5 % bovine serum albumin. The prelabeled cells were washed 3 times with a RPMI-salt-salution (120 mM
NaCl, 5.2 mM KCl, 5.1 mM Na_2HPO_4, 0.5 mM $MgCl_2$, 0.43 mM $CaCl_2$, 11 mM
glucose, 10 mM Hepes pH 7.2). Samples of 0.5 ml (10^6 cells/ml) were
incubated for 30 min. in a water bath at 37 C. Thereafter they were
stimulated with either 5µg/ml Con A (PHARMACIA) or 20 U/ml IL 2
(recombinant human IL 2 or IL 2 from a concanavalin A induced rat

spleen supernatant). The incubation was stopped at the times indicated by addition of 1.88 ml chloroform/methanol (1:2). The inositol-phosphates were extracted and separated by ionexchange chromatography as previously described[3].

RESULTS AND DISCUSSION

The increase of radiolabeled inositolphosphates in ^{3}H-inositol prelabeled cells as a response to Con A and IL 2 was measured. As shown in fig. 1a Con A provoked a sharp increase in the level of inositoltriphosphates (IP_3) up to 370 % of the control within 3 minutes after addition of Con A. The Con A induced enhancement of radioactivity in IP_2 and IP_1 was less than in IP_3 in the range of 170 and 200 % respectively (fig. 1b and 1c). For the inhibition of phos-phatases, which hydrolyse the products of the phospholipase C reaction, 5 mM LiCl was added to the samples. Without LiCl increases in the levels of inositolphosphates were lower than in the presence of LiCl (fig. 1a, 1b, 1c). As compared to bulk lymphocytes the effect of Con A was small[3]. IL 2 had no significant effect on the production of inositolphosphates under the same conditions in which the influence of Con A was tested (fig. 1d, 1e, 1f).

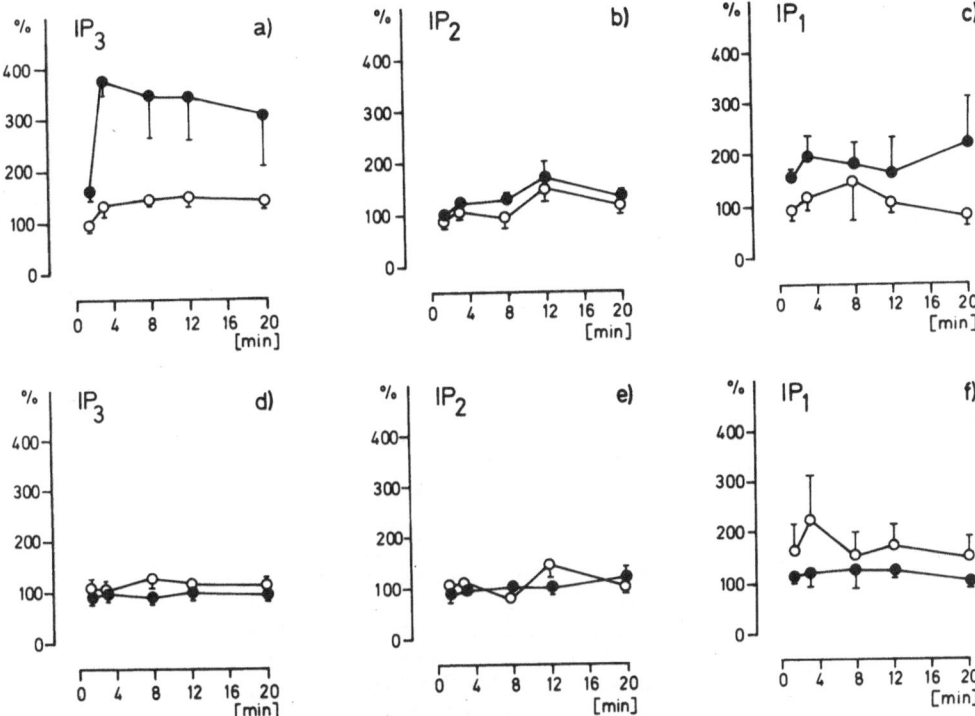

Fig. 1. RELEASE OF INOSITOLPHOSPHATES. ^{3}H-inositol prelabeled cells were incubated with either 5 µg/ml Concanavalin A (fig. 1a-1c) or 20 U/ml IL 2 (fig. 1d-1f) for the indicated time. Experi-ments were performed with (closed symbols) and without (open symbols) addition of 5 mM LiCl. Inositolphosphates were extracted and the individual inositolphosphates were separated by ion-exchange chromatography. The figures show the time course of the accumulation of radioactivity in the individual inositol-phosphates as percentage of the controls. Data repre-sent the mean +/- SEM values of 3 independent experiments.

Parallel to the examinations of the inositolphosphate release we tested the proliferation of clone II-8 as a response to Con A and IL 2. 6 to 7 days after the last restimulation clone cells could be stimulated by exogenously added IL 2. A weak proliferation could also be obtained by stimulating the clone cells with Con A (1µg/ml). The Con A-induced growth of cells, however, reached only 5 to 20 % of the maximal proliferation induced by saturating IL 2 concentrations (more than 12 U/ml).

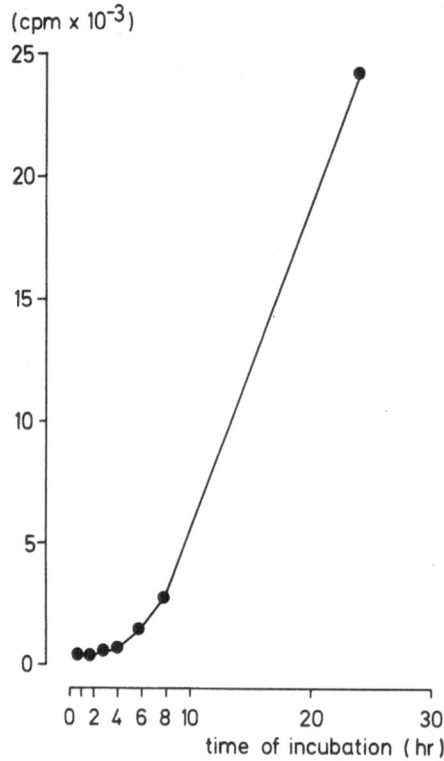

Fig. 2. DEPENDENCE OF PROLIFERATION ON THE PERIOD OF EXPOSURE TO IL 2. Clone cells were incubated with 16 U/ml IL 2 for the indicated times, washed twice, and incubated for 20 hours including the time of IL 2 incubation. Proliferation was measured by incorporation of ^{3}H-thymidine during the following 4 hours.

The strong differences between Con A and IL 2 in their potencies to provoke a phosphoinositide hydrolysis and proliferative response suggest that the PI-response is not involved in the final mitogenic signal pathway induced by IL 2. This is consistent with very recent findings by Mills et al. who could not find alterations in the phosphoinositide turnover in response to IL 2 in human or murine IL 2 sensitive lymphocytes[9]. To investigate whether or not the activated IL 2 receptor provokes an early and transient post receptor signal which is capable to induce mitosis we measured the dependence of the proliferative response on the period of exposure to IL 2. We could show that the cells had to be exposed to IL 2 more than 4 hours to increase the DNA synthesis (fig. 2). There are reports that IL 2 enhances the amount of IL 2 receptors on T-lymphocytes. Such an increase in the level of receptors or shift in the affinity state of receptors might take place during the first hours of stimulation with IL 2. If this hypothesis is right one cannot exclude that the low level of receptors at the beginning of the incubation with IL 2 is responsible for the lack of a detectable release of inositolphosphates.

REFERENCES

1. T. A. Waldmann, The structure, function , and expresssion of interleukin-2 receptors on normal and malignant lymphocytes. Science. 232:727 (1986).
2. J. B. Imboden and J. D. Stobo, Transmembrane signalling by the T cell antigen receptor. J. Exp. Med. 161:446 (1985).
3. M. V. Taylor, J. C. Metcalfe, T. R. Hesketh, G. A. Smith and J. P. Moore, Mitogens increase phosphorylation of phosphoinositides, Nature. 312:462 (1984).
4. T. R. Hesketh, J. P. Moore, D. H. Morris, M. V. Taylor, J. Rogers, G. A. Smith and J. C. Metcalfe, A common sequence of calcium and pH signals in the mitogenic stimulation of eukaryotic cells, Nature. 313:481 (1985).
5. M. Kohno, S. Kuwata, Y. Namba and M. Hanaoka, Interleukin 2 induces rapid phosphorylation in murine T-lymphocytes. FEBS. 198: 33 (1986).
6. T. Ishii, K. Sugamura, M. Nakamura and Y. Hinuma, Interleukin 2 (IL 2) rapidly induces phosphorylation of a cellular protein, pp67, in an IL 2 dependent murine cell line. Biochem. Biophys. Res. Comm. 135:487 (1986).
7. G. N. Gaulton and D. D. Eardley, Interleukin 2-dependent phosphorylation of interleukin 2 receptors and other T cell membrane proteins. J. Immunol. 136:2470 (1986).
8. W. L. Farrar and W. B. Anderson, Interleukin 2 stimulation of protein kinase C plasma membrane association. Nature. 315:233 (1985).
9. G. B. Mills, D. J. Stewart, A. Mellors and E. W. Gelfand, Interleukin 2 does not induce phosphatidylinositol hydrolysis in activated T cells. J. Immunol. 136:3019 (1986).
10. J. M. Depper, W. J. Leonard, C. Drogula, M. Krönke, T. A. Waldmann and W. C. Greene, Interleukin 2 (IL-2) augments transcription of the IL-2 receptor gene. Proc. Natl. Acad. Sci. USA. 82:4230 (1985)
11. J. Weiß, B. Schwinzer, H. Kirchner, D. Gemsa and Resch, Effects of Cyclosporin A on Functions of Specific Murine T Cell Clones: Inhibition of Proliferation, Lymphokine secretion and Cytotoxicity. Immunobiol. 171:234 (1986).

PHOSPHORYLATION OF TWO DIFFERENT FORMS OF THE CYTOSKELETAL PROTEIN

CALDESMON BY PROTEIN KINASE C

David W. Litchfield and Eric H. Ball

Department of Biochemistry
University of Western Ontario
London, Ontario, Canada. N6A 5C1

Tumour-promoting phorbol esters have been shown to cause cytoskeletal alterations in cultured cells (Rifkin et al., 1979; Schliwa et al., 1984). These compounds exert their biological effects by activation of protein kinase C (Ashendel, 1985; Nishizuka, 1986) suggesting that cytoskeletal changes may be brought about by the phosphorylation of regulatory components of the cytoskeleton.

The recently discovered actin- and calmodulin-binding protein caldesmon is a candidate for a role in the organization of the cytoskeleton (Sobue et al., 1981; Bretscher, 1986). Caldesmon was originally identified in chicken gizzard smooth muscle (Sobue et al., 1981) as a doublet composed of polypeptides of 150 kDa and 147 kDa. Immunologically related polypeptides have been subsequently identified in a number of cell and tissue types (Owada et al., 1984; Bretscher and Lynch, 1985). On the basis of molecular weight, these related polypeptides fall into two apparent classes, one of 140-150 kDa and one of 70-80 kDa. Proteins from both classes bind to actin and to calmodulin (Sobue et al., 1985) and exhibit heat stability (Bretscher and Lynch, 1985; Dingus et al., 1986). The two classes of proteins have recently been called caldesmon$_{77}$ (CD77)* and caldesmon$_{150}$ (CD150) to reflect their similar functional features while distinguishing them with respect to molecular weight differences (Sobue et al., 1985). The molecular basis of the similarities and differences between CD77 and CD150 are as yet unknown, although it is unlikely that CD77 is simply a proteolytic fragment of CD150 (Sobue et al., 1985; Dingus et al., 1986).

Smooth muscle caldesmon (CD150) has recently been identified as an _in vitro_ substrate for protein kinase C (Umekawa and Hidaka, 1985). The phosphorylated form of caldesmon could inhibit myosin light chain kinase whereas the non-phosphorylated form could not. In this report, we have extended these observations by demonstrating that both classes of caldesmon, CD77 and CD150, from bovine liver can be phosphorylated by protein kinase C. The two phosphoproteins have identical tryptic phosphopeptide maps demonstrating that there are conserved sequences in both types of caldesmon. The results

---

*The abbreviations used are: CD77, 70-80 kDa caldesmon; CD150, 140-150 kDa caldesmon; SDS, sodium dodecyl sulphate; PAGE, polyacrylamide gel electrophoresis; PKC, protein kinase C; EGTA, ethylene glycol bis (β-amino-ethyl ether)-N,N,N',N'-tetraacetic acid.

point toward a possible role for the phosphorylation of caldesmon in the changes in cytoskeletal organization which follow the treatment of different cells with tumour-promoting phorbol esters.

MATERIALS AND METHODS

Caldesmon (CD77 and CD150) was purified from fresh bovine liver by a modification of the method of Bretscher (1984). CD77 and CD150 were identified by SDS-PAGE and by their cross-reactivity with affinity-purified rabbit antibodies against chicken gizzard caldesmon.

Protein kinase C was purified from 5 grams of fresh rat brain by the method of Wolf et al. (1985). PKC was identified by SDS-PAGE and was assayed by its ability to phosphorylate histone (Sigma, type IIIS) in the presence of phosphatidylserine and calcium as previously described (Litchfield and Ball, 1986). The specific activity of PKC was 110 nmolar units/mg protein. One nmolar unit of activity is the amount of enzyme required to transfer one nmol of phosphate to histone (type IIIS) from ATP per minute at 30°C.

CD77 (3 μg) or CD150 (1 μg) was phosphorylated by PKC (0.01 nmolar units) at 30°C for 10 minutes. The reaction was conducted in a volume of 0.05 ml containing 20 mM Tris-HCl, pH 7.5, 10 mM MgCl$_2$, 10 μM [γ-^{32}P]ATP (specific activity 200 cpm/pmol) and either phosphatidylserine (50 ug/ml) and calcium (0.5 mM) or EGTA (0.5 mM). The reaction products were analysed by SDS-PAGE and autoradiography using 8% polyacrylamide gels and the buffer system of Laemmli (1970).

Two dimensional phosphopeptide mapping was done by the method of Elder et al. (1977) and phosphoamino acid analysis by the method of Cooper et al. (1983). Protein determinations were done as described by Hartree (1972).

RESULTS AND DISCUSSION

Caldesmon is found associated with actin in a variety of tissue types and cultured cell lines, suggesting a common role in cytoskeletal function or organization. Smooth muscle caldesmon (CD150) has previously been identified as an in vitro substrate for protein kinase C (Umekawa and Hidaka, 1985) and for a calcium/calmodulin-dependent protein kinase (Ngai and Walsh, 1984) suggesting that caldesmon function could be modulated by phosphorylation. To extend these observations, we have examined the phosphorylation of two forms of non-muscle caldesmon. The results indicate that CD77 (Fig. 1B, lane 1) and CD150 (Fig. 1B, lane 3) can be phosphorylated by PKC in the presence of phosphatidylserine and calcium. In addition to CD77 and CD150, an autophosphorylated band of PKC (Mr 82,000) can also be seen. The autophosphorylation of PKC, which is well documented (Wolf et al., 1985), is also demonstrated (Fig. 1B, lane 5) in the absence of CD77 or CD150.

The physiological significance of caldesmon phosphorylation cannot be established until the effects of phosphorylation on the properties of caldesmon are known, and phosphorylation has been shown to occur in vivo. In addition to its actin-and calmodulin-binding abilities, caldesmon has been shown to exert an inhibitory effect on the actin-activated myosin ATPase of smooth muscle (Ngai and Walsh, 1984). This inhibitory effect was alleviated when caldesmon was phosphorylated by the calcium/calmodulin-dependent kinase (Ngai and Walsh, 1984). It was also demonstrated by Umekawa and Hidaka (1985) that caldesmon previously phosphorylated by PKC could inhibit myosin light chain kinase. These observations have not yet been extended to the phosphorylated forms of bovine liver CD77 and CD150, nor to an <u>in vivo</u> system. As a first step in establishing the physiological relevance of caldesmon phosphorylation, we have demonstrated that caldesmon (CD77) is phosphorylated by PKC in intact platelets following treating with tumour-promoting phorbol esters (Litchfield and Ball, manuscript submitted).

Two dimensional phosphopeptide mapping reveals that CD77 (Fig. 2A) and CD150 (Fig. 2B) are both phosphorylated at multiple sites. Phosphorylation occurs exclusively on serine residues in both CD77 (Fig. 2C) and CD150 (Fig. 2D). An interesting revelation of phosphopeptide mapping is that the two proteins have the same sites of phosphorylation, each protein having the same two major and several minor phosphopeptides. These results indicate that despite the large difference in molecular weight, CD77 and CD150 do contain conserved sequences near the sites of phosphorylation. These results are not unexpected in view of the similar functional characteristics demonstrated by the two proteins, and have been verified by phosphopeptide maps from avian CD77 and CD150 (our unpublished observations). These findings also suggest that CD77 and CD150 may be subject to similar mechanisms of regulation.

The actual molecular basis of the differences between the two proteins is not yet known. Although the possibility that CD77 is a proteolytic

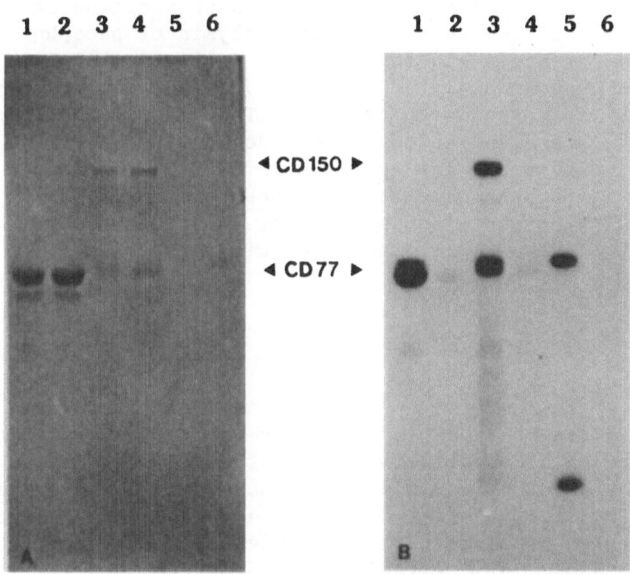

Fig. 1. Phosphorylation of different caldesmon forms by protein kinase C. CD77 (lanes 1,2) or CD150 (lanes 3,4) was incubated with PKC in the presence (lanes 1,3,5) or absence (lanes 2,4,6) of phosphatidylserine and calcium. PKC in the absence of CD77 or CD150 is also shown (lanes 5,6). The reaction mixtures were run on SDS-PAGE which was stained with Coomassie Blue (A) prior to autoradiography (B).

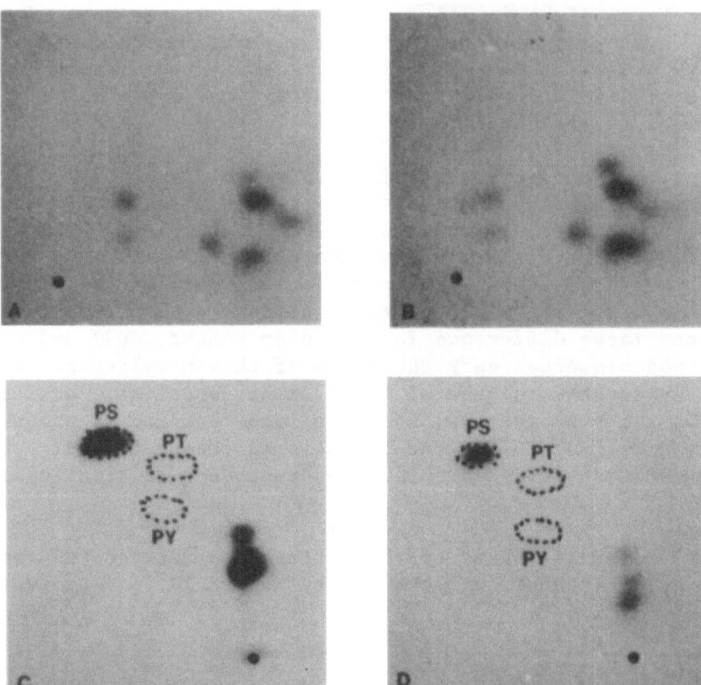

Fig. 2. Two dimensional tryptic phosphopeptide maps
and phosphoamino acid analysis of phosphorylated caldesmon forms. Following SDS-PAGE,
phosphorylated CD77 (A,C) or CD150 (B,D) was
digested with trypsin. The tryptic peptides
were subjected to peptide mapping (A,B) or
phosphoamino acid analysis (C,D). Standard
phosphoamino acids - phosphoserine (PS),
phosphothreonine (PT) or phosphotyrosine
(PY) - were visualized with ninhydrin.

fragment of CD150 cannot be excluded by our results, it has been previously
demonstrated that this is unlikely (Sobue et al., 1985; Dingus et al.,
1986). Thus, CD77 and CD150 could be products of distinct genes or could
result from alternate processing of a single gene or its product. Sobue et
al. (1985) have also proposed that the native proteins have similar molecular weights (300,000), CD150 existing as a dimer and CD77 as a tetramer.
Therefore, it is possible that CD150 may be derived by duplication of the
CD77 gene, or by covalent crosslinking of two CD77 polypeptides. Further
experimentation will be necessary to resolve this issue.

The results of this study demonstrate that two forms of caldesmon from
liver can be phosphorylated by protein kinase C. The effects of phosphorylation on the known properties of caldesmon have, however, not yet been
established. A better understanding of the physiological role of caldesmon
phosphorylation must also await an improved knowledge of the in vivo role of
caldesmon. Nevertheless, the results do indicate that caldesmon is one
candidate for the action of PKC which may be important in mediating the
effects of tumour-promoting phorbol esters on cytoskeletal organization.

ACKNOWLEDGEMENTS

This research was supported by a grant from MRC Canada. E.H.B. is the holder of an MRC Scholarship and D.W.L. is a holder of an MRC Studentship.

REFERENCES

Ashendel, C. L., 1985, The phorbol ester receptor: a phospholipid-regulated protein kinase, Biochim. Biophys. Acta, 822:219.

Bretscher, A., 1984, Smooth muscle caldesmon - Rapid purification and F-actin cross-linking properties, J. Biol. Chem., 259:12873.

Bretscher, A., 1986, Thin filament regulatory proteins of smooth- and non-muscle cells, Nature, 321:726.

Bretscher, A., and Lynch, W., 1985, Identification and localization of immunoreactive forms of caldesmon in smooth and nonmuscle cells: A comparison with the distributions of tropomyosin and α-actinin, J. Cell Biol., 100:1656.

Cooper, J. A., Sefton, B. M., and Hunter, T., 1983, Detection and quantification of phosphotyrosine in proteins, Methods Enzymol., 99:387.

Dingus, J., Hwo, S., and Bryan, J., 1986, Identification by monoclonal antibodies and characterization of human platelet caldesmon, J. Cell Biol., 102:1748.

Elder, J. H., Pickett, R. A.,II, Hampton, J., and Lerner, R. A., 1977, Radioiodination of proteins in single polyacrylamide gel slices, J. Biol. Chem., 252:6510.

Hartree, E. F., 1972, Determination of protein: A modification of the Lowry method that gives a linear photometric response, Anal. Biochem., 48:422.

Laemmli, U. K., 1970, Cleavage of structural proteins during the assembly of the head of bacteriophage T4, Nature, 227:680.

Litchfield, D. W., and Ball, E. H., 1986, Phosphorylation of the cytoskeletal protein talin by protein kinase C, Biochem. Biophys. Res. Commun., 134:1276.

Ngai, P. K., and Walsh, M. P., 1984, Inhibition of smooth muscle actin-activated myosin Mg^{2+}-ATPase activity by caldesmon, J. Biol. Chem, 259:13656.

Nishizuka, Y., Studies and perspectives of protein kinase C, 1986, Science, 233:305.

Owada, M. K., Hakura, A., Iida, K., Yahara, I., Sobue, K., and Kakiuchi, S., 1984, Occurrence of caldesmon (a calmodulin-binding protein) in cultured cells: Comparison of normal and transformed cells, Proc. Natl.Acad.Sci.U.S.A., 81:3133.

Rifkin, D. B., Crowe, R. M., and Pollack, R., 1979, Tumor promoters induce changes in the chick embryo fibroblast cytoskeleton, Cell, 18:361.

Schliwa, M., Nakamura, T., Porter, K. R., and Euteneuer, U., 1984, A tumor promoter induces rapid and coordinated reorganization of actin and vinculin in cultured cells, J. Cell Biol., 99:1045.

Sobue, K., Tanaka, T., Kanda, K., Ashino, N., and Kakiuchi, K., 1985, Purification and characterization of caldesmon$_{77}$: A calmodulin-binding protein that interacts with actin filaments from bovine adrenal medulla, Proc.Natl.Acad.Sci.U.S.A., 82:5025.

Sobue, K., Muramoto, Y., Fujita, M., and Kakiuchi, S., 1981, Purification of a calmodulin-binding protein from chicken gizzard that interacts with F-actin, Proc.Natl.Acad.Sci.U.S.A., 78:5652.

Umekawa, H., and Hidaka, H., 1985, Phosphorylation of caldesmon by protein kinase C, Biochem. Biophys. Res. Commun., 132:56.

Wolf, M., Cuatrecasas, P., and Sahyoun, N., 1985, Interaction of protein kinase C with membranes is regulated by Ca^{2+}, phorbol esters and ATP, J. Biol. Chem., 260:15718.

GUANINE NUCLEOTIDE REGULATORY PROTEIN COUPLES ANGIOTENSIN II RECEPTORS TO PHOSPHOLIPASE C IN MESANGIAL CELLS

Josef Pfeilschifter

Physiologisches Institut der Universität Zürich
Winterthurerstr. 190
CH-8057 Zürich, Switzerland

INTRODUCTION

Recently I have shown that vasoconstrictive hormones, such as angiotensin II, vasopressin and norepinephrine, caused a rapid hydrolysis of phosphatidylinositol 4,5-bisphosphate (PIP_2) in rat mesangial cells with subsequent formation of diacylglycerol (DG) and inositoltrisphosphate (IP_3) [1,2]. The activation of phospholipase C, which accounted for this degradation of PIP_2 further leads to an increased synthesis of prostaglandins (PG). The present study was carried out to gain information about the nature of the link between the occupation of the angiotensin II receptor and the activation of phospholipase C in mesangial cells.

EXPERIMENTAL PROCEDURES

Cell culture: Rat mesangial cells were cultivated as described earlier [1]. For all experiments the first passage of mesangial cells was used. The cells were grown in RPMI 1640 supplemented with 10% fetal bovine serum, penicillin, streptomycin and bovine insulin.

Measurement of $[Ca^{2+}]_i$: The intracellular free calcium concentration was measured by using the fluorescent calcium indicator quin 2 as described [3].

Determination of inositol phosphates: The cells were prelabelled for 72 h with myo-[2-^{3}H]inositol ($10 \mu Ci/ml$) in MEM. The cells were then stimulated with angiotensin II (10^{-7}M) for 10 s and the reaction was terminated by the addition of ice-cold trichloroacetic acid and the inositol phosphates separated by anion exchange chromatography as described [2].

Angiotensin II-binding assay: Binding experiments were done with dissociated cells scraped off their flasks exactly as described [3]. Brief: ^{125}I-angiotensin II was added at 0,5nM and incubated with 30-40 µg of cellular proteins per tube for 40 min. At the

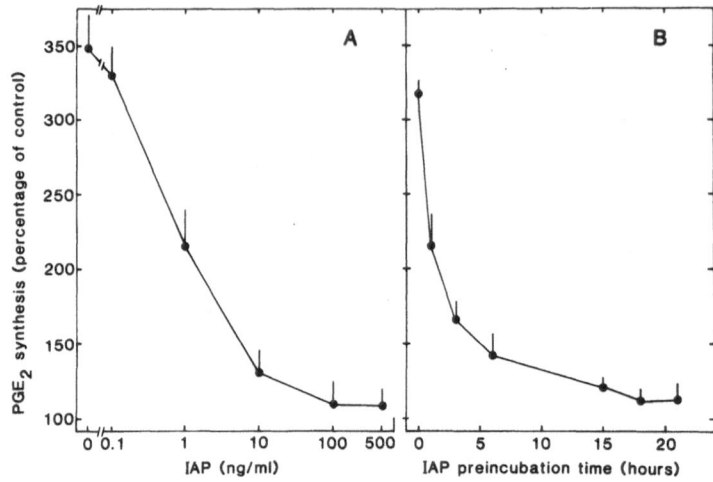

Fig. 1. Time- and dose-dependency of inhibition
of angiotensin II-induced PG synthesis
by pertussis toxin.

Mesangial cells were incubated for 5 min with
angiotensin (10^{-7}), and PG synthesis is expressed as a percentage of control (cells stimulated with vehicle alone). Values are means
$\pm$ S.E.M. for 5 experiments. a) Cells preincubated for 15 h with the indicated concentration
of pertussis toxin (islet-activating protein,
IAP). b) Cells preincubated with 100 ng/ml
for the indicated time period.

end of the incubation, bound radioactivity was separated by
filtration. Specific binding was calculated by subtracting the
binding in the presence of 5mM unlabelled angiotensin II from
total binding and expressed as fmol of bound angiotensin II/mg
of cell protein.

Prostaglandin and protein analysis: PGE_2 was determined by
radioimmunoassay and cell protein was determined by the method
of Lowry (4).

RESULTS

The basal rate of PGE_2 synthesis in mesangial cells was
1,8 ng/5 min per mg of protein. Addition of angiotensin II
(10^{-7}M) increased PGE_2 formation by 348% to 6.3 ng/5 min per
mg of protein. Preincubation of mesangial cells with pertussis
toxin dose- and time-dependently attenuated the angiotensin II
evoked PGE_2 synthesis, as shown in Fig. 1. Pertussis toxin did
not inhibit the PGE_2 synthesis stimulated by the calcium ionophore A23187 (10^{-6}M): 9.5 $\pm$ 1,2 ng/5 min per mg of protein
(A23187), 8.7 $\pm$ 1.1 ng/5 min per mg of protein (pertussis toxin)
(means $\pm$ S.E.M., n = 3).

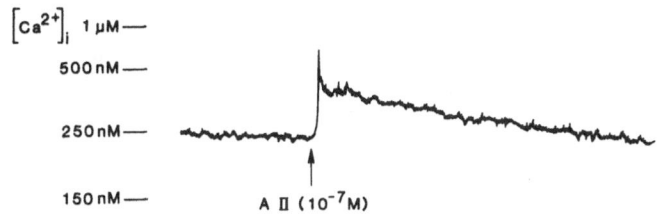

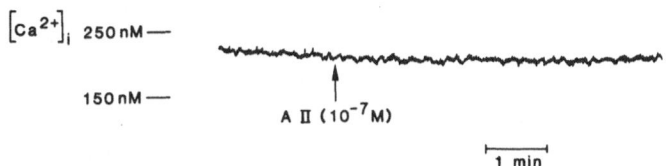

Fig. 2. $[Ca^{2+}]_i$ in mesangial cells as indicated by quin 2 fluorescence.
$[Ca^{2+}]_i$ was calculated from changes in quin
2 fluorescence. The upper part of the Figure
shows the effect of angiotensin II (10^{-7}M)
on $[Ca^{2+}]_i$. The lower part of the Figure
shows that pretreatment of mesangial cells
with pertussis toxin (100 ng/ml; 15 h)
totally abolished the angiotensin II-induced
increase in $[Ca^{2+}]_i$. Arrow indicates
addition of angiotensin II.

The basal $[Ca^{2+}]_i$ as measured with the fluorescent calcium indicator quin-2 was 199$\pm$4 nM (mean $\pm$ S.E.M., n = 17). Angiotensin
II (10^{-7}M) increased this transiently to 382$\pm$41 nM (n = 9). Pretreatment with pertussis toxin did not alter basal $[Ca^{2+}]_i$, but
completely abolished the angiotensin II-induced increase in
$[Ca^{2+}]_i$, as shown in Fig. 2.

As I have shown previously, angiotensin II activates a
phospholipase C in mesangial cells (2). In cells prelabelled
with [^{3}H] inositol, angiotensin II caused an increased formation of IP_3, inositolbisphosphate (IP_2) and inositolmonophosphate
(IP_1), and this increased formation of inositolphosphates could
be inhibited by prior treatment of mesangial cells with pertussis
toxin (Fig. 3). These data indicate that pertussis toxin is able
to prevent an activation of phospholipase C in mesangial cells
stimulated with angiotensin II.

To exclude the possibility that the pertussis toxin-induced
attenuation of the angiotensin II effects is due to a decreased
density or affinity of receptors on mesangial cells, I determined
the specific binding of ^{125}I-angiotensin II. Scatchard analysis
gave a linear curve, corresponding to one group of receptor sites.
The maximal number of binding sites (39 fmol/mg of protein) and
the apparent dissociation constant, K_d (2,1nM) were calculated.
There was no significant difference either in the number of

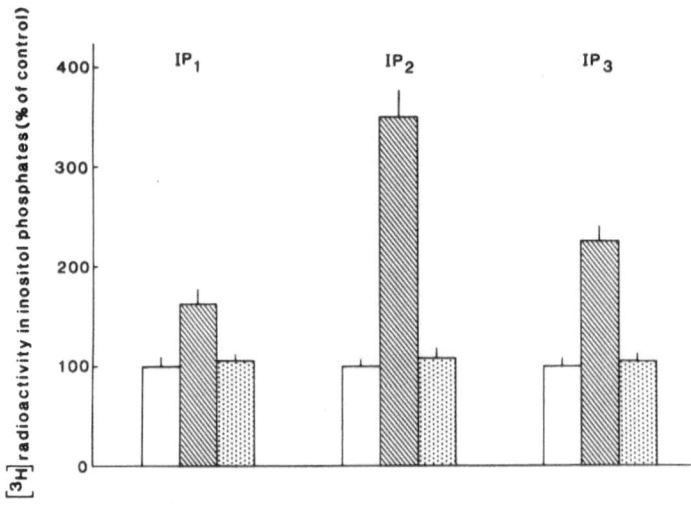

Fig. 3. Effect of pertussis toxin on angio-
        tensin II-induced inositol phosphates
        formation in mesangial cells

Mesangial cells were preincubated with pertussis
toxin (100ng/ml) or vehicle for 15 h. After
this preincubation period cells were stimulated
with angiotensin II (10^{-7}M) or vehicle for
10 s and thereafter the inositol phosphates
separated. Open bars (control values), hatched
bars (cells stimulated with angiotensin II)
and dotted bars (cells pretreated with pertussis toxin and stimulated with angiotensin II)

binding sites or in their affinities for angiotensin II between
control and pertussis toxin treated cells.

DISCUSSION

Cockcroft & Gomperts (5) have suggested an involvement of
a GTP-binding protein in coupling of receptor to phospholipase
which mediates the Ca^{2+}-dependent histamine release from mast
cells. My results showed that phospholipase C stimulation by
angiotensin II resulting in the degradation of phosphoinosition
is sensitive to pertussis toxin. In analogy to mast cells,
neutrophils and HL-60 cells (6-8), this might indicate that a
GTP-binding protein is involved in coupling of the angiotensin
receptor to phospholipase C activation in renal mesangial cell.
Our results also provide evidence that activation of phospholipase C is an essential step in the angiotensin II-induced
liberation of arachidonic acid and subsequent synthesis of PG.
These findings strengthen the hypothesis that there is a cause
relationship between agonist-induced activation of phospholipase
C and subsequent increased PG synthesis in mesangial cells.

REFERENCES

1. Pfeilschifter, J., Kurtz, A. & Bauer, C.,(1984)
   Biochem. J. 223, 855-859
2. Pfeilschifter, J., Kurtz, A. & Bauer, C.,(1986)
   Biochem. J. 234, 125-130
3. Pfeilschifter, J. & Bauer, C. (1986) Biochem. J. 236,
   289-294
4. Lowry, O.H., Rosebrough, N.J., Farr, A.L. & Randall, R.J.
   (1951) J. Biol. Chem. 193, 265-275
5. Cockcroft, S. & Gomperts, B.D. (1985) Nature (London) 314,
   534-536
6. Nakamura, T. & Ui, M. (1985) J. Biol. Chem. 260, 3584-3593
7. Brandt, S.J., Dougherty, R.W., Lapetina, E.G. & Niedel, J.E.
   (1985) Proc. Natl. Acad. Sci. USA 82, 3277-3280
8. Volpi, M., Naccache, P.H., Molski, T.F.P., Shefcyk, J.,
   Huang, C.K., Marsh, M.L., Munoz, J., Becker, E.L. & Sha'afi,
   R.I. (1985) Proc. Natl. Acad. Sci. USA 82, 2708-2712

METABOLISM OF PHOSPHATIDYLINOSITOL IN RETICULOCYTES -

ISOPROTERENOL AFFECTS THE TURNOVER OF POLYPHOSPHOINOSITIDES

Dieter Maretzki, Barbara Reimann, Evelin Schwartzer,
Milosav Kostic*) and Samuel Rapoport

Institute of Biochemistry, Humboldt University
1040 Berlin, German Democratic Republic
*)Institute of Medical Physiology, Markovic University
34000 Kragujevac, Yugoslavia

SUMMARY

Pulse-chase experiments with rabbit reticulocytes and erythrocytes prelabelled with inorganic ^{32}P-phosphate revealed a high turnover of the phosphomonoester groups of polyphosphoinositides in reticulocytes which declines strongly during maturation. Reticulocytes incorporated ^{3}H-inositol into phosphatidylinositol (PI), phosphatidylinositol-4-phosphate (PIP) and phosphatidylinositol-4,5-bisphosphate (PIP$_2$) in the proportion of 17 : 1 : 1.8, respectively. A slow release of ^{3}H-labelled inositol phosphates was found in reticulocytes, indicating a very low phosphodiesterase activity (Phospholipase C) in the intact cells. Determination of the ^{32}P/^{3}H-ratio in double-labelled phosphatidylinositides in reticulocytes demonstrated an about 3000-fold turnover of the phosphomonoester groups as compared with that of the phosphodiester group. (-)-isoproterenol stimulates the incorporation of ^{32}P-phosphate into the polyphosphoinositides in reticulocytes but did not influence the release of ^{3}H-inositol phosphates. A rapid response to (-)-isoproterenol (0.1 uM) with an increase of up to 60 % was found in the ^{32}P-labelling of PIP$_2$. The effect was antagonized by alprenolol. The stimulation of the polyphosphoinositide phosphate turnover might be mediated by an elevated concentration of cyclic AMP in reticulocytes.

INTRODUCTION

Several membrane functions are lost during the terminal maturation from the reticulocyte to the erythrocyte (Rapoport, 1986). Whereas the breakdown of inositol-containing phospholipids is an irreversible process in erythrocytes, the reticulocytes maintain the ability to synthesize them 'de novo'. PI, PIP and PIP$_2$ are interconvertible in red blood cell membranes by the action of two specific inositide kinases and two phosphatases (Palmer, 1985). In a previous study (Maretzki et al., 1986) we reported on the strong decline in the rates of ^{32}P-phosphate incorporation into the polyphosphoinositides during reticulocyte maturation which may be due to higher activities of these enzymes in reticulocytes or to the formation of 1,2-diacylglycerols and inositol trisphosphate which are now considered as second messengers (Berridge, 1983).

Here we present results of experiments on ^{32}P-phosphate "pulse-chase", the incorporation of ^{3}H-inositol into phosphatidylinositides and their labelling with both radioactive precursors as well as the effects of the ß-adrenergic agonist (-)-isoproterenol on the phosphoinositide metabolism in reticulocytes. Isoproterenol was found recently (Kostic et al., 1986) to increase the lactate formation in reticulocytes more than 2-fold accompanied by a strong accumulation of cyclic AMP. ATP-consuming processes like the polyphosphoinositide turnover (Reimann et al., 1981) could be involved in the increased lactate production which generates ATP.

RESULTS AND DISCUSSION

Incorporation of ^{32}P-phosphate into phosphatdiylinositides

Table 1 shows the specific radioactivities of the phosphomonoester groups of the polyphosphoinositides and of ATP after 30 minutes of incubation of rabbit erythrocytes and reticulocytes with ^{32}P-P$_i$ (70 µCi/ml cells). On the basis of equal labelling of the 4- and 5-positions in the inositol ring of PIP$_2$ the specific radioactivity was 13-fold higher in reticulocytes compared with that of mature erythrocytes. Assuming the predominant labelling of the 5-P of PIP$_2$, as recently reported by Hawkins et al. (1984), for human erythrocytes the specific radioactivity of the 5-position may actually approach that of cellular γ-ATP in reticulocytes. The specific radioactivity of PIP was about 6-fold higher in the reticulocytes as compared with erythrocytes. PI was labelled to a minor extent even in reticulocytes in the range of 5 % of the incorporation into the polyphosphoinositides.

The endogeneous ^{32}P-ATP pools in reticulocytes and erythrocytes were used after prelabelling for 15 minutes with ^{32}P-P$_i$ to chase the incorporated radioactivity of the polyphosphoinositides by addition of inorganic phosphate at 1 mM and 10 mM to the extracellular medium (Fig. 1). The rationale for this quasi "pulse-chase" experiment is that the rate limiting step for the ^{32}P-labelling of the ATP pools and subsequently the polyphosphoinositides is the phosphate transport into the red blood cells (Maretzki et al., 1986).

Table 1: Specific radioactivities (cpm/nmole phosphate) of polyphosphoinositides and γ-ATP of rabbit red blood cells (average of four experiments)

	PI-4,5-P$_2$ (4-and 5-P)	5-P*	PI-4-P	γ-ATP
Erythrocytes	525	725	816	11 670
Reticulocytes	6773	9347	4993	12 350

*) calculated on the basis of a 5-P to 4-P ratio of 69 versus 31 % according to Hawkins et al., 1984.

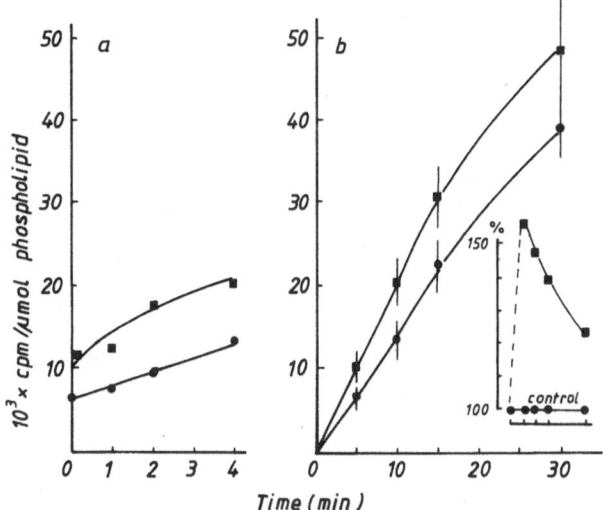

Fig. 1
Time course of cellular γ-^{32}-ATP (left panel), of ^{32}-PIP$_2$ (middle) and of ^{32}P-PIP (right panel) in "pulse-chase" experiments with intact cells.
After prelabelling with ^{32}P-P$_i$ (100 µCi/ml) for 15 min the cells were reincubated in fresh medium without tracer but containing 1 mM and 10 mM inorganic phosphate as indicated.
Filled symbols: Reticulocytes; Open Symbols: Erythrocytes.

The data in Figure 1 confirm the greater turnover of the phosphomonester groups in the PIP$_2$ in reticulocytes which decreases much more strongly than that of PIP during maturation. An explanation for the discrepancy in the specific radioactivities between ATP and the individual polyphosphoinositides could be the existence of different pools of phosphatidylinositides in the membrane and that only a portion is in a rapid exchange with the ATP pool. Our experiments indicate a high activity of the phosphoinositide-5-kinase/phosphatase cycle in reticulocytes.

^{3}H-inositol and ^{32}P-incorporation into phosphoinositides

In mature erythrocytes incubated with ^{3}H-inositol PI was poorly labelled and no radioactivity associated with PIP or PIP$_2$ was detectable. In reticulocytes ^{3}H-inositol was most extensively incorporated into PI indicating an active PI cycle. The rates of incoproration into PIP and PIP$_2$ (Table 2) suggest that both polyphosphoinositides represent constant minor pools with slow exchange of the inositol moiety with the larger PI pool.

Table 2: Time course of ^{3}H-inositol incorporation into phosphatidylinositides of rabbit reticulocytes

	0.5 h	1 h	1.5 h	2 h	3 h
	dpm x 10^{-4}/ml cells/ μCi added				
PI	2.05	4.2	6.5	9.6	17.2
PIP	0.12	0.26	0.37	0.45	0.95
PIP$_2$	0.28	0.41	0.63	0.96	1.9

A slow release of ^{3}H-labelled inositol phosphates, the breakdown
products of phosphatidylinositides, was observed during the time
course of 3 hours incubation with ^{3}H-inositol.

Reticulocytes had a higher content of released ^{3}H-labelled inositol
phosphates as compared to that of erythrocytes, expecially of
inositol trisphosphate and bisphosphate. The release of inositol
phosphates by reticulocytes was not enhanced in the presence of 10
mM Li$^+$ and 1.5 mM Ca^{2+} in the incubation medium (data not shown).

The very low incorporation of ^{32}P-phosphate into PI in comparison to
the high phosphomonoester group turnover of the polyphosphoinositides in reticulocytes was analyzed by the use of double-labelling
of the phosphatidylinositides. Reticulocytes that had been labelled
to isotopic equilibrium with ^{3}H-inositol (10 μCi/ml) for 3 hours
prior to removal of extracellular tracer were incubated and at time
zero ^{32}P-P$_i$ was added at 50 μCi/ml (Table 3). Taking into account
the relative proportion of the ^{3}H-inositol label into PI : PIP :
PIP$_2$ as 17 x 1 x 1.8 (Table 2) as a measure for the amount of each
of the subclass of the phosphatidylinositides and their ^{32}P/^{3}H-ratio
(Table 3) it becomes evident that reticulocytes exhibit about 3000
fold higher exchange rates of the phosphomonoester groups in the
inositol ring compared with that of the phosphodiester group in the
inositollipids. This feature seems to be specific for immature red
blood cells which express pesumable a low phosphodiesterase
activity.

Stimulation of the poly-PI-phosphate turnover

The ß-adrenergic stimulation by (-)-isoproterenol of the adenylate
cyclase of reticulocytes induced a large initial rate of cyclic AMP
foramtion (Kostic et al., 1986) and a rapid increase of the ^{32}P-
incorporation into polyphosphoinositides as shown for PIP$_2$ in Figure
2.

Table 3: Time course of the $^{32}P/^3H$-ratios of phosphatidylinositides in rabbit reticulocytes. The relative amounts of ^3H-label incorporated into PI, PIP and PIP_2 remained constant at 86, 5 and 9 %.

Time	PI	PIP	PIP_2
5 min	0.001	0.15	0.23
10 min	0.0016	0.29	0.45
15 min	0.0022	0.39	0.71

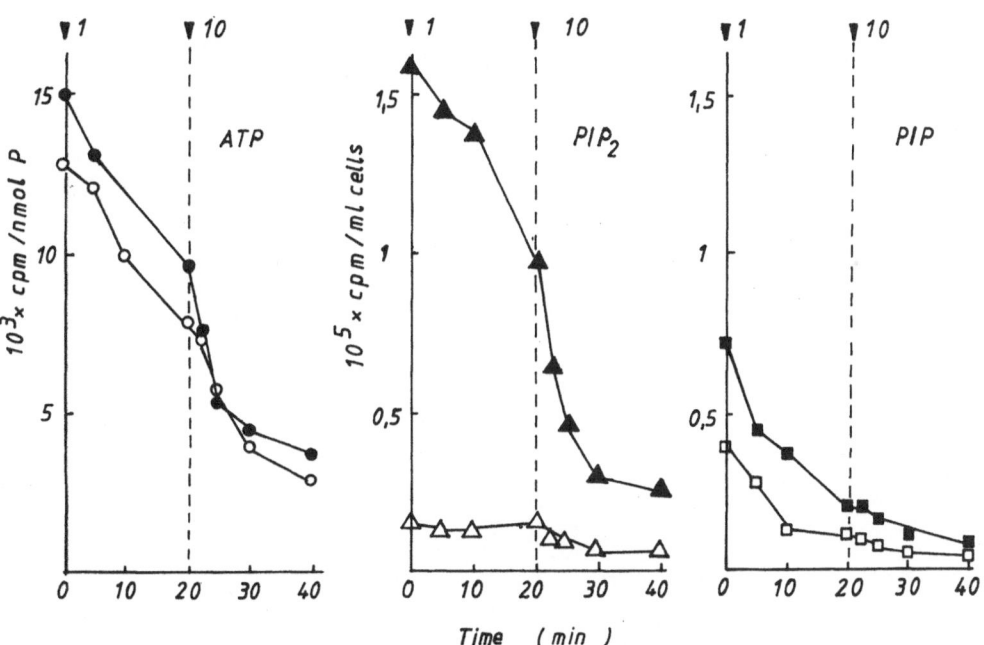

Fig. 2
Changes of ^{32}-phosphate incorporation into PIP in rabbit reticulocytes incubated with 0.1 μM (a) or 10 μM (b) (±)-isoproterenol. In (a) the cells were prelabelled for 5 min with $^{32}P_i$ (100 μCi/ml). Control (●); Addition of isoproterenol (■).

REFERENCES

Berridge, M.J., 1984, Inositol trisphosphate and diacylglycerol as second messenger, Biochem. J., 220:345.

Hawkins, P.T., Michell, R.H. and Kirk, C.J., 1984, Analysis of the metabolic turnover of the individual phosphate groups of phosphatidylinositol-4-phosphate and phosphatidylinositol-4,5-bisphosphate, Biochem. J, 218:785.

Kostic, M., Müller, M., Maretzki, D., Krause, E.G. and Rapoport, S.M., 1986, Metabolic effects of (-)-isoprenalin stimulation of adenylate cyclase in reticulocytes, Biomed., Biochim. Acta, 45:973.

Maretzki, D., Kostic, M., Reimann, B., Schwarzer, E and Rapoport, S.M., 1986, Maturation of rabbit reticulodytes: Strong decline of the turnover of polyphosphoinositides, Biomed. Biochim. Acta, 45:1227.

Palmer, F.B.St.C., 1985, Polyphosphoinositide metabolism in ageing erythrocytes, Can. J. Biochem. Cell. Biol., 63:927.

Rapoport, S.M., 1986, "The Reticulocytes", CRC Press, Boca Ration, Florida.

Reimann, B., Klatt, D., Tsamaloukas, A.G. and Maretzki, D., 1981, Membrane phosphorylation in intact human erythrocytes, Acta Biol. Med. Germ., 40:487.

ROLE OF CYTOPLASMIC pH FOR STIMULUS-INDUCED

CA^{2+} MOBILIZATION IN HUMAN PLATELETS

Winfried Siffert, Peter Scheid, and Jan W.N. Akkerman*

Inst. für Physiol., Ruhr-Universität, 4630 Bochum, FRG
*Department of Haematology, University Hospital
3511 GV Utrecht, The Netherlands

SUMMARY

Stimulation of human platelets by thrombin resulted in cyto-plasmic alkalinization due to activation of Na^+/H^+ exchange. Inhibition of this Na^+/H^+ exchange suppressed the thrombin-induced Ca^{2+} mobilization. Artificial cytoplasmic alkalinization, on the other hand, markedly enhanced the stimulus-induced rises in the cytosolic free Ca^{2+} concentration. It is concluded that activation of Na^+/H^+ exchange is needed for Ca^{2+} mobilization and thus constitutes an essential step in platelet activation.

INTRODUCTION

A variety of cells respond to pharmacologic stimulation with a rapid hydrolysis of phosphatidylinositol 4,5 bisphosphate (PIP_2) and subsequent formation of 1,2-diacylglycerol (DG) and inositol 1,4,5-trisphosphate (IP_3). Whereas IP_3 mobilizes Ca^{2+} from internal storage sites, DG induces Na^+/H^+ exchange via activation of protein kinase C (Pk-C).[1,2] The resultant elevation of intracellular pH (pH_i) appears to be intimately linked to stimulus-response coupling although the mechanisms by which an increase in pH_i can promote cellular functions have remained obscure.[3] Stimulation of human platelets by thrombin also increases pH_i due to activation of Na^+/H^+ exchange.[4,5] Inhibition of this process markedly affects platelet activation in that it prevents aggregation and the release of granule contents.[6-8] This study aimed at further characterizing the mechanisms which induce Na^+/H^+ exchange in platelets and at clarifying its role in cell activation.

MATERIAL AND METHODS

Human platelets were prepared from freshly drawn blood according to standard procedures.[9] - Cytoplasmic pH was measured using the fluorescent pH_i indicator 2,7-biscarboxyethyl-5-(6)-carboxyfluorescein (BCECF) by uptake of its membrane-permeable tetraacetoxymethylester (BCECF-am).[10] - The concentration of the cytosolic free Ca^{2+} concentration, $\{Ca^{2+}\}_i$, was measured in human

platelets loaded with the fluorescent Ca^{2+} indicator quin 2.[11] –
The final medium for platelet suspension consisted of (in mM): 145
NaCl, 5 KCl, 1 $MgSO_4$, 10 HEPES (free acid), 5 glucose, pH 7.4 at
37°C. Na^+/H^+ exchange was blocked by either isotonic replacement
of NaCl with cholinechloride or by addition of ethylisopropylamiloride (EIPA) to the platelet suspensions. Cytoplasmic alkalinization was achieved by adding NH_4Cl or the Na^+/H^+ ionophore monensin
to the platelet suspensions.

RESULTS

Addition of 0.05 U/ml thrombin to platelets (Fig. 1a)
increased pH_i from 7.15 ± 0.08 (S.D.; n = 64) to 7.27 ± 0.04
(n = 12). Both lowering extracellular Na^+, Na_o^+, to below 1 mM and
pre-incubating platelets with 50 μM EIPA prevented the thrombin-
induced increase in pH_i suggesting that cytoplasmic alkalinization
is mediated by Na^+/H^+ exchange. Stimulation of platelets by the
synthetic DG analogue 1-oleoyl-2-acetylglycerol (OAG), a potent
activator of Pk-C,[1] also induced Na^+-dependent, EIPA-sensitive
cytoplasmic alkalinization (Fig. 1b). It can thus be concluded
that activation of Na^+/H^+ exchange occurs via a Pk-C-dependent
pathway.

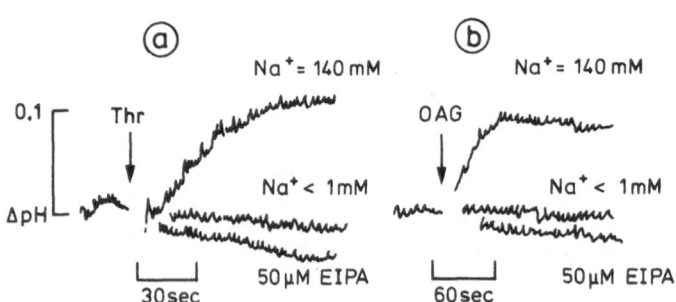

Figure 1   Cytoplasmic pH-changes in stimulated human platelets

Displayed are representative pH_i traces obtained in
BCECF-loaded platelets. Stimulus: a, 0.05 U/ml
thrombin (Thr); b, 50 μg/ml OAG. Na_o^+, external Na^+.

The effects of blocking Na^+/H^+ exchange on thrombin-induced Ca^{2+} mobilization are illustrated in Fig. 2. In medium containing 1 mM Ca^{2+} and 140 mM Na^+ (Fig. 2a) thrombin increased $\{Ca^{2+}\}_i$ by 800 ± 210 nM (n = 9), whereas at 0.5 mM external Na^+ only a small increase by 90 ± 20 nM (n = 7) was observed (Fig. 2b). In low-Ca^{2+} medium (presence of 2 mM EGTA) containing 140 mM Na^+, $\{Ca^{2+}\}_i$ rose by 140 ± 32 nM (Fig. 2c) (n = 7) upon addition of thrombin, but at low Na_o^+ (Fig. 2d) $\{Ca^{2+}\}_i$ remained at its resting level. The Ca^{2+} ionophore ionomycin increased $\{Ca^{2+}\}_i$ both in the presence (Fig. 2c) or absence of Na^+ (Fig. 2d) which indicates that removal of Na^+ did not cause depletion of Ca^{2+} from internal storage sites. Blockade of Na^+/H^+ exchange by EIPA also completely inhibited thrombin-induced Ca^{2+} mobilization (Fig. 2e,f), which was reversed when pH_i was artificially increased using the Na^+/H^+ ionophore monensin (Fig. 2g).

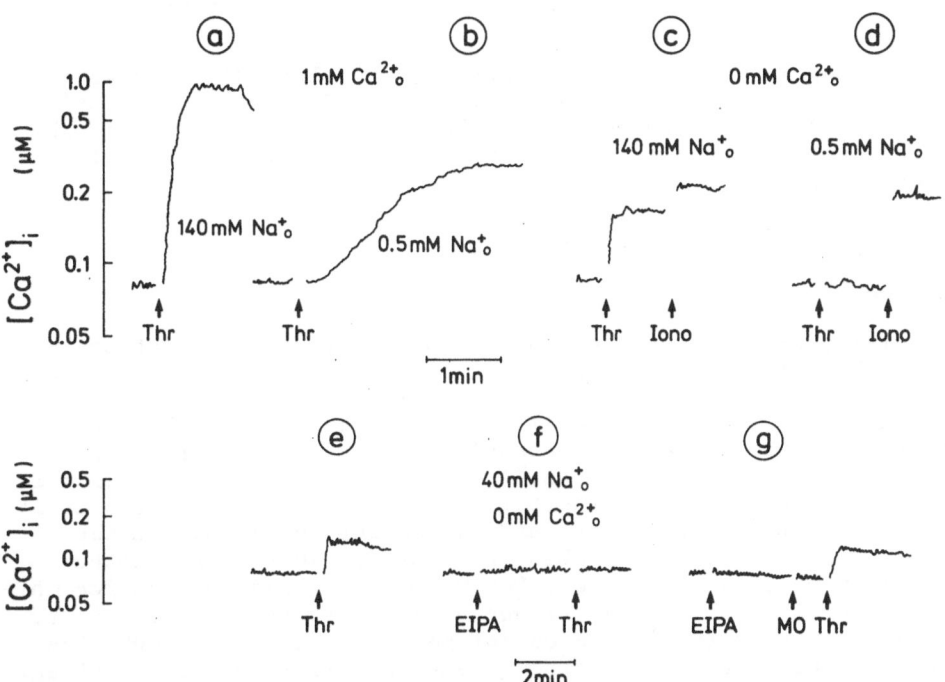

<u>Figure 2</u> Ca^{2+} mobilization in thrombin-stimulated platelets

Displayed are representative Ca^{2+} traces obtained in quin 2- loaded platelets. Stimulus: 0.05 U/ml thrombin (Thr). Na_o^+, external Na^+; Ca_o^{2+}, external Ca^{2+}; Iono, ionomycin (100 nM); EIPA, ethylisopropylamiloride (40 μM); Mo, monensin (10 μM). For details, see text.

The significance of cytoplasmic alkalinization for Ca^{2+} mobilization could also be demonstrated in experiments in which pH_i was increased by NH_4Cl or monensin prior to stimulation of platelets by thrombin (Table 1). Cytoplasmic alkalinization alone did not increase ${Ca^{2+}}_i$ in unstimulated platelets but distinctly enhanced thrombin-stimulated Ca^{2+} mobilization.

Table 1. Effect of cytoplasmic alkalinization on thrombin-induced Ca^{2+} mobilization

Additions	$\Delta{Ca^{2+}}_i$ (nM)	n
Thrombin	100 ± 15	8
NH_4Cl or monensin	0	23
NH_4Cl + thrombin	212 ± 60	8
Monensin + thrombin	180 ± 40	7

The changes in ${Ca^{2+}}_i$ upon stimulation of platelets by 0.05 U/ml thrombin were measured in quin 2 loaded platelets at 0 mM Ca^{2+}_o (2 mM EGTA). NH_4Cl and monensin were applied at 10 mM and 10 μM, respectively. Values represent means ($\pm$ S.D.); n = number of experiments.

DISCUSSION

The present results reconfirm and extend previous observations that stimulation of human platelets induces Na^+/H^+ exchange and thereby increases pH_i.[4,5] The finding that OAG could elicit the action of thrombin (Fig. 1) suggests that protein kinase C is involved in the activation process, and similar, observations have been reported for other cells.[12] The most striking result of this study was, hovever, that inhibition of Na^+/H^+ exchange also inhibited Ca^{2+} mobilization. This effect was mot pronounced in the absence of external Ca^{2+}, when increases in ${Ca^{2+}}_i$ reflected release from internal stores. The most likely explanation for this phenomenon is, that the IP_3-induced Ca^{2+} release is pH-sensitive as recently reported.[13]- We do not yet know how pH_i affects the influx of Ca^{2+} across the plasma membrane. Platelets lack voltage-dependent Ca^{2+} channels[14] and Na^+/Ca^+ exchange[15] and the processes which increase the permeability of the plasma membrane for Ca^{2+} remain to be determined.

Our study also provides evidence, that a rise in pH_i alone is insufficient for Ca^{2+} mobilization but provides an essential 'co-signal' in stimulus-induced Ca^{2+} mobilization. Besides Ca^{2+} mobilization additional intracellular events may depend on increases in pH_i[3] but determination of these processes was beyond the scope of this study.

ACKNOWLEDGEMENTS

This study was supported by the Deutsche Forschungsgemeinschaft, grant Sche 46/5-1. Participation of W.S. in this
conference was supported by the Boehringer Ingelheim Fonds.

REFERENCES

1. M. Berridge, Inositol trisphosphate and diacylglycerol as
   second messengers, Biochem. J. 220:345 (1984)
2. M. Berridge and R.F. Irvine, Inositol trisphosphate, a novel
   second messenger in cellular signal transduction, Nature 312:
   315 (1984)
3. W.B. Busa and R. Nuccitelli, Metabolic regulation via
   intracelluar pH, Am. J. Physiol. 246:R409 (1984)
4. W.C. Horne, N.E. Norman, D.B. Schwartz, and E.R. Simons,
   Changes in cytoplasmic pH and in membrane potential in
   thrombin-stimulated human platelets, Eur. J. Biochem. 120:295
   (1981)
5. W. Siffert, G. Fox, K. Mückenhoff, and P. Scheid, Thrombin
   stimulates Na^+/H^+ exchange across the human platelet plasma
   membrane, FEBS Lett. 172:272 (1984)
6. W.C. Horne and E.R. Simons, Effects of amiloride on the
   response of human platelets to α-thrombin, Thromb. Res. 13:599
   (1978)
7. T.M. Connolly and L.E. Limbird, Removal of extraplatelet Na+
   eliminates indomethacin-sensitive secretion from human
   platelets stimulated by epinephrine, ADP, and thrombin. Proc.
   Natl. Acad. Sci. USA 80:5320 (1983)
8. W. Siffert, S. Gengenbach and P. Scheid, Inhibition of
   platelet aggregation by amiloride. Thromb. Res., in press.
9. L.A. Harker and T.S. Zimmerman, Measurements of platelet
   function, Churchill Livingstone, New York (1983)
10. T.J. Rink, R.Y. Tsien, and T. Pozzan, Cytoplasmic pH and free
    Mg^{2+} in lymphocytes, J. Cell Biol. 95:189 (1982)
11. T.J. Rink, A. Sanchez, and T.J. Hallam, Diacylglycerol and
    phorbol ester stimulate secretion without raising cytoplasmic
    free calcium in human platelets, Nature 305:317 (1983)
12. S. Grinstein and A. Rothstein, Mechanisms of regulation of the
    Na^+/H^+ exchanger. J. Membrane Biol. 90:1 (1986)
13. L.F. Brass and K. Joseph. A role for inositol trisphosphate in
    intracellular Ca^{2+} mobilization and granule secretion in
    platelets, J. Biol. Chem. 260:15172 (1985)
14. V.M. Doyle and U.T. Rüegg, Lack of evidence for voltage-
    dependent calcium channels on platelets. Biochem. Biophys.
    Res. Commun. 127:161 (1985)
15. L.F. Brass, The effect of Na^+ on Ca^{2+} homeostasis in
    unstimulated platelets, J. Biol. Chem. 259:12571 (1984)

PHOSPHATIDYLINOSITOL PHOSPHORYLATION IN FAST SKELETAL MUSCLE MEMBRANES

G. Behle, M. Varsanyi, R. Thieleczek and L.M.G. Heilmeyer,Jr.

Institut für Physiologische Chemie
Abteilung für Biochemie Supramolekularer Systeme
Ruhr-Universität Bochum
Universitätsstraße 150
4630 Bochum 1
West-Germany

Phosphatidylinositol 4-phosphate (PIP) associated with the isolated Ca^{2+} transport ATPase from rabbit skeletal muscle sarcoplasmic reticulum (SR) is formed when the enzyme is incubated at high membrane protein concentration with ATP/Mg^{2+} (1). This observation indicates that an endogenous phosphatidylinositol kinase (PI kinase) is present in these membranes. The PIP formation rate and the steady state level of PIP is enhanced by phosphorylase kinase, showing that this protein kinase expresses also lipid kinase activity (2,3). It is still unclear what kind of relation exists between the membrane associated PI kinase and the PI kinase present, as side activity, in phosphorylase kinase.

SR membrane associated PI kinase activity can be solubilized by the nonionic detergent Triton X 114 and transferred to the aqueous phase to a degree of 40%. Simultaneously phosphorylase kinase activity can be solubilized from these membranes and is coextracted to the same extend. These two kinase activities present in the aqueous phase were not separable by classical protein purification methods. Activity of the SR associated PIP monoesterase, the enzyme which catalyses the dephosphorylation of PIP (4), can also be solubilized by Triton X 114 to a degree of 70%.

Monoclonal antibodies to α - and ß-subunits of phosphorylase kinase (anti-α, anti-ß) have been used to study the correlation between PI kinase and phosphorylase kinase. By means of the immunoblot technique, the presence of antigenic structures identical or related to the α- and ß-subunits of

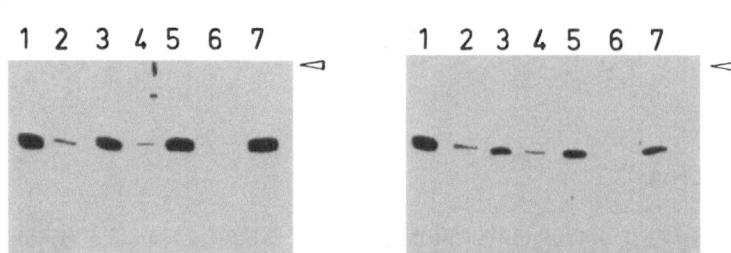

Fig. 1. Identification of antigenic structures of phosphorylase kinase in the aqueous phase of Triton X
114 solubilized SR in the immunoblot, using monoclonal antibodies to the α-subunit (A) and the ß-
subunit (B) of phosphorylase kinase. The numbers
indicate the slots of the microgels. The samples
A4-A7 and B4-B7 were preincubated with purified
monoclonal anti-α antibody (diluted 1:10 (A4, A5,
B4, B5) and non-diluted (A6,A7, B6, B7)) and
subsequently immunoprecipitated by Staphylococcus
aureus cells. Slots A1 and B1 are loaded with 1 µg
and 0.5 µg phosphorylase kinase respectively. A2,
A3 and B2, B3 contain the non-precipitated aqueous
phase. The arrowheads mark the start and dye front
of the microgels.

phosphorylase kinase can be directly demonstrated in the aqueous phase after
Triton X 114 solubilization of the SR membranes (5) (Fig. 1A and 1B). Both
subunits, α and ß, are removed from the aqueous phase by immunoprecipitation
with increasing concentrations of the purified anti-α antibody.

The supernatants after immunoprecipitation have been examined for
phosphorylase kinase and PI kinase activities (Table 1). After the first
immunoprecipitation with purified anti-α the supernatant contains ca. 70%
of the residual phosphorylase kinase activity whereas the PI kinase activity
remains unchanged. A second immunoprecipitation causes only a small further
reduction of phosphorylase kinase activity whereas the PI kinase activity is
reduced to an extend of 50%.

The corresponding pellets to the immunoprecipitation described in
Table 1 are analyzed by SDS-polyacrylamide gel electrophoresis (Fig. 2), and
the protein bands are visualized by the silver staining method according to
Merril et al. (6).

The fact that all three subunits of phosphorylase kinase can be detected in the immunoprecipitates after SDS-gel electrophoresis indicates that

Table 1. Immunoprecipitation of phosphorylase kinase and PI kinase activities by purified anti-α from the aqueous phase of solubilized SR membranes

sample no.	assay	purified anti-α (µg/ml assay)	protein (µg/ml sample)	phosphorylase kinase activity			phosphatidylinositol kinase activity		
				U/ml sample	resid. activity in % relative to buf. con.	prec. con.	U/ml sample	resid. activity in % relative to buf. con.	prec. con.
1	APB/PBS	0	462	358	100		83	100	
2	APB/STA	0	451	392	100		67	100	
3	APA/PBS	6	458	415	116	100	66	80	100
4	APA/STA	6	471	284	72	68	66	98	100
5	SS4/AS/STA	6	408	254	65	61	31	46	47

The assay conditions are described using the following abbreviations: AB (antibody buffer), AS (antibody solution = AB containing purified anti-α), PBS (phosphate buffered saline), APB (aqueous phase mixed with AB (antibody-free)), APA (aqueous phase mixed with AS (antibody containing)), STA (Staphylococcus aureus cell suspension), SS4 (supernatant of sample 4). The supernatants of the assays after centrifugation are named samples. The supernatant of the first immunoprecipitation (sample 4, SS4) was subjected to a second immunoprecipitation yielding assay SS4/AS/STA. Residual activities (resid. activity) were normalized to corresponding buffer conditions (buf. con.; assay 1 and 2) or pre-precipitating conditions (prec. con.; assay 3). U is given in pmoles per min.

the decrease in phosphorylase kinase activity is due to a removal of the holoenzyme (α β γ). Obviously, a part of the phosphorylase kinase activity is insensitive to immunoprecipitation. This might be due to substitution of the α-subunit in some of the holoenzymes for a modified α-subunit, α^*, which would prevent detection of the α^* β γ-complex by anti-α. Otherwise it might be possible that the insensitive residual phosphorylase kinase activity is related to dissociated γ-subunits, which would not be precipitated by the anti-α antibodies. The isolated γ-subunit is known to express phosphorylase kinase activity (7).

The observation that the PI kinase activity is not affected by the first immunoprecipitation but is reduced to 50% by the second one cannot be due to low-affinity antibody binding sites in the latter case. The introduced antibody concentration was sufficient to saturate also such sites during the first immunoprecipitation. The precipitation of PI kinase activity by anti-α suggests the presence of epitopes referable to PI kinase which are structurally similar to those of phosphorylase kinase. However, their detection by anti-α depends on the previous precipitation of phosphorylase kinase activity.

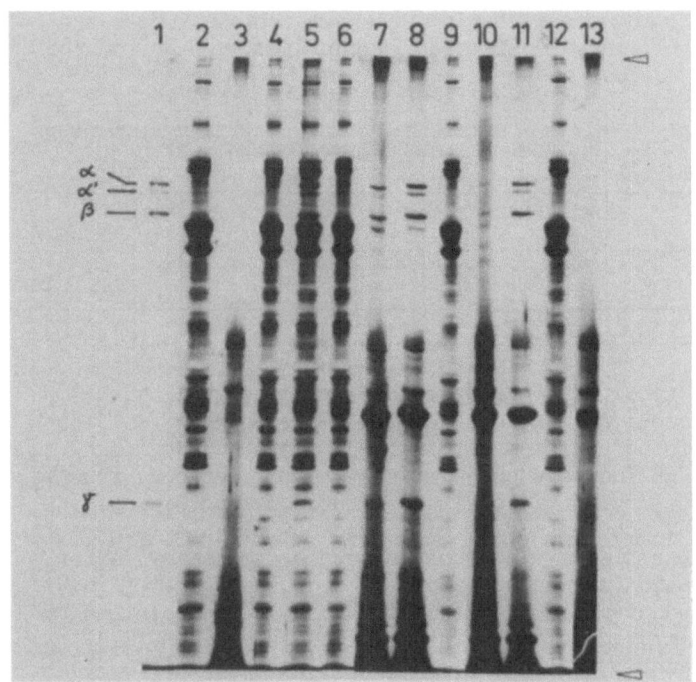

Fig. 2. SDS-polyacrylamide gel electrophoresis
of the aqueous phase of Triton X 114
solubilized SR. Protein bands were
visualized by silver staining (7). The
numbers indicate the slots of the gel
which contains the following samples
(PLAK, phosphorylase kinase; S, supernatant; RP, redissolved immunoprecipitate): 1, 0.1 µg of PLAK; 2, 1.5 µg
APB/PBS-S; 3, 15 µl APB/STA-RP; 4, 1.5
µg APB/STA-S; 5, sample 4 and 50 ng
PLAK; 6, 1.5 µg APA/STA-S; 7, 15 µl
APA/STA-RP; 8, sample 7 and 50 ng PLAK;
9, 1.5 µg S of (sample 6/AS/STA)-S
after 2nd precipitation; 10, 15 µl S of
(sample 6/AS/STA)-RP after 2nd precipitation; 11, sample 10 and 50 ng PLAK;
12, 1.5 µg S of (sample 6/STA)-S after
2nd precipitation; 13, 15 µl S of
(sample 6/STA)-RP after 2nd precipitation.
The arrowheads point to the start and the
dye front of the gel. The greece letters
mark the PLAK-subunits $\propto$, $\propto'$(isoform of
$\propto$ from red muscle), ß and γ.

References

1. M. Varsanyi, H. G. Tölle and L. M. G. Heilmeyer, Jr., R. M. C.
    Dawson. R. F. Irvine, Activation of sarcoplasmic reticular Ca2+
    transport ATPase by phosphorylation of an associated phospha-
    tidylinositol, The EMBO J. 2:1543 (1983).
2. Z. Georgoussi and L. M. G. Heilmeyer, Jr., Evidence that phos-
    phorylase kinase exhibits phosphatidylinositol kinase activity,
    Biochemistry 25:3867 (1986).

3. M. Varsanyi, G. Behle and M. Schäfer, Stimulation of phosphatidyl-inositol phosphorylation in the sarcoplasmic reticular Ca2+ transport ATPase by vanadate, Z. Naturforsch. 41c:310 (1986).

4. M. Schäfer, "Untersuchungen zur Dephosphorylierung des Phosphatidyl-inositolphosphates in isolierter Ca2+-Transport ATPase des Sarkoplasmatischen Retikulums", PH.D. Thesis, Bochum (1986).

5. C. Bordier, Phase separation of integral membrane proteins in Triton X-114 solution, J. Biol. Chem. 256:1604 (1981).

6. C. R. Merril, M. L. Dunau and D. Goldman, A rapid sensitive silver stain for polypeptides in polyacrylamide gels, Anal. Biochem. 110:201 (1981).

7. S. M. Kee and D. G. Graves, Isolation and properties of the active γ-subunit of phosphorylase kinase, J. Biol. Chem. 261:4732 (1986).

THE ROLE OF PHOSPHORYLATION IN GROWTH CONTROL AND MALIGNANT TRANSFORMATION

Tony Hunter

Molecular Biology and Virology Laboratory
Salk Institute, P.O. Box 85800
San Diego, California 92138

Growth control

Experiments with cultured cells suggest that the growth of mammalian cells is regulated largely by growth factors and inhibitors acting at the G_0/G_1 border in cell cycle. Growth factors are mostly polypeptide in nature (e.g. EGF, PDGF, FGF, IGF-1, bombesin) and range in mass from 2,000 to 70,000 daltons. Cells responsive to a given growth factor have specific cell surface receptors, usually numbering between 10^3 and 10^6 per cell, with a high affinity (K_d 10^{-8} to 10^{-11}) and specificity for their ligand. Occupancy of these receptors results in the transduction of a signal across the membrane to the cytoplasm, the nature of which will be discussed below. Bona fide growth inhibitors have only recently been purified to homogeneity (e.g. TGF-β, interferon), although their existence had been surmised for many years. Their mode of action is not understood, but, like the polypeptide growth factors, growth inhibitors have specific high affinity surface receptors, which presumably transduce signals to the cytoplasm. Another external component essential for cell growth is the extracellular matrix, comprised of proteins like fibronectin, laminin, and collagen, and the glycosaminoglycans. Adherent cells will not grow in the absence of an extracellular matrix, but most cells are capable of elaborating their own matrix. There are specific surface receptors for these matrix proteins, but it is not yet known whether the binding of ligand to this type of receptor causes a signal transduction event.

Growth factors often interact in a synergistic manner to elicit a growth response. For some cell types, such as Balb/c 3T3, growth factors which cooperate can be divided into two categories. The first category consists of "competence" factors which render cells capable of receiving a signal from a second type of factor, called a "progression" factor. At low doses competence factors are not mitogenic, but require the secondary addition of a progression factor in order to trigger movement through the cell cycle. PDGF and FGF are competence factors, whereas EGF and IGF-1 are progression factors. For most growth factors less than 100% receptor occupancy is needed for a mitogenic response, but receptors must be occupied for a certain time, normally between 1 and 6 hours, before the cells are "committed" to divide. The synergistic interactions between different growth factors (and negative modulation by growth inhibitors) are presumably necessary to provide fool-proof safeguards against accidental growth stimulation by transient exposure to a growth factor.

The cellular response to growth factors can coveniently be divided into three phases. There is a series of early effects occurring between 0 and 15 minutes after growth factor treatment. These include an increase in ion flux (e.g. Na^+ influx), an alkalinization of the cytoplasm due to elevated H^+ export, a stimulation of phosphatidylinositol turnover giving rise to diacylglycerol (DAG) and IP_3, an increase in nutrient import (e.g. glucose and amino acids), an enhanced phosphorylation of proteins and a number of rapid morphological changes (e.g. surface ruffling). There is an increase in cytosolic Ca^{2+} as a result of the generation of IP_3, which stimulates Ca^{2+} release from intracellular vesicle systems. The enhanced protein phosphorylation is due in part to the rise in DAG, which activates protein kinase C. Although these early events have not been presented here in any particular order, they are likely to be linked by a chain of causality, which, as will be discussed below, is initiated by a primary signal from the growth factor receptor.

Intermediate events occurring between 15 and 60 minutes following growth factor treatment include a 3-5 fold rise in the rate of protein synthesis, an induction of the transcription of specific genes (e.g. the c-*myc* and c-*fos* genes) and the translation of their mRNAs. The number of primary genes induced by mitogens in quiescent cells appears to be somewhere between 20 and 100. Late events happening between 1 - 24 hours would encompass all the processes classically associated with the cell cycle, culminating in DNA synthesis and cell division. The progression through the cell cycle must involve an ordered series of steps, which includes gene expression cascades. There is good evidence for a number of decision or restriction points during the G_1 phase. One possible explanation for the continued requirement for growth factors to obtain commitment is that one or more of the early events, or perhaps the primary receptor signal itself, may be needed a second time at these decision points.

Table 1. Growth factor receptors

A. Receptors which have intrinsic protein-tyrosine kinase activity

EGF receptor	FGF receptor?
PDGF receptor	Bombesin receptor?
CSF-1 receptor	neu(c-*erb*-B-2) protein?
Insulin receptor	c-*ros* protein?
IGF-1 receptor	*trk* protein?
	met protein?
	c-*kit* protein?

B. Receptors which lack intrinsic protein kinase activity

T cell antigen receptor
Immunoglobulin antigen receptor
IL-2 receptor
IL-3 receptor
TGF-β receptor
IGF-2 receptor

To understand this chain of events we need to know the nature of the signal delivered by growth factor receptors when they bind their ligands. Growth factor receptors can be divided into two categories based on their function (Table 1). Over half the characterized receptors have ligand-stimulated protein-tyrosine kinase activities, and treatment of cells with the relevant growth factors can be shown to induce increased tyrosine phosphorylation of the receptor itself and of cellular proteins[1,2]. Several members of this class of receptor have been cloned and a common structure has emerged[3-8]. They all possess a large glycosylated disulphide-bonded external ligand-binding domain, a single transmembrane domain, and a cytoplasmic domain, which contains a 270 amino acid sequence recognizable as a protein kinase catalytic domain. The PDGF, EGF and CSF-1 receptors are found as monomers, whereas the insulin and IGF-1 receptors exist as disulphide-linked dimers. The mechanism of signal transduction is not fully understood, but there is increasing evidence that this requires dimerization. Ligand-induced dimerization of the external domains of two receptor molecules would bring together their cytoplasmic domains, and the interaction of the two protein kinase domains could increase catalytic activity through an allosteric effect. Another mode of regulation for the growth factor receptor protein-tyrosine kinases, secondary to ligand binding, is the stimulatory effect of ligand-dependent autophosphorylation on enzyme activity. The autophosphorylation sites are either in the catalytic domain (e.g the insulin receptor), or in a separate regulatory domain (e.g. the EGF receptor). The effect of these twin regulatory systems is to maintain the unoccupied receptor in a state of very low protein kinase activity, and yet allow a 10-20 fold activation upon ligand binding.

Included in the protein-tyrosine kinase receptor category in Table 1 are several other less well characterized proteins. The FGF[9] and bombesin[10] receptors have been reported to possess ligand-induced protein-tyrosine kinase activity, but the molecular structure of these receptors is poorly defined. The other proteins in this category are the cellular homologues of a series of oncogenes, whose products all have protein-tyrosine kinase activity. The structures of the oncoproteins are reminiscent of those of the protein-tyrosine kinase growth factor receptors. Indeed $p185^{neu}$ is closely related to the EGF receptor[11,12], but so far no ligand has been identified for this protein. Likewise no ligands have been identified for the other putative growth factor receptors.

The second class of growth factor receptors tend to have very small cytoplasmic domains, and the isolated proteins do not display protein kinase activity (Table 1). This might suggest that these receptors transmit a different type of signal, such as that used by the adrenergic receptors which involves the activation of G proteins. Nevertheless, because many of these receptors are specifically associated with other membrane or cytoplasmic proteins, one could propose as a unifying hypothesis that they act as regulators of associated protein-tyrosine kinases. Indeed there is evidence that occupancy of the T cell antigen receptor leads to tyrosine phosphorylation of an associated subunit.

The fact that the majority of known growth factor receptors have ligand-stimulated protein-tyrosine kinase activity strongly suggests that tyrosine phosphorylation can be utilized to initiate a mitogenic response. Formal proof for this assertion, however, is lacking, and it remains possible that these receptors all have some other property which is instrumental in transmembrane signalling. Evidence for an essential role for tyrosine phosphorylation could be obtained by identifying substrates of the growth factor receptor protein-tyrosine kinases whose activity is changed in a meaningful way upon phosphorylation. At present, however, there are no intracellular substrates for the growth factor receptors whose function is known (see Table 5). An alternative proof would be to show that mutant protein kinase-deficient receptors have a correlated defect in transmitting a mitogenic signal. The availability of cDNAs for several growth factor receptors now makes it possible to use this approach. A number of site-directed mutations in the EGF receptor abolish protein kinase activity. When introduced into cells lacking EGF receptors, such mutant receptors, unlike the normal EGF receptor, fail to render cells mitogenically responsive to EGF[13].

Transformation

Compared to normal cells transformed malignant cells display a number of characteristic changes. Transformed cells typically show dramatic morphological changes usually concomitantly with disruption of cytoskeletal filament organization. Transformed cells have diminished serum or growth factor requirements for growth in monolayer, and there is a good correlation between tumorigenicity and anchorage-independent growth. Transformation also leads to increases in nutrient transport, protease secretion, growth factor production and glycolytic flux.

The discovery of the role of oncogenes in viral and carcinogen-induced tumorigenesis leads one to the expectation that a molecular explanation for the altered phenotype of the transformed cell should be attainable. For this reason intensive efforts have been made to determine the functions and subcellular locations of oncogene products. As a result it is now possible to classify oncoproteins based on their likely functions (Table 2)[14,15].

Table 2. Functions of cell-derived oncogenic proteins

ONCOGENE	TRANSFORMING PROTEIN	LOCATION	FUNCTION
Class 1 - Protein-tyrosine kinases			
src (V)	pp60^{v-src}	Plasma membrane	Tyrosine kinase
yes (V)	P90$^{gag-yes}$	Plasma membrane?	Tyrosine kinase
fgr (V)	P70$^{gag-fgr}$	?	Tyrosine kinase
lck (T)	p56^{lck}	Plasma membrane	Tyrosine kinase
fps (V)	P140$^{gag-fps}$	Cytoplasm	Tyrosine kinase
fes (V)	P85$^{gag-fes}$	Cytoplasm	Tyrosine kinase
abl (V)	P160$^{gag-abl}$	Plasma membrane	Tyrosine kinase
ros (V)	P68$^{gag-ros}$	?	Tyrosine kinase
erb-B (V)	gp68/74^{erb-B}	Plasma and cytoplasmic membranes	EGF receptor tyrosine kinase domain
neu (T)	p185^{neu}	Plasma membrane	Tyrosine kinase
fms (V)	gP180$^{gag-fms}$	Plasma and cytoplasmic membranes	CSF-1 receptor tyrosine kinase
kit (V)	P80$^{gag-kit}$	?	Tyrosine kinase?
trk (T)	P70^{trk}	?	Tyrosine kinase
met (T)	p65^{met}	?	Tyrosine kinase
sea (V)	gP155$^{env-sea}$	Plasma membrane	Tyrosine kinase
ret (T)	p96^{ret}	?	Tyrosine kinase?
- Protein-serine kinases			
raf (V)	P90$^{gag-raf}$	Cytoplasm	Serine kinase?
mil (V)	P100$^{gag-mil}$	Cytoplasm	Serine kinase?
mos (V)	P37^{v-mos}	Cytoplasm	Serine kinase?
pim-1 (T)	p36^{pim-1}	?	Serine kinase?
Class 2 - Receptors lacking protein kinase activity			
erb-A (V)	P75$^{gag-erb-A}$	Cytoplasm?	Steroid receptor related
mas (T)	p35^{mas}	?	Adrenergic receptor related

ONCOGENE	TRANSFORMING PROTEIN	LOCATION	FUNCTION
Class 3 - Growth factors			
sis (V)	p28$^{env-sis}$	Secreted	PDGF-like growth factor
Class 4 - GTP binding/GTPase proteins			
H-ras (V/T)	p21^{H-ras}	Plasma membrane	GTP binding/GTPase
K-ras (V/T)	p21^{K-ras}	Plasma membrane	GTP binding/GTPase
N-ras (T)	p21^{N-ras}	Plasma membrane	GTP binding/GTPase
mel (T)	?	?	?
Class 5 - Nuclear proteins			
myc (V/T)	P110$^{gag-myc}$	Nucleus	DNA binding
N-myc (T)	p66^{N-myc}	Nucleus	DNA binding
L-myc (T)	p64^{L-myc}	Nucleus	DNA binding?
myb (V)	P48$^{gag-myb-env}$	Nucleus	DNA binding
fos (V)	p55^{v-fos}	Nucleus	DNA binding?
ski (V)	P125$^{gag-ski}$	Nucleus	?
p53 (T)	p53	Nucleus	?
rel (V)	p59^{v-rel}	Nucleus	?
Unclassified			
ets (V)	P135$^{gag-myb-ets}$	?	?
dbl (T)	?	?	?
bcl-1 (T)	?	?	?
bcl-2 (T)	p38^{bcl-2}	?	?
int-1 (T)	gp40^{int-1}	Membranes/secreted?	?
int-2 (T)	p34^{int-2}	?	?
jun (V)	?	?	?

When there are several isolates of an oncogene a typical example is listed. The protein products of the oncogenes are given according to convention with the number representing the size of the protein in kDa and the following superscript giving the genetic loci that make up the protein in order from N to C terminus. (V) indicates viral and (T) tumor oncogene.

Class 1: There are several cellular genes closely related to the oncogenic protein kinase genes. The syn (also known as slk) and lyk genes are highly homologous to the src gene; the PDGF receptor gene is related to the fms gene; the insulin and IGF-1 receptor genes are related to the met and trk genes. All of these cellular protein-tyrosine kinase genes may be proto-oncogenes. The neu gene is also known as the c-erb-B-2 gene. The fes and fps oncogenes represent homologous cat and avian genes. Likewise the raf and mil genes are homologous mouse and avian genes.

Class 4: Other related cellular genes include the rho and R-ras genes.

Class 5: Other related cellular genes include the R-myc and r-fos genes.

Unclassified: These are oncogenes about which little is known. All of them have been molecularly cloned, and although sequences can be predicted for several of their products, they are not homologous to known proteins.

About half the 41 oncogenes listed in Table 2 encode proteins with proven or suspected protein kinase activity. The majority of these are protein-tyrosine kinases. All of these proteins contain a sequence of about 270 residues which is recognizable as the catalytic domain[1,16]. Comparative analysis of these sequences with those of normal cellular protein-tyrosine kinases allows one to divide the protein-tyrosine kinases into subfamilies.

Table 3. Protein-tyrosine kinase subfamilies

A. *src*

$pp60^{c-src}$
$pp62^{c-yes}$
$p56^{lck}$
c-*fgr* protein
$p59^{syn}$
$p56^{lyk}$

B. *fps/fes*

$p92^{c-fes}/p98^{c-fps}$
NCP94

C. *abl*

$p150^{c-abl}$ - 4 different N-termini

D. **Growth factor receptors**

EGF receptor (c-*erb*-B protein)
c-*erb*B-2 protein

PDGF receptor
CSF-1 receptor (c-*fms* protein)
c-*kit* protein?

Insulin receptor
IGF-1 receptor
c-*ros* protein
trk protein
met protein
c-*sea* protein
ret protein?

There are other less well-defined protein-tyrosine kinase activities, including a liver enzyme (P75) and a brain enzyme (P120)

Cells transformed by oncogenic proteins derived from the protein-tyrosine kinases in classes A-C have a 5-10 fold greater level of phospho-tyrosine in cellular protein than untransformed cells (phosphotyrosine normally accounts for 0.05% of phosphate linked to protein-hydroxyl groups)[16]. In contrast oncoproteins derived from the growth factor receptor related protein-tyrosine kinases cause very little if any increase in the level of phosphotyrosine in cellular proteins. Why should this be the case? It is becoming clear that cellular protein-tyrosine kinases are normally tightly negatively regulated. As is true for most oncogenes, activation of protein-tyrosine kinase genes into oncogenes requires mutations leading to structural alterations. In the case of the protein-tyrosine kinase genes the activating mutations generally result in the loss of negative regulatory domains, while the catalytic domain remains intact.

The oncogenic activation of $pp60^{c-src}$ and the EGF receptor are the two best understood examples. $pp60^{c-src}$ is negatively regulated by tyrosine phosphorylation at Tyr 527, which lies 10 residues from the C-terminal end of the protein kinase domain and just six residues from the C-terminus of the protein[17-20]. Tyr 527, which is phosphorylated by an unknown protein-tyrosine kinase, is normally fully phosphorylated in the cell, and it has been proposed that the phosphotyrosine at position 527 may inhibit pp60[c-src] protein kinase activity by acting as a product analogue[20]. The oncogenic pp60[v-src] differs primarily from its normal cellular homologue in that the 19 C-terminal amino acids of pp60[c-src] have been replaced by 12 different amino acids in pp60[v-src], resulting in the loss of Tyr 527. In consequence the protein kinase activity of pp60[v-src] is about 10-fold higher than that of pp60[c-src]. As might be predicted, the simple replacement of the codon for Tyr 527 with one for phenylalanine converts the c-src gene into a transforming gene. All of the src gene family have a tyrosine embedded in a similar sequence close to the C-terminus of the protein (Pro/Gly.Gln.**Tyr**.Gln.Pro/Gln). Two other independent v-src gene isolates also lack the region containing Tyr 527, and diverge from the c-src sequence at residues 519 and 525 respectively. The v-yes and v-fgr genes also diverge from their cellular counterparts just upstream of the tyrosine equivalent to Tyr 527. Although there is no direct proof, it seems likely that every member of the src gene family will be negatively regulated by phosphorylation of this C-terminal tyrosine, and that oncogenic activation can be achieved by elimination of this domain resulting in unregulated and increased protein-tyrosine kinase activity.

Table 4.   The C-termini of normal and oncogenic protein-tyrosine kinases

End of protein kinase domain ↓

```
                                            Tyr 527
             W        RP F        *           ↓
   c-src     WRRDPEERPTFEYLQAFLEDYFTSTEPQYQPGENL(533)
   v-src     --K--------K----QLLPACVLEVAE(526)

   c-yes     WKKDPDERPTFEYIQSFLEDYFTATEPQYQPGENL(543)
   v-yes     -----------------------A--SGY(812)

   syn       WKKDPEERPTFEYLQSFLEDYFTATEPQYQPGENL(537)

   c-fgr     WRLDPEERPTFEYLQSFLEDYFTSAEPQYQPGDQT(?)
   v-fgr     ---------------------NGPQQN(663)

   lck       WKERPEDRPTFDYLRSVLDDFFTATEGQYQPQP(508)

   lyk       WKEKAEERPTFDYLQSVLDDFYTATEGQYQQQP(512)
```

The C-termini of the cellular protein-tyrosine kinases of the src gene family are shown in the one letter amino acid code. Where they exist, the sequence of the corresponding oncogene products are given. Identities are indicated by (-); residues where the oncogenic proteins diverge from the normal are bolded. Conserved residues in the protein kinase catalytic domain are given above the c-src sequence. The (*) indicates a conserved hydrophobic residue which marks the C-terminal end of the catalytic domain.

The other well understood example of oncogenic activation is the EGF receptor gene, which has given rise to several isolates of the v-erb-B oncogene. All the v-erb-B proteins have a very similar structure, being truncated at both ends with respect to the EGF receptor[3,21]. They lack the majority of the external EGF-binding domain, but leave a stub exposed on the outside of the cell. The cytoplasmic region contains an intact protein kinase domain, but is missing one, or two, of the three C-terminal auto-phosphorylation sites. In the unstimulated EGF receptor the auto-phosphorylation sites are believed to occupy the active center of the catalytic domain thus blocking access by other substrates. Phosphorylation would then cause this regulatory region to move out of the active site. A priori the absence of the ligand-binding site, in combination with the loss of autophosphorylation sites might have caused a dramatic activation of the v-erb-B protein-tyrosine kinase. In fact the v-erb-B proteins have rather poor protein kinase activity[22,23]. This may be because the loss of the external domain has eliminated most of the contact sites necessary for dimer formation and allosteric activation. Consequently, the v-erb-B proteins show only the marginal increase in activity over the EGF receptor that one would anticipate from the loss of the autophosphorylation sites. In this case the transforming ability of the v-erb-B proteins may lie not so much in increased protein kinase activity, but in chronic low level unregulated activity.

From these two examples it becomes apparent that the lack of the elevation in cellular phosphotyrosine in cells transformed by the growth factor receptor-related oncogenes compared to that induced by the other protein-tyrosine kinase oncogenes may in part be explained by the different degree of enzyme activation afforded by the alterations of their regulatory domains. An additional possibility is that the growth factor receptor protein-tyrosine kinases display a rather different substrate specificity.

Given that 40% of known oncogenes encode protein-tyrosine kinases, how good is the evidence that tyrosine phosphorylation is the molecular basis of transformation by these oncogenes? Analysis of transformation defective mutants shows that in general there is an excellent correlation between the protein kinase activity of these oncoproteins and their ability to transform, but, as discussed below, few if any cellular proteins which might be critical substrates have yet been identified. Based on the precedents with serine phosphorylation, it is reasonable to suppose that tyrosine phosphorylation of proteins can alter their activities, and that the protein-tyrosine kinases, like the protein-serine kinases, have pleiotropic substrate specificities. The oncogenic protein-tyrosine kinases are therefore likely to have multiple primary substrates. Relevant target proteins could exist among the substrates for the growth factor receptor protein-tyrosine kinases, which would enable the oncogenic enzymes to elicit a mitogenic response. In addition, since tyrosine phosphorylation appears to be involved in the normal regulatory processes governing cell shape, other substrates might exist in these pathways.

Substrates of protein-tyrosine kinases

The search for substrates of the oncogenic protein-tyrosine kinases has been hampered by the paucity of phosphotyrosine in proteins, compared to phosphoserine and phosphothreonine, even in transformed cells. This is compounded by the fact that the available methods for detecting phospho-tyrosine-containing proteins are not very sensitive[24]. One popular technique requires two-dimensional gel analysis of ^{32}P-labeled cells followed by alkali treatment of the gel to enrich for phosphotyrosine-containing proteins. The drawback to this method is that many proteins do not focus well in the first dimension or are resolved as a series of spots, thus diminishing the sensitivity of detection. A second technique involves immunoprecipitation of candidate substrates from ^{32}P-labeled cells followed by phosphoamino acid analysis of the gel band. This approach obviously requires not only that one has suitable immunological reagents, but also that one can guess the nature of substrates. A third technique, which is gaining in popularity, is the use of anti-phosphotyrosine antibodies for immunoprecipitation or immunoblotting. Increasingly, Na_3VO_4 is being used as an adjunct to identifying phosphotyrosine-containing proteins, since it inhibits phosphotyrosine phosphatases when used to treat intact cells, and thereby increases the level of phosphotyrosine in proteins. Although the search for substrates of many protein-serine kinases has been aided by in

vitro studies, the promiscuity of the protein-tyrosine kinases in vitro
largely vitiates this approach. In vitro phosphorylation experiments,
however, can be useful in indicating whether proteins are able to serve as
direct substrates for the protein-tyrosine kinase in question. In this
context it should be borne in mind that it is very difficult to prove that
phosphotyrosine-containing proteins detected in the cell are primary
substrates for a particular protein-tyrosine kinase, rather than substrates
for another protein-tyrosine kinase which is activated secondarily.

Using these techniques a number of phosphotyrosine-containing proteins
have been detected in intact cells either transformed by oncogenic protein-
tyrosine kinases or treated with growth factors whose receptors are
protein-tyrosine kinases (Table 5)[24].

Table 5. Substrates for protein-tyrosine kinases

Substrate	Retroviral protein-tyrosine kinases	Growth factor receptor protein-tyrosine kinases
Talin	+	?
Fibronectin receptor	+	?
Vinculin	+	-
p81 (ezrin)	+	+
p50	+	?
p42	(+)	+
Enolase	+	-
p36 (calpactin I)	+	(+)
p35 (calpactin II)	-	+
LDH	+	-
PGM	+	-
Calmodulin	+	-

All protein-tyrosine kinases autophosphorylate, and most of them contain
phosphotyrosine when isolated from cells. A 185 kDa phosphotyrosine-
containing protein has been detected in insulin-treated cells. A 21 kDa
protein associated with the T cell receptor is phosphorylated on tyrosine
upon antigen treatment. Numerous phosphotyrosine-containing proteins have
been detected by immunoblotting with anti-phosphotyrosine antibodies, but
little is known about these proteins except their molecular masses.

In cells transformed by oncogenes corresponding to the protein-
tyrosine kinase subfamilies A-C, there proves to be considerable overlap in
the spectrum of phosphotyrosine-containing proteins detectable[24]. The more
general approach of immunoblotting with anti-phosphotyrosine antibodies
also indicates a great degree of substrate similarity. As noted above the
level of phosphotyrosine-containing proteins in cells transformed by
oncogenes corresponding to subfamily D is considerably less, and the listed
substrates are phosphorylated to a variable extent. In contrast there is
little overlap between the phosphotyrosine-containing proteins in the
relevant virally-transformed cells and growth factor treated cells,
although fewer substrates have been detected in the latter cells.

In general the apoproteins of the identified substrates are relatively
abundant cellular proteins (0.05-0.3% of total cell protein). Moreover the
stoichiometry of tyrosine phosphorylation for these proteins is commonly
between 1 and 10%. This suggests that in many cases the phosphorylation is
spurious to the transformed phenotype and occurs as a consequence of the
presence of a disregulated protein-tyrosine kinase. This is likely to be
true for the three glycolytic enzymes among the substrates. Such gratuitous
tyrosine phosphorylations are presumably tolerated by the cell due to the

activity of phosphotyrosine phosphatases. Six of the listed proteins form part of a submembraneous cortical skeleton -- talin, fibronectin receptor, vinculin, p81, calpactins I and II. The first three are enriched in specialized adhesion structures known as adhesion plaques. Since most of the oncogenic protein-tyrosine kinases are localized to the membrane, all these skeletal proteins are in a position to be phosphorylated directly, which in turn could affect membrane architecture. However, it is not yet clear whether tyrosine phosphorylation of these proteins is involved in the morphological changes characteristic of transformed cells.

The major phosphotyrosine-containing protein detected in cells transformed by the retroviral protein-tyrosine kinases is the heavy chain of calpactin I (p36)[24]. Calpactin I is a Ca^{2+}/phospholipid binding protein[25,26], which can also associate with F-actin and spectrin. Thus calpactin I can in principle act as a cross-linker between the membrane and the cytoskeleton. Phosphorylation of the heavy chain might affect its ability to bind to phospholipids or the cytoskeleton or both. Recently another function has been suggested for calpactin I, and the closely related calpactin II[27]. Both proteins have been found to be capable of inhibiting phospholipase A_2 in vitro. Activation of phospholipase A_2 leads to the formation of arachidonic acid, which is converted into prosta-glandins and leukotrienes that are both cellular activators. Inhibitors of phospholipase A_2 are known collectively as lipocortins, and the anti-inflammatory effect of glucocorticoids is believed to be a result of the induced secretion of lipocortin. If the calpactins are lipocortins, then in principle phosphorylation might neutralize their inhibitory activity and lead to activation of phospholipase A_2. In practice it seems unlikely that calpactins I and II act as lipocortins in the cell. First, they are very abundant proteins, which are not detectably secreted even wtih gluco-corticoid treatment. Second, it appears that the inhibitory effects are due to their ability to sequester phospholipid substrate, rather than to a direct interaction with phospholipase A_2. Third, the phospholipid, actin and spectrin-binding properties of the calpactins suggest an eminently reasonable alternative function for these proteins.

The major phosphotyrosine-containing protein detected in quiescent cells treated with EGF or PDGF is a 42 kDa protein[24], although p42 may not be a primary substrate for these receptors. In contrast to most of the proteins phosphorylated by the retroviral enzymes, p42 is a minor cellular protein (~0.002% of total cellular protein), which is rapidly and nearly stoichiometrically phosphorylated in response to growth factors[28]. In large part due to its scarcity, rather little is known about p42 except that it is well-conserved evolutionarily and fractionates as a cytosolic protein. Interestingly, in mitogenically responsive cells tyrosine phosphorylation of p42 is also induced by tumor promoters, which do not themselves activate a protein-tyrosine kinase, but rather stimulate the protein-serine kinase, protein kinase C. Clearly protein kinase C must be able to activate one or more protein-tyrosine kinases, thus constituting a protein kinase cascade. (Conversely both the oncogenic and growth factor receptor protein-tyrosine kinases are able to stimulate a protein-serine kinase(s), which phospho-rylates the ribosomal protein S6.) This suggests that tyrosine phospho-rylation of p42 could be essential for the mitogenic response, although the rapidity of its phosphorylation would also be consistent with a role in an early response to growth factors. p42 is not detectably phosphorylated on tyrosine in every relevant transformed cell, however, which implies that p42 is not one of the substrates usurped by the oncogenic protein-tyrosine kinases through whose phosphorylation a continuous growth state is induced.

There are good reasons for believing that there are many other proteins phosphorylated by the transforming protein-tyrosine kinases, as well as by the growth factor receptors. Where should one look for these substrates and how can one tell whether they are critical targets? One clue to the location of targets for the retroviral enzymes comes from a study of mutants. pp60^{v-src} is normally associated with cellular membranes by virtue of a covalently-attached N-terminal myristoyl group. Non-myristoylated mutant pp60^{v-src}'s, which are unable to associate with membranes but are fully active as protein kinases, prove to be non-transforming[29]. This suggests that many of the critical substrates for transformation are either integral membrane or membrane-associated proteins. Given the plasma membrane location of growth factor receptors, they too may phosphorylate primarily membrane proteins. Table 6 lists candidate substrates for protein-tyrosine kinases, many, but not all of which, are plasma membrane-associated. These proteins were chosen because changes in their activities following tyrosine phosphorylation would have profound consequences for cellular physiology. As immunological reagents become available it will be possible to test whether any of these proteins are in fact targets.

Table 6. Candidate substrates for oncogenic protein-tyrosine kinases

Phospholipases C and A$_2$
Protein kinase C
G proteins
c-ras proteins
Phosphatidylinositol kinases
IP$_x$ phosphatases
Ion channels
Transport proteins
Extracellular matrix receptors
Protein-serine kinases
cNMP phosphodiesterases
Protein phosphatases
Nuclear regulatory proteins

Protein kinase C

As mentioned earlier a common response to growth factor treatment is the stimulation of phosphatidylinositol turnover, leading to the generation of two second messengers -- DAG and IP$_3$. DAG is an activator of the Ca^{2+} and phospholipid-dependent protein-serine kinase, protein kinase C[30]. As a consequence of activation by DAG a number of protein kinase C mediated phosphorylation events are detected following mitogen treatment. Tumor promoters, such as TPA, can substitute for DAG in the activation of protein kinase C. Indeed the only identified receptor for tumor promoters is protein kinase C. Protein kinase C is present in most cell types, implying that it has a critical cellular function. Recent evidence suggests that it exists in an inactive state in an intimate but reversible association with cellular membranes by virtue of its Ca^{2+}-dependent phospholipid binding activity. The generation of DAG or addition of TPA leads to activation of protein kinase C and a Ca^{2+}-independent association of the enzyme with the membrane.

Recent molecular cloning studies have shown that protein kinase C is actually a family of at least four enzymes derived from three genes[31-33]. It is not yet known exactly how these enzymes differ, but their expression is modulated in a tissue specific fashion. It seems probable that these protein kinase C isozymes will have subtly different substrate or activator specificities. The fact that activation of protein kinase C is a common response to almost all mitogens, and that tumor promoters are mitogens for certain cell types has led to an intensive search for substrates for protein kinase C. A number of bona fide substrates have been identified for protein kinase C, using the criteria that the phosphorylation of the

protein should be increased in cells treated with TPA, and that the
purified protein can be phosphorylated in vitro at the same site by
purified protein kinase C (Table 7)[34].

Table 7. Protein kinase C substrates

SUBSTRATE	RESIDUE	FUNCTIONAL EFFECT
EGF receptor	Thr 654	Causes ↓ in EGF binding and tyrosine kinase activity and ↓ internalization
pp60^{c-src}	Ser 12	None known
Transferrin receptor	Ser 24	Causes ↑ internalization
IL-2 receptor	Ser 247	None known
Insulin receptor	Ser	May cause ↓ tyrosine kinase activity
IGF-1 receptor	Ser	None known
p47	Ser	↑ IP$_3$ 5' phosphatase activity
p80	Ser	Function unknown
Vitronectin receptor	Ser	Untested
Vinculin	Ser/Thr	Untested
Caldesmon	Ser	Inhibition of MLCK
Calpactin I (p36)	Ser 25	Untested

Many other characterized proteins show increased phosphorylation in TPA-
treated cells e.g. 28 kDa stress proteins, ribosomal protein S6, p56^{lck},
α and β adrenergic receptors, glycogen synthase, tyrosine hydroxylase and
myosin light chain. Some of these, plus a multitude of other proteins of
interest, are phosphorylated by protein kinase C in vitro e.g. guanylate
cyclase (activation), G$_i$ (inhibition), HMGCoA reductase (inhibition),
troponins I and T, filamin, talin, HMG 14 and 17, histones, myelin basic
protein, topoisomerase II (activation), DNA methylase (activation).

Given the location of the active enzyme, it is not surprising that
many of these substrates are membrane-associated proteins. In contrast to
the substrates for the activated protein-tyrosine kinases, many of the
protein kinase C substrates are phosphorylated to a high stoichiometry, and
in addition there are demonstrable functional consequences of
phosphorylation. Interestingly, however, most of the functional effects
characterized so far are negative in nature. Thus the phosphorylation of
the EGF receptor leads to decreased responsiveness to EGF[35]. This may form
part of the circuitry involved in the integration of the cellular response
to multiple external stimuli. This particular feedback may be the cause of
the diminished binding of EGF (a progression factor) in cells treated with
PDGF (a competence factor). The phosphorylation and activation of an IP$_3$
5' phosphatase (p47) is presumably part of a feedback loop restricting the
release of Ca^{2+} in response to phosphatidylinositol turnover. There is
also evidence that protein kinase C inhibits phospholipase C activity, but
the molecular mechanism is unknown.

There are several cytoskeletal substrates for protein kinase C, and these could be instrumental in the rapid shape changes documented in TPA- or growth factor-treated cells. The only identified substrate, however, which might play a positive role in the mitogenic response is p80. Unfortunately very little is known about p80 except that it is membrane associated[36]. In looking further for protein kinase C substrates involved in mitogenesis, the list of candidate substrates for the protein-tyrosine kinases in Table 6 would be a good starting point. Indeed there is already evidence for in vitro phosphorylation of a G protein, and the Na^+/H^+ antiporter is very likely to be a substrate for protein kinase C.

The rapid activation of protein kinase C following growth factor treatment raises the question of which of the early events are directly induced by the growth factor receptor protein-tyrosine kinases and which by protein kinase C. In addition, the mitogenic activity of TPA and the nearly universal stimulation of phosphatidylinositol turnover by growth factors might prompt one to ask whether activation of protein kinase C is the sole mitogenic pathway. There are several reasons for thinking that this is not the case. First, cells in which protein kinase C activity has been down-regulated by long term treatment with TPA still respond mitogenically to growth factors. Second, TPA non-responder mutant cell lines are responsive to growth factors. Third, some growth factors, such as EGF, do not activate phosphatidylinositol turnover significantly in mitogenically responsive cells. Therefore the protein kinase C and protein-tyrosine kinase signals are likely to form parallel synergistic mitogenic pathways, arranged so that there are multiple crossover points.

ACKNOWLEDGEMENTS

I thank Jill Meisenhelder for her constructive criticism, and the island of Spetsai for the inspiration needed to write the lectures which are summarized here.

REFERENCES

1. Hunter, T., and Cooper, J. A. Protein-tyrosine kinases. Ann. Rev. Biochem. 54:897 (1985).

2. C.-H. Heldin, and B. Westermark, Growth factors: mechanisms of action and relation to oncogenes, Cell 37:9 (1984).

3. A. Ullrich, L. Coussens, J. S. Hayflick, T. J. Dull, A. Gray, A. W. Tam, J. Lee, Y. Yarden, T. A. Libermann, J. Schlessinger, J. Downward, E. L. V. Mayes, N. Whittle, M. D. Waterfield, and P. H. Seeburg, Human EGF receptor cDNA sequence and aberrant expression of the amplified gene in A431 epidermoid carcinoma cells, Nature 309:418 (1984).

4. Y. Ebina, L. Ellis, K. Jarnagin, M. Edery, L. Graf, E. Clauser, J.-H. Ou, F. Masiarz, Y. W. Kan, I. D. Goldfine, R. A. Roth, and W. J. Rutter, The human insulin receptor cDNA; the structural basis for hormone-activated transmembrane signaling, Cell 40:747 (1985).

5. A. Ullrich, J. R. Bell, E. Y. Chen, R. Herrera, L. M. Petruzelli, T. J. Dull, A. Gray, L. Coussens, M. Isubokawa, A. Mason, P. H. Seeburg, C. Grunfeld, O. M. Rosen, and J. Ramachandran, Human insulin receptor and its relationship to the tyrosine kinase family of oncogenes, Nature 313, 756 (1985).

6. Y. Yarden, J. A. Escobedo, W. J. Kuang, T. L. Yang-Feng, T. O. Daniel, P. M. Tremble, E. Y. Chen, M. E. Ando, R. N. Harkins, U. Francke, V. A. Fried, A. Ullrich, and L. T. Williams, Structure of the receptor for PDGF helps define a family of closely-related growth factor receptors, Nature 323:226 (1986).

7. A. Ullrich, A. Gray, A. W. Tam, T. L. Yang-Feng, M. Tsubokawa, C. Collins, W. Henzel, T. LeBon, S. Kathuria, E. Chen, S. Jacobs, U. Francke, J. Ramachandran, and Y. Fujita-Yamaguchi, Insulin-like growth factor 1 primary structure: comparison with insulin receptor suggests structural determinants that define functional sepcificity, EMBO J. 5:2503 (1986).

8. L. Coussens, C. Van Beveren, D. Smith, E. Chen, R. L. Mitchell, C. M. Isacke, I. M. Verma, and A. Ullrich, Structural alterations of viral homologue of proto-oncogene *fms* at carboxyl terminus, Nature 320:269 (1986).

9. S. S. Huang, and J. S. Huang, Association of bovine brain-derived growth factor receptor with protein-tyrosine kinase activity, J. Biol. Chem. 261:9568 (1986).

10. D. A. Cirillo, G. Gaudino, L. Naldini, and P. M. Comoglio, Receptor for bombesin with associated tyrosine kinase activity, Mol. Cell. Biol. 6:4641 (1986).

11. C. I. Bargmann, M.-C. Hung, and R. A. Weinberg, The *neu* oncogene encodes an EGF receptor-related protein, Nature 319:226 (1986).

12. T. Yamamoto, S. Ikawa, T. Akiyama, K. Semba, N. Nomura, N. Miyajima, T. Saito, and K. Toyoshima, Similarity of protein encoded by the human c-*erb*-B-2 gene to EGF receptor, Nature 319:230 (1986).

13. R. Prywes, E. Linveh, A. Ullrich, and J. Schlessinger, Mutations in the cytoplasmic domain of the EGF receptor affect EGF binding and and receptor internalization, EMBO J. 5:2179 (1986).

14. J. M. Bishop, Viral oncogenes, Cell 42:23 (1986).

15. T. Hunter, Oncogenes and proto-oncogenes: how do they differ? J. Natl. Canc. Inst. 73:773 (1984).

16. T. Hunter, and J. A. Cooper, Viral oncogenes and tyrosine phosphorylation, The Enzymes XVII:191 (1986).

17. S. A. Courtneidge, Activation of $pp60^{c-src}$ kinase by middle T binding or by dephosphorylation, EMBO J. 4:1471 (1985).

18. P. M. Coussens, J. A. Cooper, T. Hunter, and D. Shalloway, Restriction in vitro and in vivo tyrosine protein kinase activities of $pp60^{c-src}$ relative to $pp60^{v-src}$, Mol. Cell. Biol. 5:2753 (1985).

19. J. A. Cooper, K. L. Gould, C. A. Cartwright, and T. Hunter, Tyr^{527} is phosphorylated in $pp60^{c-src}$: implications for regulation, Science 231:1431 (1986).

20. J. A. Cooper, and C. S. King, Dephosphorylation or antibody binding stimulates $pp60^{c-src}$, Mol. Cell. Biol. 6:4467 (1986).

21. T. Yamamoto, T. Nishida, N. Miyajima, S. Kawai, T. Oi, and K. Toyoshima, The *erb*-B gene of avian erythroblastosis virus is a member of the *src* family, Cell 35:71 (1983).

22. R. M. Kris, I. Lax, W. Gullick, M. D. Waterfield, A. Ullrich, M. Fridkin, and J. Schlessinger, Antibodies against synthetic peptides as a probe for the kinase activity of the avian EGF receptor and v-*erb*-B protein, Cell 40:609 (1986).

23. T. Gilmore, J. E. DeClue, and G. S. Martin, Protein phosphorylation at tyrosine induced by the v-*erb*-B protein, Cell 40:619 (1986).

24. J. A. Cooper, and T. Hunter, Regulation of cell growth and transformation by tyrosine specific protein kinases: the search for important cellular substrate proteins, Curr. Top. Microbiol. Immunol. 107:125 (1983).

25. J. R. Glenney, Jr., Two related but different forms of the 36,000 M_r tyrosine kinase substrate (calpactins) which interact with phospholipid and actin in Ca^{2+}-dependent manner, Proc. Natl. Acad. Sci. USA 83:4258 (1986).

26. C. J. M. Saris, B. F. Tack, T. Kristensen, J. R. Glenney, Jr., and T. Hunter, The cDNA sequence for the protein-tyrosine kinase substrate p36 (calpactin I heavy chain) reveals a multidomain protein with internal repeats. Cell 46:201 (1986).

27. K. S. Huang, B. P. Wallner, R. J. Mattaliano, R. Tizard, C. Burne, A. Frey, C. Hession, P. McGray, L. K. Sinclair, E. P. Chow, J. L. Browning, K. L. Ramachandran, J. Tang, J. E. Smart, and R. B. Pepinsky, Two human 35 kd inhibitors of phospholipase A_2 are related to substrates of pp60^{v-src} and of the EGF receptor kinase, Cell 46:191 (1986).

28. J. A. Cooper, and T. Hunter, Major substrate for growth factor-activated protein-tyrosine kinases is a low abundance protein, Mol. Cell. Biol. 5:3304 (1985).

29. M. P. Kamps, J. E. Buss, B. M. Sefton, Rous sarcoma virus transforming protein lacking myristic acid phosphorylates known polypeptide substrates without inducing transformation, Cell 45:105 (1986).

30. Y. Nishizuka, Studies and perspectives of protein kinase C, Science 233:305 (1986).

31. P. J. Parker, L. Coussens, N. Totty, L. Rhee, S. Young, E. Chen, S. Stabel, M. D. Waterfield, and A. Ullrich, The complete primary structure of protein kinase C - the major phorbol ester receptor, Science 233:853 (1986).

32. L. Coussens, P. J. Parker, L. Rhee, T. L. Yang-Feng, E. Chen, M. D. Waterfield, U. Francke, and A. Ullrich, Multiple distinct forms of bovine and human protein kinase C suggest diversity in cellular signaling pathways, Science 233:859 (1986).

33. J. L. Knopf, M.-H. Lee, L. A. Sultzman, R. W. Kriz, C. R. Loomis, R. M. Hewick, and R. M. Bell, Cloning and expression of multiple protein kinase cDNAs, Cell 48:491 (1986).

34. J. R. Woodgett, T. Hunter, and K. L. Gould, Protein kinase C and its role in cell growth, in: "Cell Membranes: Methods and Reviews," E. L. Elson, W. A. Frazier, and L. Glaser, eds., Plenum Publishing Co. New York (1987).

35. C. Cochet, G. N. Gill, J. Meisenhelder, and T. Hunter, C-kinase phosphorylates the EGF receptor and reduces its EGF-stimulated tyrosine protein kinase activity, J. Biol. Chem. 259:2553 (1984).

36. P. J. Blackshear, L. Wen, B. P. Glynn, and L. A. Witters, Protein kinase C stimulated phosphorylation in vitro of a M_r 80,000 protein phosphorylated in response to phorbol eseters and growth factors in intact fibroblasts, J. Biol. Chem. 261:1459 (1986).

THE EFFECTS OF BOTULINUM NEUROTOXIN AND TETRODOTOXIN ON

PROTEIN PHOSPHORYLATION IN PURE CHOLINERGIC SYNAPTOSOMES

Xavier Guitart, Jordi Marsal and Carles Solsona

Dpt. de Biologia Cel.lular i Anatomia Patològica
Hosp. Bellvitge. Universitat de Barcelona
Casanova, 143. E-08036 Barcelona (Spain)

INTRODUCTION

Protein phosphorylation has been related to the molecular mechanisms of neurotransmitter release (1) and some phosphoproteins as Synapsin I are thought to be implicated in the exocytotic process (2). On the other hand, depolarization of isolated synaptic buttons (synaptosomes) results in the phosphorylation of specific proteins in rat brain synaptosomes (3) and in Torpedo electric organ synaptosomes (4). The electric organ of Torpedo is a useful model to study the acetylcholine release since it is innervated exclusively by cholinergic nerves (5), and presents an homologous structure with the neuromuscular junction. In contrast, in the electric organ of Torpedo the presynaptic nerve terminals represent as much as 2% to 3% of the total volume. A pure cholinergic synaptosomal fraction can be isolated from this organ (6). In these cholinergic synaptosomes the induced release of acetylcholine (7) and ATP (8) and the calcium fluxes (9) have been characterized. Acetylcholine release can be induced by several chemical depolarising agents as high external potassium concentration, or veratridine or the calcium ionophore A 23187. Botulinum toxin (BoNTx) blocks the acetylcholine release (see 10 and 11 for reviews) in several preparations. In cholinergic synaptosomes from the electric organ of Torpedo marmorata BoNTx inhibits the induced acetylcholine release whereas ATP release, calcium uptake and membrane potential are not affected by the toxin (12). We have used this pure cholinergic preparation to study the effect of botulinum neurotoxin type A under depolarising conditions on the rate of phosphorylation of synaptosomal proteins.

MATERIALS AND METHODS

Live Torpedo marmorata specimens were obtained in Mediterranean sea. Fishes were anesthetized with MS 222 (Sandoz) and electric tissue excised in several blocks and kept in a physiological saline solution containing (in mM): NaCl 280, KCL 3, $MgCl_2$ 1.8, $CaCl_2$ 3.4, HEPES/NaOH (pH 6.8) 3.6, urea 300, sucrose 100, glucose 5.5. The pH of this

solution was adjusted, after oxygenation, to 7.0 by adding
NaHCO$_3$. Cholinergic synaptosomes were isolated from these
fragments by a method described by Israel et al. (6). The
synaptosomal suspension was then diluted with calcium-free,
sucrose-free physiological medium containing 2 mM EGTA, and
then pelleted. The pellet was resuspended in the saline
solution, without calcium and with 0.1 mM EGTA, and this
fraction was used in the phosphorylation assays. Synaptosomes
were preincubated with 32-Pi (0.250 mCi/ml, final
concentration) for 45 min at room temperature and then,
aliquots were transferred into several tubes containing the
same volume of the physiological saline solution. Some of
these tubes also contained (depending on the assay): 50 mM
KCl; the calcium ionophore A 23187, 10 μM, or veratridine,
100 μM, (final concentrations).

When the effects of BoNTx or tetrodotoxin (TTX) were
tested, these toxins were added into the prelabelling tube
during the last 10 minutes of preincubation, to give a final
concentration of 1.25 nM and 40 μgr/ml respectively.
Reactions were carried out for 30 sec and finished by adding
a small volume of a SDS plus 2-mercaptoethanol containing
solution (13) and electrophoresis were run out immediately. A
discontinuous polyacrylamide slab gel was used as described
by Laemmli (14). The resolving gel consisted of a linear
gradient of 6% to 17% and was overlaid with a 4% acrylamide
stacking gel. After electrophoresis, the resolving gel was
stained with 0.2% Coomassie Blue R-250 in 52% methanol and
6.6% acetic acid and then distained. Dried gels were exposed
to KODAK X-Omat RP films at -80-C for 3 to 4 days using
intensifying screens. After developing, autoradiograms were
scanned and the 32-P labelled proteins quantified from the
densitometric scans. We measured the areas under the peaks
and these areas have been taken as proportional to the amount
of 32-P incorporated into different bands. Results of protein
phosphorylation are given as arbitrary units since the area
of the peaks depends on the protein phosphorylation, the
darkness of autoradiograms being substracted. Apparent
molecular weight of phosphoproteins was estimated by
comparison with the mobilities of known molecular weight
proteins (15).

RESULTS AND DISCUSSION

The events presented here correspond to presynaptic
events since cholinergic synaptosomal fraction was tested
morphologically and biochemically. The nicotinic
acetylcholine receptor was measured in order to detect any
postsynaptic contamination. We only found slight
contamination (about 3 % of the nicotinic receptor present in
the electric organ homogenate) by postsynaptic elements.
Synaptosomes retain their integrity and their functional
properties. Preincubation of Torpedo marmorata electric organ
synaptosomes with 32-P led to an incorporation of phosphorus
into several proteins. Our results suggest that protein
phosphorylation is linked to acetylcholine release since
depolarization with a high potassium external concentration
or veratridine results in an increase in the incorporation of
phosphorus to the 138, 108, 63, 37 and 24 KDa proteins, as
can be seen in fig. 1. On the other hand, the calcium
ionophore A 23187 showed a similar effect that depolarising
agents and little differences are found in the rate of

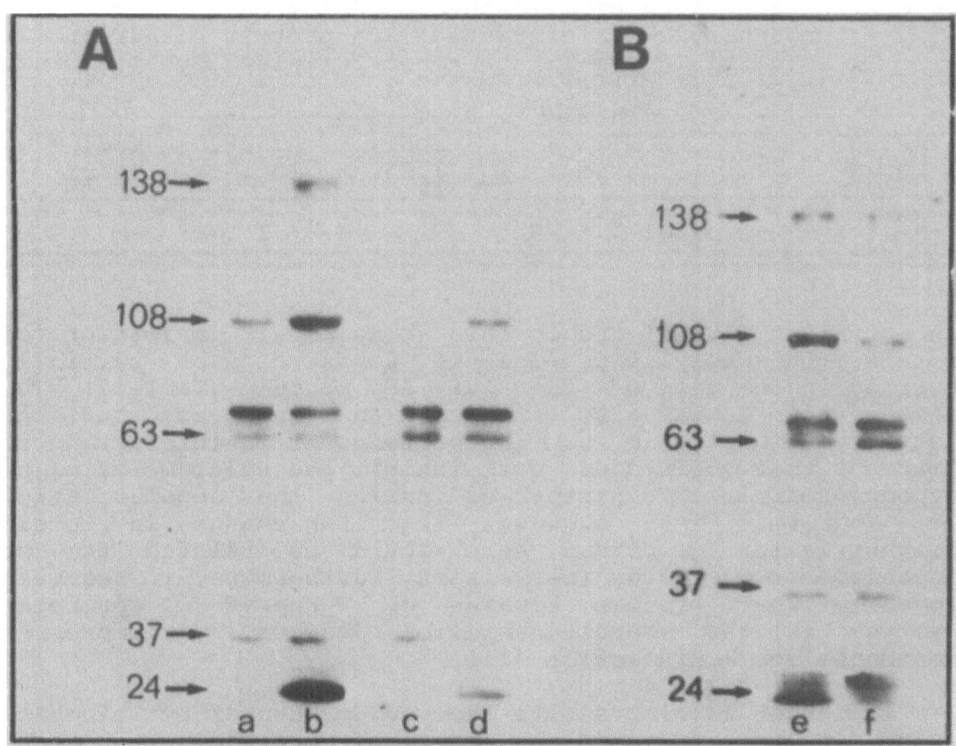

Fig. 1. Autoradiogram showing : A. The effect of
potassium induced depolarization on synaptosomal
protein phosphorylation either in botulinised
or non botulinised cholinergic synaptosomes.
Lane a : Resting condition; Lane b : KCl, 50 mM
; Lane c : Resting condition plus BoNTx;
Lane d : KCl, 50 mM plus BoNTx. B. The effects
of veratridine and TTX on synaptosomal protein
phosphorylation. Lane e : Veratridine; Lane f :
Veratridine plus TTX.

protein phosphorylation induced by the referred agents. These
differences could be related to the mechanism of action used
to trigger the acetylcholine release.

BoNTx, a strong blocker of acetylcholine release also
acts on Torpedo cholinergic synaptosomes and causes an
inhibition of the acetylcholine release up to 70% (12). This
neurotoxin inhibits depolarization induced phosphorylation of
the 138, 108 and 24 KDa proteins when intoxicated cholinergic
synaptosomes are depolarised by increasing the external
potassium concentration (fig.1). BoNTx decreases the basal
level of phosphorylation in the 37 KDa protein when
botulinised synaptosomes are depolarised by potassium. It has
been described a group of proteins isolated from the
presynaptic membrane of pure cholinergic terminals that seems
to be involved in acetylcholine release when is incorporated
to artificial membranes (14). One protein of this fraction is
a 36 KDa protein that could correspond to the 37 KDa
described here. Acetylcholine release by botulinised
synaptosomes can be partially restored by using A 23187 as a
secretagogue in presence of a high extracellular calcium

Table 1. Effect of Calcium ionophore A23187 on phosphorylation of the 108 KDa protein either in botulinised (BoNTx) or non-botulinised synaptosomes. (Arbitrary Units)

A23187 10 μM	BoNTx 1.25 nM	BoNTx + A23187 + CaCl$_2$ 1.25 nM 10 μM 10 mM
3.22 ± 0.86	0.88 ± 0.08	2.12 ± 0.15

concentration (10 mM) and in this condition, inhibition of protein phosphorylation due to BoNTx can be partially reverted in the case of the 108 KDa protein (Table 1). The presence of BoNTx also induces an increase in the phosphorylation of a 63 KDa protein. It is interesting to point out that BoNTx does not inhibit the calcium influx by Torpedo cholinergic synaptosomes during the depolarization (12) and this fact suggests that the changes in protein phosphorylation described here could be related to the mechanisms of acetylcholine release. Furthermore, it has been suggested that protein kinases or Synapsin I, proteins involved in the neurotransmission process, may represent substrates for BoNTx action (11).

TTX, that interacts with the sodium channel by blocking sodium influx, prevents the effect of veratridine, an agent that causes cell depolarization by opening the sodium channels. We have found that TTX is able to inhibit the veratridine induced protein phosphorylation of the 138, 108 and 24 KDa proteins when synaptosomes are intoxicated for 10 minutes before triggering the acetylcholine release, but has not effect on the 37 KDa protein (Fig.1). The sodium channel has been isolated from different sources (17,18) and the purified TTX-binding protein seems to be a protein of approximately 260 KDa (α-subunit) that copurifies with additional peptides depending on the source used to isolate the channel. Also, it has been described a cAMP dependent phosphorylation of the α-subunit in nerve endings (19) but we could not find any phosphorylation of such a molecular weight protein in our preparation probably due to that sodium channels are poorly represented at the nerve terminals (20). As in the case of botulinum neurotoxin, the presence of tetrodotoxin also increases the phosphorylation of a 63 KDa protein. We suggest that this similar behavior could be related to acetylcholine release inhibition since this protein is more phosphorylated when the transmitter release is inhibited. It has been found a protein of 63 KDa in the electric organ of Torpedo (21) and it has been suggested that it could be the homologous to the L-polypeptide of the mammalian neurofilament triplet. We can speculate that this protein can exist in different phosphorylation states and their different forms could be related to acetylcholine release mechanisms. Torpedo electric organ synaptosomes seem to be a good model to study the correlation between presynaptic protein phosphorylation and acetylcholine release. Acknowledgments: We thank Dr. P. Arté (I.I. Pesqueres, CSIC, Barcelona) for providinf specimens of Torpedo. We are indebted to Dra. A. Casanova for the preparation of Clostridium b. culture. This work was supported by CAICYT grant 84/2154 and CIRIT grant AR82/2-73.

REFERENCES

1. E. J. Nestler and P. Greengard, Protein phosphorylation in the brain, Nature 305:583-588 (1983).
2. F. Navone, P. Greengard and P. DeCamilli, Synapsin I in nerve terminals: selective association with small synaptic vesicles, Science 226:1209-1211 (1984).
3. B. K. Krueger, J. Forn and P. Greengard, Depolarization-induced phosphorylation of specific proteins mediated by calcium ion influx in rat brain synaptosomes, J. Biol. Chem. 252:2764-2773 (1977).
4. D. M. Michaelson and S. Avissar, Ca^{2+}-dependent protein phosphorylation of purely cholinergic Torpedo synaptosomes, J. Biol. Chem. 254:12542-12546 (1979)
5. W. Feldberg, A. Fessard and D. Nachmansohn, The cholinergic nature of the nervous supply of the electric organ of the Torpedo (Torpedo marmorata), J.Physiol. (Lond) 97:3P (1940).
6. M. Israel, R. Manaranche, P. Mastour-Frachon and N. Morel, Isolation of pure cholinergic nerve endings from the electric organ of Torpedo marmorata, Biochem. J. 160:113-115 (1976).
7. N. Morel, M. Israel, R. Manaranche and B. Lesbats, Stimulation of cholinergic synaptosomes isolated from Torpedo electric organ, Prog. Brain Res. 49:191-202 (1979).
8. N. Morel and F.M. Meunier, Simultaneous release of acetylcholine and ATP from stimulated cholinergic synaptosomes, J. Neurochem. 36:1766-1773 (1981).
9. J. Marsal, J.E. Esquerda, C. Fiol, C. Solsona and J. Tomás, Calcium fluxes in isolated pure cholinergic nerve endings from the electric organ of Torpedo marmorata, J. Physiol. (Paris) 76:443-457 (1980).
10. S. Thesleff and J. Molgó, A new type of transmitter release at the neuromuscular junction, Neurosci. 9:1-8 (1983).
11. L. C. Sellin, The pharmacological mechanism of botulism, Trends Pharmacol. Sci.,Feb:80-82 (1985).
12. J. Marsal, C. Solsona, X. Rabasseda, J. Blasi and A. Casanova, Depolarization-induced release of ATP from cholinergic synaptosomes is not blocked by botulinum toxin type A, Neurochem. Int. (in press) (1987).
13. H. Gower and R. Rodnight, Intrinsic protein phosphorylation in synaptic plasma membrane fragments from the rat. General characteristics and migration behaviour on polyacrylamide gels of the main phosphate acceptors, Biochem. Bophys. Acta 716:45-52 (1982).
14. U. K. Laemmli, Cleavage of structural proteins during the assembly of the head of bacteriophage T_4, Nature 227:680-685 (1970).
15. K. Weber and M. Osborn, The reliability of molecular weight determination by dodecyl sulfate-polyacrylamide gel electrophoresis, J. Biol. Chem. 244:4406-4412 (1969).
16. M. Israel, B. Lesbats, N. Morel, R. Manaranche, T. Gulik-Krzywicki and J. C. Dedieu, Reconstitution of a functional synaptosomal membrane possessing the protein constituents involved in acetylcholine translocation, Proc. Natl. Acad. Sci. U.S.A. 81:277-281 (1984).
17. R. L. Barchi, Protein components of the purified sodium

channel from rat skeletal muscle sarcolemma, <u>J.</u> Neurochem. <u>40:1377-1385 (1983).</u>

18. J. A. Miller, W. J. Agnew and S. R. Levinson, Principal glycopeptide of the tetrodotoxin/saxitoxin binding protein from Electrophorus electricus: isolation and partial chemical and physical characterization, <u>Biochemistry</u> 22:462-470 (1983).

19. M. R. C. Costa and W. A. Catterall, Cyclic AMP-dependent phosphorylation of the ∝-subunit of the sodium channel in synaptic nerve endings particles, <u>J.</u> Biol. Chem. <u>259:8210-8218 (1984).</u>

20. J. L. Brigant and A. Mallart, Presynaptic currents in mouse motor endings, <u>J. Physiol. (Lond)</u> 333:619-637 (1982).

21. J. H. Walker, C. M. Bousteaud, V. Witzemann, G. Shaw, K. Weber and M. Osborn, Cytoskeletal proteins at the cholinergic synapse: distribution of desmin, actin, fodrin, neurofilaments and tubulin in Torpedo electric organ, <u>Eur. J. Cell Biol.</u> 38:123-133 (1985).

PROTEIN PHOSPHORYLATION IN MEMBRANES OF NICOTIANA TABACUM

Klaus Palme, Jorge E. Mayer and Jeff Schell

Max Planck-Institut für Züchtungsforschung

D-5000 Koln 30, FRG

We have identified growth related changes in phosphoproteins from Nicotiana tabacum membranes. Changes in the phosphorylation patterns suggest that different substrates and different phosphorylation sites in substrates are used. Most remarkably, a change in the phosphorylation of a 33.000 Da phosphoprotein was correlated with the presence of cytokinin-like growth hormones in the culture media. A membrane bound protein kinase with an apparent Mr of 16.500 was isolated from Nicotiana tabacum and purified to homogeneity.

INTRODUCTION

Kinases catalyzing the transfer of the γ-phosphate from ATP to serine, threonine or tyrosine residues in protein substrates occur in a wide variety of organisms. Recent investigations have emphasized the role of protein phosphorylation in growth regulation of normal cells as well as in the establishment and maintenance of cell transformation (1,2). Furthermore, analysis of yeast cell division control genes demonstrates the role of protein kinases in cell cycle specific control events (3). However, only limited information is available about the phosphorylation of proteins in plant cells. We now report evidence for specific changes in the phosphoprotein pattern according to its status of differentiation and the modulation of activity of a purified membrane protein kinase by platelet activating factor (PAF).

MATERIALS AND METHODS

Cell cultures. Cultures of Nicotiana tabacum (vr White Burley) were cultivated in liquid Linsmeyer-Bednar and Skoog medium supplemented with 88 mM sucrose. Stock suspension cultures were subcultivated every week at the end of the linear growth phase by inoculating 0.3 l medium with 90 g cell suspension and incubated at $21^{o}C$ on a horizontal shaker (100 rpm, Braun-Melsungen) with a light period of 16 h.

<u>Extraction</u>. Cells (10 g) were disrupted for 10 min with a polytron homogenizer in 25 ml buffer A (100 mM Tris-HCl pH 8.0, 5 mM $MgCl_2$, 1 mM EDTA, 10 mM 2-mercaptoethanol, 1 mM leupeptin, 3,5 mg/l aprotinin, 5 mM ascorbic acid, and reextracted once. After centrifugation at 5.000 x g and filtration through a 132 µM nylon net, the particulate fraction was isolated by centrifugation at 100.000 x g for 30 min and washed twice in buffer A. Preparation of extracts for protein purification will be described in detail elsewhere.

<u>Protein kinase assay</u>. Membranes corresponding to 20 µg protein were incubated in 30 µl of 10 mM Pipes (piperazine-N,N´-bis(2-ethanesulfonic acid) pH 6.8, 10 mM NaCl, 1 mM $MgCl_2$, 0.2 mM EGTA, 20 mM $MnCl_2$, 0.05 mM ATP, 10 uCi $(\gamma^{32}P)$-ATP per ml at $4^{\circ}C$. The reaction was stopped by addition of 7% TCA, and the precipitates were analysed by electrophoresis in sodium dodecylsulfate/ polyacrylamide gels ($NaDodSO_4$ gel electrophoresis) containing 14% acrylamide and 0.5% N,N´-diallyltartardiamide.

RESULTS

Marked changes in the protein kinase activity level were observed in <u>Nicotiana tabacum</u> suspension cells. The enzyme levels progressively increased to the mid logarithmic growth phase and then decreased in the stationary phase (data not shown). To investigate developmental changes of endogenous substrate proteins particulate fractions were isolated from growing and fully differentiated leaves (Fig.1). Although the overall protein pattern was very similar, the phosphorylation patterns of specific proteins changed during differentiation. Prominent substrates in growing and fully differentiated cells are proteins with Mr 140.000, 38.000, 13.500 and 150.000, 35.000, 24.000, 18.000, respectively. After treatment of gels with alkali some changes in phosphorylation, otherwise masked, were detected (Fig.1B). Phosphoaminoacid analysis revealed both phosphoserine and phosphothreonine; however, phosphotyrosine was not detected. To assess the contribution of growth hormones to the phosphorylation of membrane proteins, <u>Nicotiana tabacum</u> suspension cultures were incubated in media with different growth hormones for 6 days and the particulate fraction isolated (Fig.2). Phosphorylation of a Mr 33.000 protein was markedly increased in the presence of cytokinin type growth hormones. pp33 was strongly phosphorylated on serine and only weakly on threonine.

To identify a membrane associated protein kinase in <u>Nicotiana tabacum</u> cells, suspension cultures of <u>Nicotiana tabacum</u> were lysed after 6 days of growth, the crude membrane fraction isolated by ultracentrifugation and the resulting 100.000 x g pellet resuspended in 50 mM Tris-HCl pH 7.5, 100 mM NaCl, 2% NP40, and 50 mM sucrose. After extraction for 60 min at $4^{\circ}C$ and centrifugation, the supernatant was concentrated with ammonium sulfate and chromatographed on Sephacryl S200. Protein kinase activities were detected using histone III as a substrate. Fig.3a shows the proteins eluting in the peak fraction. After in vitro phosphorylation with ^{32}P-ATP the major phosphorylated protein has a molecular weight of 16.500 Da after $NaDodSO_4$ gel electrophoresis

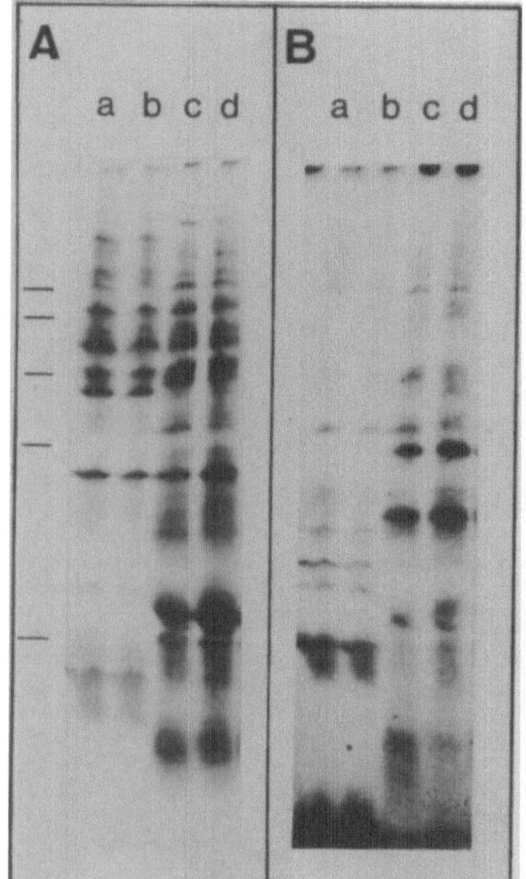

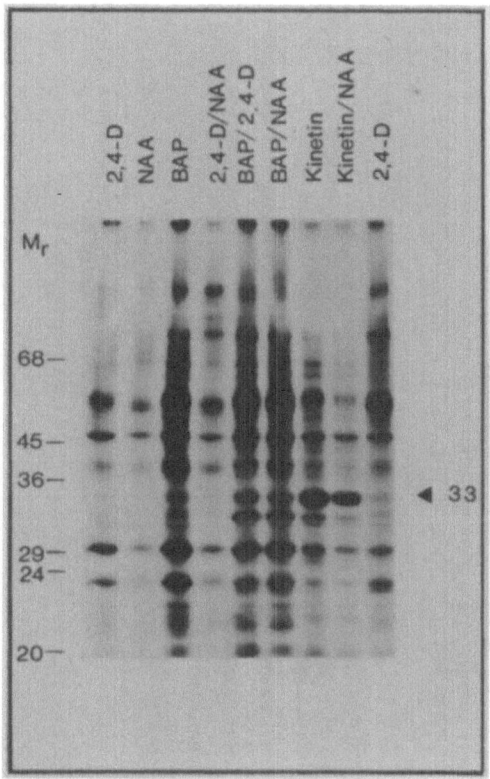

Fig.1   Fig.2

Fig.1. Leaves from <u>Nicotiana tabacum</u> plants vr W38 (lane a,c) and SR1 (lane d) were homogenized, particulate fractions isolated and analysed after incubation with $(\gamma^{32}P)ATP$ by gel electrophoresis. Growing tissue: lane a,b; fully differentiated tissue: lane c,d. B shows the same gel after treatment with 1 M KOH at $55^{\circ}C$ for 2 h.

Fig.2. <u>Nicotiana tabacum</u> suspension cells (W38) were grown in LS medium in the presence of 0.5 mg/l of the indicated growth hormones. Cells were harvested after 5 days, the particulate fraction isolated and analysed after incubation with $(\gamma^{32}P)ATP$ by gel electrophoresis.

2,4D   2,4-Dichlorophenoxyacetic acid; NAA   1-Naphtylacetic acid;
BAP   6-benzyl-aminopurine; Kinetin   6-furfurylaminopurine.

(Fig.3b). The protein kinase activity was further purified by chromatography on Cibacron F3GA Sepharose, A-Ser-Sepharose, Hydroxyapatite, and Heparin-Sepharose. With an overall yield of approximately 34%, 20 ug of homogenous protein can be obtained per

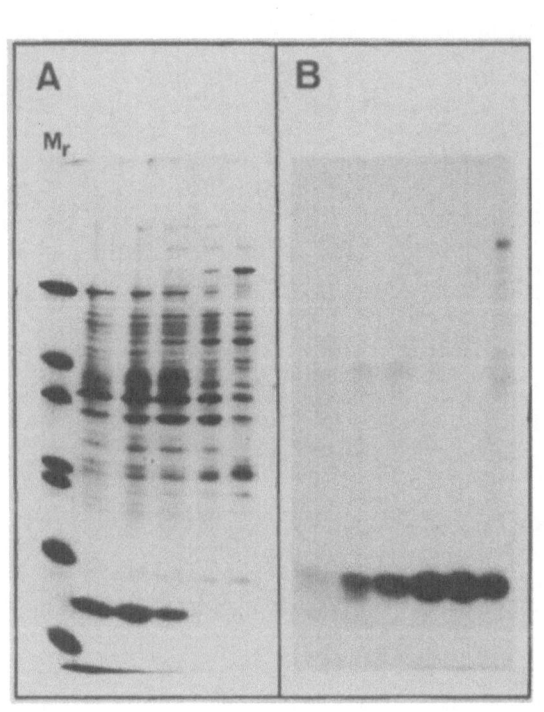

Fig.3   Fig.4

Fig.3. Elution profile of the protein kinase fraction from Sephacryl S200. The fractions were incubated with (γ ^{32}P)ATP, and analysed by NaDodSO$_4$ gel electrophoresis and stained with Coomassie blue (A) or autoradiographed (B).

Fig.4. Purified pK16. 1 µg of the purified enzyme was incubated with γ ^{32}P-ATP for 20 min at 30°C, analysed by gel electrophoresis and stained with Coomassie blue (lane a). The gel was autoradiographed for 12 h (lane b).

kg starting tissue. The homogenous protein kinase has a M_r of 16.500 Da after NaDodSO$_4$ gel electrophoresis (Fig.4, lane a). The autoradiogram in Fig.4, lane b demonstrates that the protein kinase strongly autophosphorylates. The major phosphoaminoacid in substrate proteins detected after partial hydrolysis and two-dimensional TLC is phosphoserine. The protein kinase (pK16) uses both, histones and casein as substrates with a Km of 84 and 64 uM respectively. Only ATP, not GTP can serve as a phosphate donor with an apparent Km of 16.1 uM. 5 mM Mg^{2+} is required for maximal activity, but Mn^{2+} will support phosphorylation at a lower level. The enzyme activity is not affected by cyclic AMP. pK16 shows a broad optimum at pH 7-8 and 30°C. The activity of pK16 is modulated by 1-alkyl-2-acetyl-sn-glycero-3-phosphocholine (platelet activating factor, PAF) by a factor of 2 , but not by phosphocholine or other lipids (Table 1). The stimulation is most prominent in the presence of EGTA, whereas rising the Ca^{2+} concentration in the presence of PAF nearly completely abolishes the enzymatic activity.

Phospholipid Specificity

Phospholipid	Protein kinase activity (units/mg)	relative activity
none	58.8	1.0
phosphocholine	61.74	1.05
1-alkyl-2-acetyl -sn-glycero-3- phosphocholine (PAF)	126.42	2.15
phosphatidylserine	67.62	1.15

Phospholipid Specificity

Phospholipid	Protein kinase activity (units/mg)	relative activity
none	58.8	1.0
phosphocholine	61.74	1.05
1-alkyl-2-acetyl -sn-glycero-3- phosphocholine (PAF)	126.42	2.15
phosphatidylserine	67.62	1.15

DISCUSSION

The findings presented here reveal certain developmental changes in the Nicotiana tabacum protein phosphorylation pattern. Some of the phosphorylation activities in tobacco cell suspension cultures were dependent on the presence of cytokinins in the medium, indicating the developmental relevance. A major protein kinase activity (pK16) was purified from these cultures. It is known that some protein kinases like protein kinase C or phosphorylase kinase are modulated by lipids. Upon testing of several lipids it was found that platelet activating factor (PAF) stimulated the pK16 protein kinase activity. PAF is known to have multiple functions in animal cells, e.g. the stimulation of Ca2+ influx and phospho-inoside breakdown (4-8). It was independently found that a lipid such as PAF also stimulates the proton transport in plant microsomes (9). It is therefore tempting to speculate that the influence of PAF on proton transport is mediated by its modulation of pK16 protein kinase activity.

REFERENCES

1. Cohen, P., 1985, Eur. J. Biochem, 151:439.
2. Hunter, T., and Cooper, J.A., 1985, Ann. Rev. Biochem., 54:897.
3. Nurse, P., 1985, Trends in Genetics, 1:51
4. Vargafting, B.B., and Benveniste, J., 1983, Trends in Pharm. Sci., 4:341.
5. Lee, T.-C., Malone, B., Blank, M.L., and Snyder, F., 1981, Biochem. Biophys. Res. Comm. 102:1062.
6. Shukla, S.D., and Hanahan, D.J., 1982, Biochem. Biophys. Res. Comm., 106:697.
7. Fisher, R.A., Shukla, S.D., Debuysere, M.S., Hanahan, D.J., and Olson, M.S., 1985, J. Biol. Chem., 259:8685.
8. Morre, D.J., Gripshover, B., Monroe, A., and Morre, J.T., 1984, J. Biol. Chem., 259:15364.
9. Scherer, G.F.E., 1985, Biochem. Biophys. Res. Comm., 133:1160

INTERLEUKIN 1 INDUCES TYROSINE PHOSPHORYLATION IN PLASMA MEMBRANES

OF THE TUMOR CELL LINES K 562 AND BW 5147

Michael Martin, Urte Kyas, and Klaus Resch

Div. of Molecular Pharmacology, Medical School Hannover
Konstanty-Gutschow-Str. 8, D 3000 Hannover 61, FRGermany

INTRODCUTION

Interleukin 1 (IL-1) is a protein produced mainly by activated mononuclear phagocytes which mediates the accessory cell requirement in the activation cascade of T-lymphocytes. The purification and molecular cloning of at least two different genes for IL-1 made larger amounts of the molecule available and thus it became apparent that the IL-1s have a multitude of functions besides the central role in the immune system, which can be summarized in their regulatory function in the inflammatory response (for a review see (1)). Despite this pivotal role of IL-1 in regulating the metabolism and controlling the growth of a variety of target cells, very few data are available concerning the ways of signal transduction involved.

Recently we reported that the growth of the human leukaemic tumor cell line K 562 could be inhibited by IL-1 (2). This observation initiated the search for the IL-1 receptor and the mode of signal trans-duction in these cells, which can easily be grown in numbers large enough to perform biochemical studies. The fact that Il-1 behaves like a growth factor in some systems like fibroblasts (3) or mesangial cells (4) prompted us to investigate whether protein phosphorylation was involved in the signal transduction pathway of IL-1.

Here we report that IL-1 induces tyrosine specific phosphorylation of a single protein in plasma membranes of the two tumor cells K 562 and BW 5147. Further data will be discussed, suggesting that this protein is part of a receptor complex for IL-1 or the receptor itself.

MATERIALS AND METHODS

K 562 or BW 5147 cells were disrupted by nitrogen cavitation as described for lymphocytes (5). Purified plasma membranes were prepared by differential ultracentrifugation at 4°C and membranes were stored in liquid nitrogen.

Recombinant human IL-1 alpha was obtained from Genzyme (Munich, FRG). Natural murine IL-1 was produced in P388D1 using a superinduction protocol (6) and purified by chromatography and chromatofocusing steps.

The fractions exhibiting the highest IL-1 activities (pI 4.8 to 5.0)
were used for the assays. The stimulation of murine thymocyte proliferation (C3H/HeJ mice) in the presence of submitogenic concentrations
of phytohemagglutinin was used to quantitate IL-1. The units were
defined as the reciprocal of the dilution at which half maximal incorporation of thymidine was measured.

The phosphorylation assays were performed at 0°C in a final volume of
50 µl containing: 10 µg plasma membrane protein , 20 mM Hepes, pH 7.4,
5 mM $MnCl_2$, 60 mM NaCl, 0.1 % bovine serum albumin, 15 µM ATP, with 4 µCi
gamma $32P$-ATP (Amersham-Buchler, Braunschweig, FRG). Phosphoprotein
patterns were analyzed after denaturing SDS-polyacrylamide gel electrophoresis (7) with 1 mm thick slab gels.

RESULTS AND DISCUSSION

An in vitro assay was established in order to investigate the
mechanism(s) whereby the IL-1 signal is transduced through the plasma
membrane. Protein phosphorylation could be shown to be involved in the
in vitro assay consisting of purified plasma membranes and IL-1.
Plasma membrane vesicles were obtained by using a protocol yielding
highly purified and enriched "right side out" vesicles without
cytosolic contamination, as could be shown by enzymatic profiles (5).
The incubation of these membranes with natural or recombinant IL-1 in
the presence of gamma-32P-labeled ATP revealed that IL-1 was able to
induce the specific phosphorylation of a 41 kDa protein in the human K
562 line (8) (Fig. 1) and a 43 kDa protein in the murine BW 5147
line (U.Kyas, M.Martin, and K.Resch, manuscript submitted).

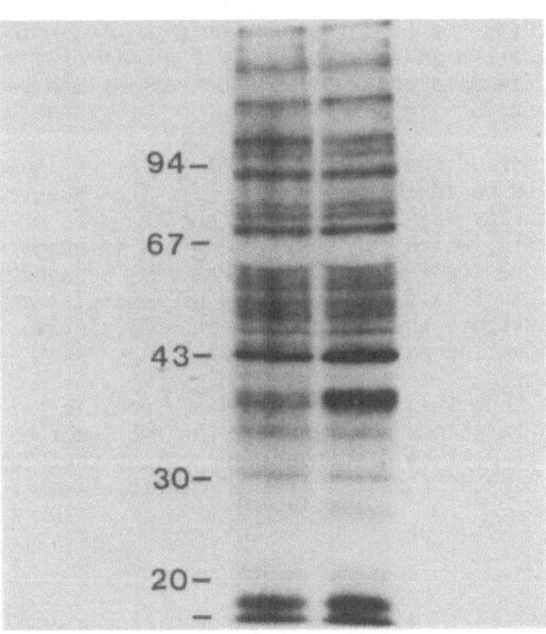

Fig.1. Induction of phosphorylation of the 41 kDa protein in plasma
membranes from K 562. 10 µg of protein were incubated for
30 minutes at 0°C. (Left lane: without IL-1; right lane:
1 unit of natural murine IL-1).

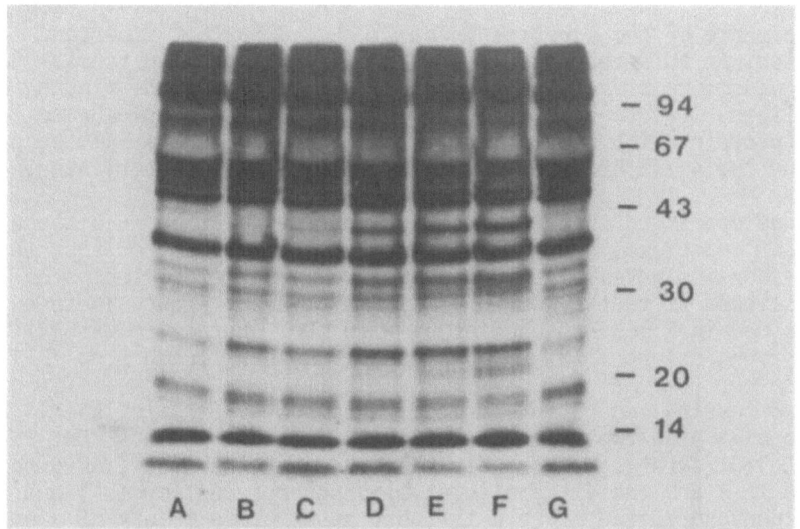

Fig.2. Concentration dependence of the induction of phosphorylation
by recombinant human IL-1 alpha in plasma membranes from
the murine tumor cell BW 5147. (A and G: without IL-1;
B: 0.1 U; C: 0.2 U; D: 0.5 U; E: 1U; F: 2 U of IL-1).

The differences in molecular weights may be explained by the different
species from which these tumors were obtained. The induction of phosphorylation of these proteins was specific for IL-1; lipopolysaccharide or
phorbolesters had no effect on the phosphorylation in the assays, the
latter indicating that no proteinkinase C was involved or present. The
induction of phosphorylation was dependent on the concentration of IL-1
in the assays, less than 0.5 U/10 µg protein being effective (Fig. 2).
This phosphorylation was also dependent on the time of incubation with
IL-1, BW 5147 being more rapidly phosphorylated than K 562 at 0°C (30
seconds compared to 5-10 minutes). The assays were carried out at 0°C
where kinases were very active. Phosphatases were inactive at that
temperature. This could be shown by dephosphorylation experiments, which
were performed in a pulse chase manner. The pulse was made with radioactive ATP and I1-1 at 0°C for 30 minutes, whereas the chase was made at
37°C in the presence of exogenous amounts of unlabeled ATP. A very rapid
dephosphorylation took place, which could be totally inhibited by
zincate, not by fluoride, but in a time dependent way by p-nitrophenyl-
phosphate and most strikingly by ortho vanadate, reported to be a
specific inhibitor for phosphatases with tyrosine phosphate specificity.
Alkali treatment of the gels and analysis of the phosphoamino acid
composition proved that tyrosine phosphate was present in the 41 kDa/43
kDa protein.

Performance of the experiments at 0°C had a further important implication. At this temperature practically no lateral diffusion takes place in plasma membranes, one therefore has to expect the binding domain for the ligand IL-1 to be tightly in contact with a transmembrane spanning protein which itself must be in contact with the induced protein kinase and the 41/43 kDa protein, ultimately being phosphorylated.

Thus we concluded that the receptor for IL-1 must be in a close spacial and functional association with these proteins, or that these different domains have to be in one entity. We therefore iodinated IL-1 and crosslinked it to the membranes with a bivalent agent. These experiments yielded in K 562 one band at 68 - 70 kDa, whereas in BW 5147 one band was found at about 75 kDa, besides of the non-linked free IL-1.

These results can be interpreted in two ways. Firstly, the 41 kDa or 43kDa proteins bind two molecules of IL-1 (natural IL-1:14 kDa, recombinant IL-1: 17 kDa). Or secondly that the proteins crosslinked to iodinated IL-1 are the extracellular binding proteins for IL-1 and the tyrosine phosphorylated protein is the transmembrane moiety of a protein kinase performing autophosphorylation. Drawing parallels to other receptors for growth factors, it seemed promising to look for a kinase in this region of 41/ 43 kDa in the membranes. A necessary but not exclusive prerequisite for a kinase is an ATP-binding and cleaving site. We thus tried to identify such a site at 41/43 kDa respectively, by using a radioactive affinity label. Indeed we could show a strong radioactive band in this region by incubating the membranes with 14C-labeled 5-p-fluorosulfonylbenzoyladenosine. The final decision whether this band is the phosphorylated protein itself or a second different protein with the same molecular weight, cannot be made at present and remains to be elucidated, e.g. by the use of monoclonal antibodies, purification etc. Experiments to clarify these issues are performed at present, and hopefully will help to clarify this controversary interpretation.

The summation of the data presented here show that tyrosine specific phosphorylation is involved in the signal transduction triggered by IL-1 in plasma membrane vesicles. The 41/ 43 kDa protein phosphorylated and the protein kinase are most likely one molecule. The binding domain for Il-1 is either on this protein itself or on a closely associated but not covalently bound protein, yet to be identified in our systems. Our studies and the method of analysis used do not exclude a dimeric structure or a complex structure with additional proteins. Membranes of a subclone of K 562 which did not respond to Il-1 by growth inhibition , did not show an induction of the phosphorylation of the 41 kDa protein. These data strongly suggest, that in vivo IL-1 may exert its biological function in a similar way to the in vitro assay via a common way of signaltransduction, i.e. the tyrosine phosphorylation of a putative regulatory protein in the cytosol of the responder cell, mediated by a plasma membrane receptor protein kinase.

REFERENCES

1. J.J.Oppenheim, E.J.Kovacs, K.Matsushima, and S.K.Durum, There is more than one Interleukin 1, Immunology Today 7:45 (1986)
2. D.H.Lovett, B.Kozan, M.Hadam, K.Resch, and D.Gemsa, Macrophage Cytotoxicity: Interleukin 1 as a mediator of tumor cytostasis, J.Immunol. 136:340 (1986)
3. J.A.Schmidt, S.B.Mizel, D.Cohen, and I.Green, Interleukin 1, a potential regulator of fibroblast proliferation, J.Immunol. 128:2177 (1982)

4. D.H.Lovett, J.L.Ryan, and R.B.Sterzel, Stimulation of rat mesangial cell proliferation by macrophage interleukin 1, J.Immunol. 131:2830 (1983)

5. V.Kaever, M.Szamel, M.Goppelt, and K.Resch, Characterization and subcellular localization of nucleotide cyclase in calf lymphocytes, Biochim.Biophys.Acta 776:133 (1984)

6. S.B.Mizel and D.Mizel, Purification to apparent homogeneity of murine interleukin 1, J.Immunol. 126:834 (1981)

7. U.K.Laemmli, Cleavage of the structural proteins during assembly of the head of bacteriophage T4, Nature 227: 680 (1970)

8. M.Martin, D.H.Lovett, and K.Resch, Interleukin 1 induces specific phosphorylation of a 41 kDa plasma membrane protein from the human tumor cell line K 562, Immunobiol. 171:165 (1986)

THE REGULATION OF cAMP-DEPENDENT PROTEIN KINASES IN NORMAL AND TRANSFORMED L6 MYOBLASTS

Ian A. J. Lorimer and B. D. Sanwal

Department of Biochemistry
University of Western Ontario
London, Ontario, Canada. N6A 5C1

L6 myoblasts are a permanent cell line which differentiates in culture to form multinucleate myotubes containing large numbers of the proteins characteristic of mature muscle (Yaffe, 1968). There is some evidence which suggests a role for the cAMP-dependent protein kinases in the regulation of this process (Ball and Sanwal, 1980; Curtis and Zalin, 1981). In this report we have compared the cAMP-dependent protein kinases of L6 myoblasts and a spontaneously-transformed L6 cell line which is unable to differentiate in culture. In particular we have studied the type I cAMP-dependent protein kinase as we have found previously (Rogers et al., 1984) that this isozyme is regulated during differentiation, suggesting that it is somehow involved in the control of this process. We show that the regulatory subunit of this isozyme (RI)* is degraded rapidly in L6 myoblasts in the presence of cAMP, and that this behaviour is altered in the spontaneously-transformed cell line. We also present some data which suggest a physiological role for changes in RI levels during the process of withdrawal from the cell cycle.

MATERIALS AND METHODS

L6 cells were grown in α-MEM supplemented with 10% horse serum as described previously (Cates et al., 1984). The spontaneously-transformed L6 cell line, JRU5, has also been described previously (Seth et al., 1983).

Antibody was raised in rabbits against rat RI purified by the method of Dills et al.(1975). An affinity column of RI coupled to sepharose was used to purify the antibody from whole serum.

For Western blotting, 15 ug of total L6 protein per lane was electrophoresed on 9% SDS-polyacrylamide gels and then transferred to nitrocellulose using a Bio-Rad Trans blot apparatus. Blots were blocked with 4% BSA in TBS for 1 h and then incubated in the same buffer containing

---

*The abbreviations used are: RI, the regulatory subunit of type I cAMP-dependent protein kinase; TBS, Tris-buffered saline (20 mM Tris-HCl, 146 mM NaCl, pH 7.4); SDS, sodium dodecyl sulphate; Bt_2cAMP, N^6,2-0-dibutyryl adenosine 3:5-cyclic monophosphate; 8-Br-cAMP, 8-Bromoadenosine 3:5-cyclic monophosphate

1 ug/ml of affinity-purified antibody. Blots were then washed with 0.05%
Triton X-100 in TBS four times for ten minutes each and incubated in 4% BSA
in TBS containing 0.1 uCi/ml ^{125}I-goat anti-rabbit antibody. Blots were
washed again, dried and autoradiographed.

Half-lives of RI were determined by labelling L6 cells for 6 h with
100 uCi/ml of ^{35}S-methionine in methionine-free medium and then chasing
with regular medium containing 4 mM unlabelled methionine. RI was immunoprecipitated from samples isolated at different periods of chase as
described previously (Rogers et al., 1984). Levels of labelled RI remaining
were then quantitated by scanning densitometry of the autoradiograms.

RESULTS AND DISCUSSION

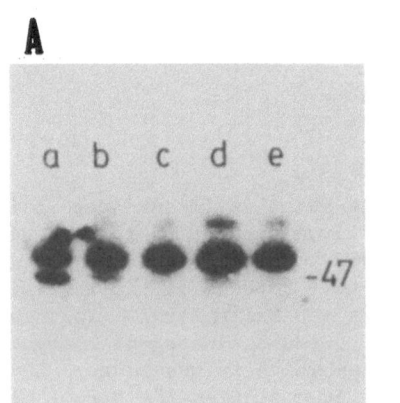

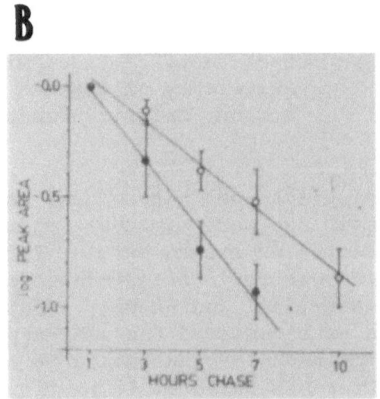

Fig. 1. (A) Western blot of L6 myoblasts after treatment with
different agents which raise intracellular cAMP. The
molecular weight of RI (47,000) is indicated on the
right. The upper band is IgG heavy chain from the
culture media. Lane a, untreated; b, 0.25 mM Bt$_2$cAMP;
c, 0.25 mM 8-Br-cAMP; d, 10 uM forskolin; e, 0.25 mM
Bt$_2$cAMP and 0.25 mM 8-Br-cAMP. 0.1 mM 3-isobutyl-1-
methylxanthine was included in all treatments which
were for 40 h. (B) Half-life of RI in L6 myoblasts. O,
untreated; ●, treated with 0.25 mM Bt$_2$cAMP.

Figure 1(A) shows that in L6 myoblasts treatments with different cAMP
analogues or forskolin all resulted in a decrease in total RI levels. After
the 40 h exposure to Bt$_2$cAMP the level of RI was decreased about three-
fold. RI was barely detectable after treatment with the other analogues or
forskolin. To determine the mechanism by which cAMP decreased RI levels we
measured the rates of turnover of RI in the presence or absence of Bt$_2$cAMP.
Figure 1(B) shows the results of these experiments. In untreated myoblasts
radiolabelled RI levels decreased exponentially with a half-life of 3.1 h.
Treatment of myoblasts with cAMP resulted in a significant increase in the
rate of degradation of RI. The half-life in the presence of 0.25 mM Bt$_2$cAMP
was calculated to be only 1.9 h. It therefore appears that the decreases in
RI levels are due to an increase in rate of degradation of RI in the
presence of cAMP.

Figure 2(A) shows the effect of different cAMP analogues and forskolin
on the levels of RI in the non-differentiating L6 cell line JRU5. By
comparison with figure 1A it can been seen that RI levels are similar in
untreated L6 and JRU5 cells. However, treatment of JRU5 with Bt$_2$cAMP caused

A

B

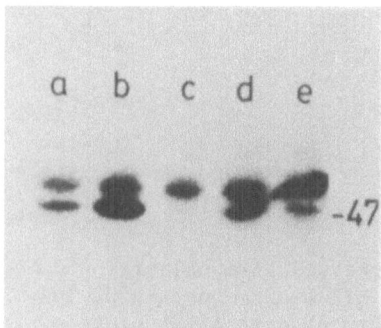
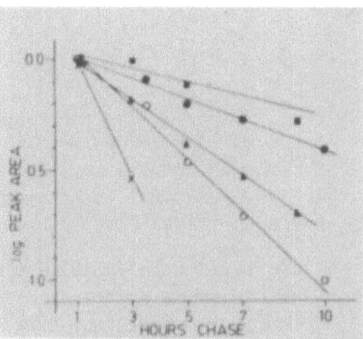

Fig. 2 (A) Western blot of JRU5 treated with different agents
which raise intracellular cAMP. Lanes are as described
in Fig. 1(A). (B) Half-life of RI in JRU5. O,untreated;
●, 0.25 mM Bt_2cAMP; ✗, 0.25 mM 8-Br-cAMP;
Δ, 10 uM forskolin; ■, 0.25 mM N^6-benzoyl-cAMP.

a three-fold increase in RI levels, whereas the same treatment in L6 caused
a decrease in RI levels. The increase in RI is only seen with some
analogues. 8-Br-cAMP was found to decrease RI levels in JRU5, as it did in
L6. Treatment with forskolin did not cause any change in levels of RI and
treatment with a combination of Bt_2cAMP and 8-Br-cAMP caused only a small
decrease in RI levels. These discrepancies in RI behaviour with different
cAMP analogues are also seen when half-life values are measured (figure
2B). The half-life of RI in untreated JRU5 was about 2.2 h. Treatment with
Bt_2cAMP or N^6-benzoyl-cAMP increased the half-life of RI to 6.4h and 9.8 h
respectively. The half-life decreased to about 1.1 h in L6 treated with 8-
Br-cAMP whereas forskolin did not change the half-life significantly (3.4
h). Therefore all the changes in RI levels observed by Western blotting are
apparently due to changes in the rates of degradation of RI.

The differences in the behaviour of RI seen with different analogues
make it difficult to interpret what the actual effect of endogenous changes
in cAMP levels is in JRU5. It has been shown that different cAMP analogues
show specificity for the different cAMP binding sites of RI (Rannels and
Corbin, 1980). Analogues which are derivatized at the 8 position of the
adenine ring selectively bind to the cAMP binding site designated as site
I. Analogues derivatized at the 6 position are selective for site II.
Therefore there is apparently a correlation between the site specificity of
the analogues and their effect on RI half-life (this has been suggested
previously by Steinberg and Agard, 1981). Two 6-substituted analogues
(Bt_2cAMP and N^6-benzoyl cAMP) were stabilizing, whereas a 8-substituted
analogue (8-Br-cAMP) was destabilizing. Since cAMP itself will occupy both
binding sites, we feel that the forskolin treatment or the combination of
two analogues with different site specificities will most accurately
reflect the in vivo response. With these treatments there was a distinct
decrease in RI levels in L6, but little or no decrease in JRU5.

We have not as of yet determined what the actual cause of the change
in RI behaviour is in JRU5. A comparison of RI immunoprecipitated from L6
and JRU5 showed there were no differences in the isoelectric point or two-
dimensional tryptic maps of the subunit (not shown). This suggests that a
change in some specific protease, rather than a change in the subunit
itself, may be responsible for the altered rates of degradation of RI.

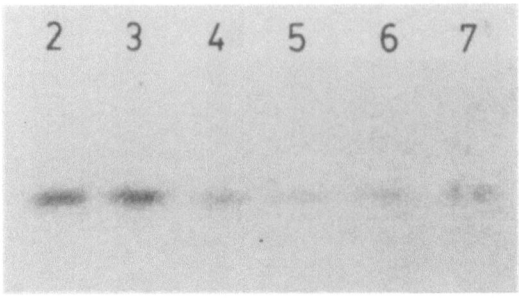

Fig. 3. RI levels in confluence-arrested L6 myoblasts.
Cells were grown in Ham's F12 medium containing 10%
fetal calf serum. The number of days after plating
is indicated along the top.

Figure 3 shows the regulation of RI during changes in the growth state
of L6 myoblasts. L6 cells were grown under medium conditions in which they
withdraw from the cell cycle upon reaching confluence but do not
differentiate (Pinset and Whalen, 1984). In the proliferative stage (days 2
and 3) RI levels are the same as in L6 myoblasts grown in regular media.
Once the cells reach confluence at around day 3, however, the levels of RI
decrease to almost undetectable levels. The levels of RI then remain low,
apparently for as long as the cells remain quiescent. We have shown
previously (Rogers et al., 1984) that RI levels remain constant throughout
the growth of JRU5 in regular media.

These results suggest that the regulation of RI levels may be
important physiologically during the processes of cell cycle withdrawal in
L6 myoblasts. Since we have shown in this report that RI levels can be
regulated by exogenously added cAMP, we are currently performing
experiments to determine whether or not this is the mechanism by which RI
levels are regulated during the transition from the proliferating state to
growth arrest at confluence. If this should prove to be the case, it will
give added significance to the changes in RI behaviour we have described in
the spontaneously-transformed L6 cell line, JRU5.

ACKNOWLEDGEMENTS

This work was supported by a grant from the Medical Research Council of
Canada. I. A. J. L. is the recipient of a Steve Fonyo Studentship from the
National Cancer Institute of Canada.

REFERENCES

Ball, E. H. and Sanwal, B. D. (1980) A synergistic effect of
glucocorticoids and insulin on the differentiation of myoblasts, J. Cell
Physiol., 102:27.

Cates, G. A., Brickenden, A. M. and Sanwal, B. D. (1984) Possible
involvement of a cell surface glycoprotein in the differentiation of
skeletal myoblasts, J. Biol. Chem., 259:2646.

Curtis, D. H. and Zalin, R. J. (1981) Regulation of muscle differentiation:
stimulation of myoblast fusion in vitro by catecholamines, Science,
214:1355.

Dills, W. L., Beavo, J. A., Bechtel, P. J. and Krebs, E. G. (1975) Purification of rabbit skeletal muscle protein kinase regulatory subunit using cyclic adenosine 3:5-monophosphate affinity chromatography, <u>Biochem. Biophys. Res. Comm.</u>, 62:70.

Pinset, C. and Whalen, R. G. (1984) Manipulation of medium conditions and differentiation in the rat myogenic cell line L6. <u>Dev. Biol.</u>, 102:269

Rannels, S. R. and Corbin, J. D. (1980) Two different intrachain cAMP binding sites of cAMP-dependent protein kinases, <u>J. Biol. Chem.</u>, 255:7085

Rogers, J. E., Narindrasorasak, S., Cates, G. A. and Sanwal, B. D. (1984) Regulation of protein kinase and its regulatory subunits during skeletal myogenesis, <u>J. Biol. Chem.</u> 260:8002.

Seth, P. K., Rogers, J., Narindrasorasak, S. and Sanwal, B. D. (1983) Regulation of cyclic adenosine 3:5-monophosphate phosphodiesterases: altered pattern in transformed myoblasts, <u>J. Cell Physiol.</u>, 116:336.

Steinberg, R. A. and Agard, D. A. (1981) Turnover of regulatory subunit of cyclic AMP-dependent protein kinase in S49 mouse lymphoma cells, <u>J. Biol. Chem.</u>, 256:10,731.

Yaffe, D. (1968) Retention of differentiation potentialities during prolonged cultivation of myogenic cells, <u>Proc. Natl. Acad. Sci. USA</u>, 67:477

PROTEIN PHOSPHORYLATION IN RESPONSE TO DIVERSE MITOGENIC AGENTS IN DOG THYROID CELLS

L. Contor, F. Lamy, R. Lecocq and J.E. Dumont

Institute of Interdisciplinary Research (IRIBHN).
Free University of Brussels, Campus Erasme
Route de Lennik 808
B - 1070 Brussels, Belgium

INTRODUCTION

Among the variety of changes induced upon addition of growth factors to quiescent cells are changes in the phosphorylation of cellular proteins which occur rapidly. The receptor molecules for epidermal growth factor (EGF), and for platelet derived growth factor (PDGF) and insulin, have been shown to contain an inherent, growth factor-dependent tyrosine protein kinase activity that was first identified in association with certain viral transforming proteins. In addition, protein kinase C has been reported as a receptor protein for tumor promoters, such as tetradecanoyl-phorbol-acetate (TPA), that stimulate the kinase activity. A rapid tyrosine-specific phosphorylation of two proteins of 42K is observed in fibroblasts of several species stimulated to proliferate in a variety of ways : growth factor induction, trypsin treatment, exposure to phorbol esters and ASV transformation. It has been suggested that phosphorylation of these 42K proteins might be important in the pathway leading to the mitogenic response.

The aim of the present study was to investigate in primary culture of dog thyroid cells[1], protein phosphorylation in the beginning of the prereplicative phase of the cell cycle and to determine whether mitogenic stimuli acting by different intracellular signals induced tyrosine-specific phosphorylation of these two 42K proteins. We have used serum-free, defined culture conditions, in which dog quiescent thyroid cells, which have not proliferated in vitro before stimulation, can be induced to proliferate by different mitogenic agents : thyrotropin (TSH) acting via cyclic AMP, EGF and TPA acting through a cyclic AMP independent mechanism. In thyroid cells TSH also induces differentiation expression while EGF and TPA inhibit this process[1]. We have studied the effects of these agents on the pattern of protein phosphorylation using 2D gel electrophoresis.

METHODS

Cell culture and protein phosphorylation

Primary cultures of dog thyroid cells were realized as described by Roger et al.[2]. After 4 days of culture the cells were incubated for 8 hours in MEME (Dulbecco's Minimum Essential Medium Eagle) containing 9.10^{-6}M cold phosphate and 1 or 3 mCi/ml [^{32}P]phosphate. The mitogenic agents were added 15 min before the end of the incubation period. In some experiments the cells were incubated for 20h instead of 8h in the presence of [^{32}P]phosphate in order to reach isotopic equilibrium.

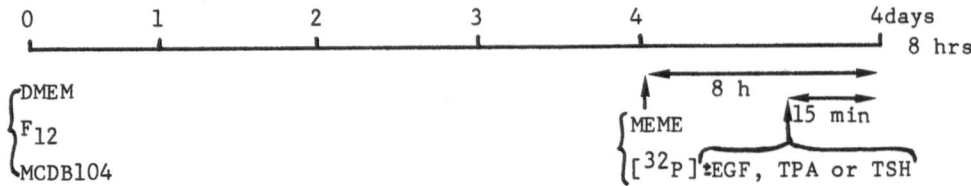

To control the mitogenic stimulation by the different agents tested, the proliferation assay was realized after 4 days and 8 hours of culture as shown below :

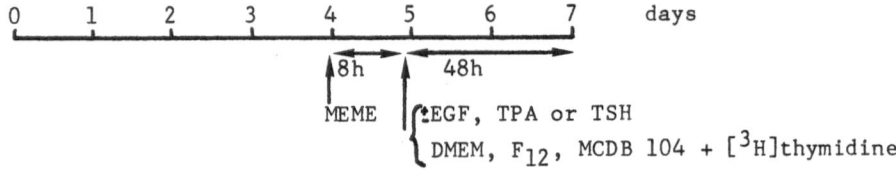

The agents tested were present during the last 48 hours of the incubation period together with 10 μCi/ml of [^{3}H]thymidine. DNA synthesis was measured by the frequency of the [^{3}H]thymidine nuclei estimated by autoradiography.

Electrophoretic techniques

Preparation of the sample and two-dimensional gel electrophoresis of total cellular proteins was performed according to O'Farrell. Electrophoretic transfer of proteins on membranes was realized following NEN's procedure.

Alkali treatment of proteins separated by 2D gel electrophoresis

Phosphotyrosine represents less than 1% of the total phosphate incorporated in the three acid-stable phosphorylated amino acids and is thus masked by phosphoserine and phosphothreonine on autoradiographs of 2D-gels. Taking advantage of the fact that phosphoserine is more labile to alkali hydrolysis than phosphothreonine and phosphotyrosine, a technique based on alkali treatment of proteins separated by 2D gel electrophoresis has been developed to decrease the proportion of phosphoserine in the phosphorylated peptides[3]. As proteins are soluble in sodium hydroxide we suspected that alkali treatment of a gel would lead to partial diffusion of proteins out of the gel. To evaluate the importance of this phenomenon, proteins labelled with [^{35}S] methionine were subjected to 2D gel electrophoresis. The gel was first dried and autoradiographed. It was then reswelled and treated with NaOH 1N for two hours at 55°C. It was dried again and exposed for a second time. To obtain autoradiographs of overall equal densities it was necessary to expose the alkali treated gel twice the time needed for the untreated gel. This suggests that during alkali treatment diffusion of proteins occurs. To overcome this drawback, the proteins after being subjected to 2D gel electrophoresis were transferred by electrophoresis to nylon membranes[4]. Nitrocellulose membranes could not be used for alkali treatment as they are soluble in NaOH 1N. The nylon membranes were then submitted to alkali treatment in NaOH at 55°C for two hours. Equal overall densities of the autoradiographs of the nylon membrane before and after alkali treatment were obtained using the same exposure time.

These results indicate that no loss nor diffusion of proteins occured during alkali treatment of the nylon sheet.

Phosphoamino acid analysis

After 2D gel electrophoresis separation, the proteins were transferred to nitrocellulose membranes, then excised and eluted from the membranes by overnight incubation at 37°C in 50% pyridine 0.1M ammonium acetate pH 8.9 [5]. The phosphopeptides were hydrolyzed in HCl 6N at 110°C for one hour. Carrier phosphoserine, phosphothreonine and phosphotyrosine were added to the sample and the phosphoamino acids were separated by two dimensional electrophoresis on cellulose plates at pH1.9 in the first dimension and at pH 3.5 in the second dimension. The carriers are revealed by ninhydrin staining and the [32P] marked phosphoamino acids by autoradiography.

RESULTS

EGF, TPA and TSH are mitogenic agents for dog thyroid cells in culture. Each of them induces DNA synthesis in these cells after a lag period of about 18 hours [1]. The effect of these agents on protein phosphorylation has been studied during the beginning of the prereplicative phase of the cell cycle.

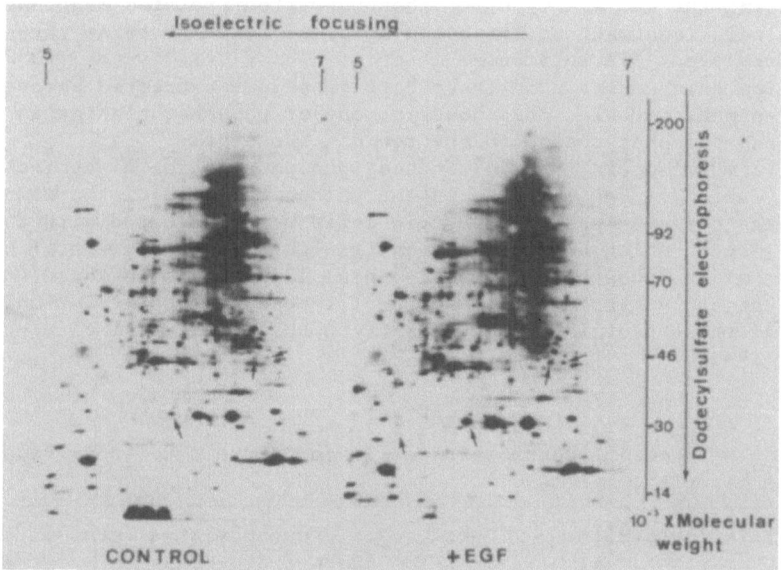

Fig. 1.   Autoradiographs of 2D gel electrophoresis of total proteins extracted from quiescent thyroid cells or from thyroid cells exposed to EGF.
Proteins incubated with [32P]phosphate for 8 hours and with EGF (50 ng/ml) for the last 15 min of the incubation period were subjected to 2D gel electrophoresis as described in Methods. The proteins whose phosphorylation is increased by EGF are indicated with an arrow.

Figure 1 shows that EGF (50 ng/ml during 15 min) stimulates the phosphorylation of 4 proteins whose molecular weights are 27K, 28K and 45K (twice). After alkali treatment of the proteins first separated by 2D gel electrophoresis and then transferred to nylon membranes, the

phosphate linked to the 28K protein was not detectable any more, suggesting an absence of phosphorylated tyrosine in this substrate. We
submitted the 4 substrates, extracted before alkali treatment, to
phosphoamino acid analysis : phosphoserine and/or phosphothreonine were
present in all 4 proteins while phosphotyrosine was only present in the
two 45K proteins.

As shown in table 1, the more acidic form of the 45K pair contained
phosphotyrosine, phosphothreonine and phosphoserine, while the basic
form contained phosphotyrosine and phosphoserine.

Table 1.

Characteristics of the two phosphotyrosine containing proteins of 45K.

	pH_i	Phosphorylated amino acids
p45A	7.2	TYR + SER + THR
p45B	7.3	TYR + SER

The pattern of protein phosphorylation induced by TPA (50 ng/ml
during 15 min) showed an increased phosphorylation of 20 proteins (table
2), including the two 45K tyrosine phosphorylated proteins mentioned
above. Alkali treatment of the nylon membrane fixed proteins revealed
8 alkali resistant TPA substrates.
Phosphoamino acid analysis of these 8 polypeptides extracted before
alkali treatment revealed phosphoserine and/or phosphothreonine in all
proteins and phosphotyrosine in the two 45K proteins.

TSH (1 mU/ml during 15 min) induces phosphorylation of at least 12
proteins (Table 2). An increase in the phosphorylation of the same
proteins was observed when dog thyroid cells were incubated with forskolin
(5.10^{-5}M), agent which enhances the intracellular accumulation of cyclic
AMP. None of the phosphate linked to these 12 proteins resisted to
alkali treatment suggesting an absence of tyrosine phosphorylation.
Only one of these proteins (28K) has its phosphorylation also increased
by EGF or TPA.

Table 2.

Stimulation of protein phosphorylation in dog thyroid cells in response
to diverse mitogenic agents.

Phosphorylated substrates	Agent	Phosphorylated amino acids	
		Ser and/or Thr	Tyr
28K	EGF or TPA or TSH	+	-
27K	EGF or TPA	+	-
two 45K	EGF or TPA	+	+
14K, two 38K, two 48K 49K, 50K, two 59K, 61K 65K, 68K, two 69K, two 76K	TPA	+	-
17K, 18K, two 21K, 23K two 42K, 46K, 57K, 65K, 80K	TSH	+	-

DISCUSSION

We have studied the effect of TSH, EGF and TPA on protein phosphorylation in dog thyroid cells in the beginning of the prereplicative phase
of the cell cycle. We have identified two common substrates of 45K
phosphorylated on tyrosine residues in response to EGF and TPA. These
two proteins are similar (isoelectric points, approximate molecular
weight, composition in phosphorylated amino acids) to the two 42K
proteins described in other systems and which have been suggested to be
important for some event in the mitogenic response to diverse agents.
We have observed that phosphorylation of the pair of 45K proteins on
tyrosine residues is not a common characteristic of mitogen treated
thyroid cells : it is induced by EGF and TPA but not by TSH or forskolin,
a cyclic AMP enhancer. We observed no overlap in the patterns of protein
phosphorylation induced by TSH on one hand and by EGF and TPA on the
other hand, except for one protein, 28K, whose phosphorylation is
enhanced but not induced by all agents. Another work performed in our
laboratory by Lamy et al.[1] showed no common proteins whose synthesis is
induced by these different mitogenic stimuli during the prereplicative
phase of the cell cycle : TSH specifically stimulated the synthesis of 8
proteins while EGF and TPA specifically increased the synthesis of one
other protein. Thus, obviously two different phenomenologies could be
involved in the proliferation response to TSH through cyclic AMP on the
one hand and epidermal growth factor and phorbol ester, presumably
through protein tyrosine phosphorylation on the other hand.

REFERENCES

1. Lamy, F., Roger, P.P., Lecocq, R. and Dumont, J.E., Differential
   protein synthesis in the induction of thyroid cell proliferation
   by thyrotropin, epidermal growth factor or serum. Eur. J.
   Biochem. 155: 265-272 (1986).
2. Roger, P.P., Servais, P. and Dumont, J.E., Stimulation by thyrotropin
   and cyclic AMP of the proliferation of quiescent canine thyroid
   cells cultured in a defined medium containing insulin. FEBS Lett.
   157: 323-329 (1983).
3. Cooper, J.A. and Hunter, T., Changes in protein phosphorylation in
   Rous Sarcoma Virus-transformed chicken embryo cells. Mol. Cell.
   Biol. 1: 165-178 (1981).
4. Contor, L., Lamy, F. and Lecocq, R., Use of electroblotting to detect
   and analyse phosphotyrosine containing peptides separated by
   two-dimensional gel electrophoresis. Anal. Biochem. (in press).
5. Parekh, B.S., Mehta, H.B., West, M.D. and Montelaro, R.C.,
   Preparative elution of proteins from nitrocellulose membranes
   after separation by sodium dodecyl sulfate-polyacrylamide gel
   electrophoresis. Anal. Biochem. 148: 87-92 (1985).

CONTROL OF PROTEIN SYNTHESIS INITIATION FACTOR eIF-2α

PHOSPHORYLATION BY INTERFERON

George Thomas

Friedrich Miescher-Institut
Postfach 2543
CH-4002 Basel
Switzerland

INTRODUCTION

This paper is intended to cover the main points of the lecture which was presented on interferon and eIF-2α phosphorylation as part of the Spetsai Summer School Course on "Signal Transduction and Protein Phosphorylation". It was impossible to cover this field in the time allotted, therefore this manuscript should be viewed as a limited introduction to the field. The goal of the talk was 3-fold. First, to briefly describe the effect of viral infection on protein synthesis and how pretreatment with interferon protects the cell against this response. Secondly, to review experiments which led to the finding that pretreatment of cells with interferon induces the synthesis of a protein kinase which phosphorylates the α subunit of protein synthesis initiation factor eIF-2. Finally, to discuss recent results leading to the identification of the molecular components of the kinase complex and the role that each of these components may play in the kinase response.

VIRAL INFECTION AND INTERFERON

Viral infection of many vertebrate cell types leads to an overall reduction in protein synthesis, an almost complete shutoff of host cell mRNA translation and the selective synthesis of viral proteins.[1] A number of mechanisms have been proposed to explain the control of protein synthesis following viral infection. These include intrinsic differences in the secondary structure of viral encoded mRNA as compared to cellular mRNA,[2] thus affecting their relative efficiencies in translation, the synthesis of viral gene products which may be important for viral protein synthesis,[3-5] and the production of large amounts of viral mRNA which in simple numbers may overwhelm their cellular counterparts.[6]

The infection process is also accompanied by the induction of a host cell gene family collectively referred to as the interferons. The type of interferon produced depends on the nature of the inducer and the producing cell. In response to

interferon the producing cells and other cells in the population are stimulated to synthesize a number of enzymes which
protect the cell from viral infection.[ ] At the level of translation, pretreatment of cells with interferon and subsequent
viral infection still leads to a reduction of overall protein
synthesis.[ ] However, in this case, depending on the dose of
interferon, the reduction in protein synthesis is not as severe,
there is an absence of the appearance of viral proteins, and
there is no subsequent cell death.[ ]

    To identify the step in protein synthesis where interferon
was exerting its effect, it was necessary to first develop
protein synthesis systems which mimicked the in vivo effects
of interferon on protein synthesis. In general, cell lysates
were prepared from parallel cultures of cells which had been
pretreated with interferon followed by infection with virus or
heat-inactivated virus (mock infection). The resulting lysates
prepared from mock-infected cells were found to be more active
than those prepared from infected cells. Furthermore, if the
ability of these lysates to utilize initiator [^{35}S]-methionyl
tRNA was tested a very similar response was observed, arguing
that the lesion was at the initiation stage of protein synthesis.[8,9] Further experiments showed that this lesion resided in
the initiation factor fraction. Each of these factors and the
role which they play in the initiation process are outlined in
Figure 1. In general, the approach then was to examine these
partial reactions and determine which were affected by viral
infection. The results showed that the lesion was in the formation of the ternary complex between GTP·tRNA·eIF-2, pointing
to an alteration in eIF-2.[9] Indeed, readdition of eIF-2 to
whole cell extracts led to restoration of protein synthetic
activity.

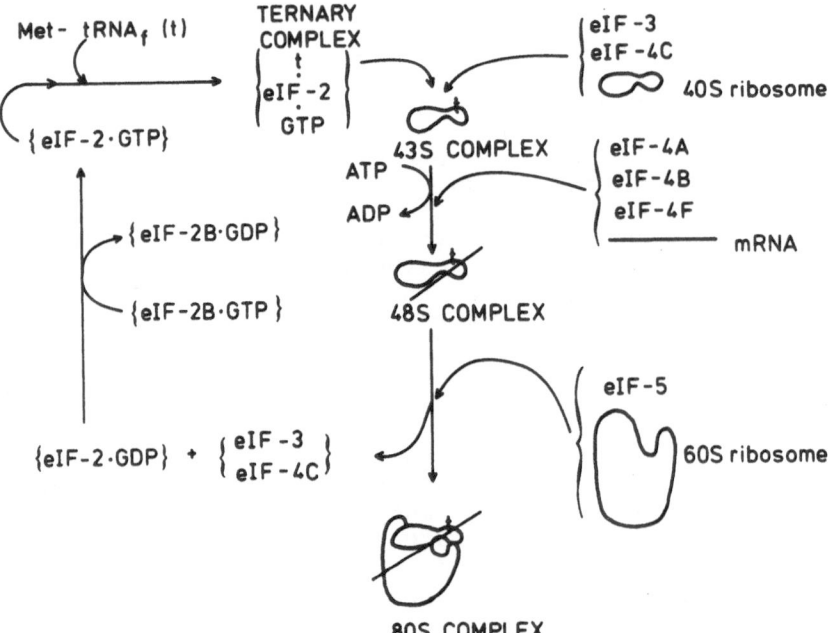

Figure 1.   Initiation of Protein Synthesis

From previous studies it was known that treatment of cells with interferon resulted in the induction of a number of genes.[1] Constitutive levels of these genes were present in varying amounts, usually quite low, in a number of cell types. One of these gene products was a synthetase which synthesized oligomers of 2'-5' adenylate, which have been collectively referred to as 2-5 A. In turn, 2-5 activates a latent cellular endo-RNAase which cleaves mRNA and rRNA.[7] A second product was shown to be a protein kinase which copurified with a protein of molecular weight 68,000 daltons (68 K).[10] Both of these proteins are found in cell lysates in an inactive form and are activated by the addition of double-stranded RNA (dsRNA), which is thought to mimic viral RNA during the infection process. In cell lysates it was also found that treatment with dsRNA, along with increasing the phosphorylation of the 68 K protein, also led to phosphorylation of another protein of 35,000 daltons.[10] It was already known from studies in reticulocytes that the α subunit of eIF-2, which also has a molecular weight of 35,000 daltons, became phosphorylated during hemin deprivation and that this phenomenon was thought to be responsible for the lysate losing protein synthetic activity <u>in vitro</u>.[11] Thus it was not long before it was shown that the <u>primary</u> substrate of the interferon-induced and dsRNA-activated kinase was the same α-subunit of eIF-2. It should also be noted that the kinases activated by hemin deprivation and by dsRNA are distinct, but that the sites of phosphorylation appear identical.

The extent to which eIF-2α becomes phosphorylated in the whole cell can be monitored by its shift on IEF-two-dimensional polyacrylamide gels using a monoclonal antibody and $[I^{125}]$-labeled protein A.[12] The phosphorylated form is readily visible by its acidic shift on the gel. Employing this method investigators have recently followed the fate of eIF-2α phosphorylation in control cells or cells pretreated for 20 hours with two concentrations of interferon followed by viral infection. In the controls there is no change in the level of eIF-2α phosphorylation, whereas in cells pretreated with interferon the basal level of eIF-2α phosphorylation increases from 20% to 40%. This increase takes place around 6 to 12 hours postinfection, a time when protein synthesis begins to decrease.[7]

Though results like those described above agree in kinetic terms, a number of observations have not fit the general hypothesis that eIF-2α phosphorylation is directly responsible for the inhibitions observed in protein synthesis.[11] First, though protein synthesis can be inhibited by greater than 90% the amount of eIF-2α which becomes phosphorylated is only 30% to 40%, the change only being in the order of 20%. Secondly, <u>in vitro</u> partial reactions show that phosphorylated eIF-2 is utilized about as efficiently as nonphosphorylated eIF-2α in which only stoichiometric reactions are followed. Finally, the addition of GTP to lysates can restore translation, without affecting the phosphorylation state of eIF-2α.[11] The first clue to how eIF-2α phosphorylation may affect translation came from the observation that the affinity of eIF-2 was much higher for GDP than GTP. Indeed, this finding together with the known intracellular concentrations of GDP and GTP would have argued that over 90% of eIF-2 should be in an inactive eIF-2·GDP complex. These results then suggested the existence of a GDP-

GTP exchange factor, analogous to the elongation factors of protein synthesis. Such a factor was eventually identified by a number of laboratories. In the presence of this factor, eIF-2β, the affinity of eIF-2 is much higher for GTP than GDP such that at intracellular concentrations of GDP and GTP, over 90% of eIF-2 would be expected to be in an active eIF-2·GTP complex.[11]

How then does phosphorylation of eIF-2α affect its interaction with eIF-2β? Initially it was thought that this event would prevent interaction with eIF2-β, thus blocking the exchange reaction. However, because there is an apparent excess of eIF-2 over eIF-2β in the cell, if only 30% -40% is in the phosphorylated form, 60%-70% of eIF-2 would still be free to interact with eIF-2β. A more attractive possibility which has recently been proposed is that in vitro, at salt concentrations similar to those in the cell, it appears that the phosphorylated form of eIF-2 has a much higher affinity for eIF-2β than the unphosphorylated form. Thus only a small change in eIF-2α phosphorylation could shift all of eIF-2β into the inactive eIF-2β·eIF-2· GDP complex. Under these conditions the remaining eIF-2 would shift into the inactive eIF-2·GDP complex leading to overall inhibition of protein synthesis.[11,12]

dsRNA-ACTIVATED PROTEIN KINASE

What is known concerning the molecular makeup of the dsRNA-activated protein kinase? As mentioned above, it was initially found that this enzyme copurified with a protein which became phosphorylated when dsRNA was added to lysates of normal or interferon-treated cells.[10] To determine whether this protein was indeed the kinase, Hovanessian and colleagues recently prepared monoclonal antibodies to the kinase.[13] To prepare these antibodies they took advantage of the high affinity of the kinase for poly(A)·poly(U) Sepharose. The sepharose beads containing the kinase were used as antigen to raise monoclonal antibodies. In this way they were able to raise monoclonal antibodies which recognized the 68 K protein on Western Blots. These antibodies were then attached to Sepharose and their ability to immunoprecipitate the kinase from cell extracts was tested. The results showed that these monoclonal antibody-linked Sepharose beads (Mab-Seph) contained dsRNA-stimulatable kinase activity.[14] Indeed, these same beads could also use eIF-2α as well as histone as substrate. The surprise was that when the beads were analyzed, that along with the 68 K protein there was a second protein present of molecular weight 48,000 daltons (48 K). As with the 68 K protein interferon also induced the synthesis of the 48 K protein.[14]

The problem then arose which of these two molecules is the kinase? Initial experiments suggested the 48 K protein was the kinase, since the amount of dsRNA kinase activity from different preparations appeared to closely follow the amount of 48 K protein present. Therefore experiments were designed to selectively remove the 48 K protein from the Mab-Seph complex. Using a combination of 1M KCl and ethylene glycol they found they could selectively remove the 48 K protein and leave the 68 K protein bound to beads. However, kinase activity was recovered neither in the unbound 48 K protein nor in the Mab-Seph-bound 68 K protein.[14] Taking advantage of the ability to remove the 48 K protein from the complex a second approach was taken to study the interaction between the two proteins. The Mab-Seph protein complex was first incubated with ATP, then the 48 K protein was either removed or left associated with the complex, and each sample was incubated with histone and ATP. In both cases the complex was capable of phosphorylating the histone as well as the endogenous 68 K protein. These results led to the proposal that the induction of $eIF-2\alpha$ phosphorylation by dsRNA was a three step process (Figure 2).[14] The first step was the activation of 48 K protein kinase by dsRNA. Next, the activated 48 K protein kinase phosphorylates and activates the 68 K protein kinase. Finally, in this activated-phosphorylated state this enzyme phosphorylates $eIF-2\alpha$ leading to an overall inhibition of protein synthesis.

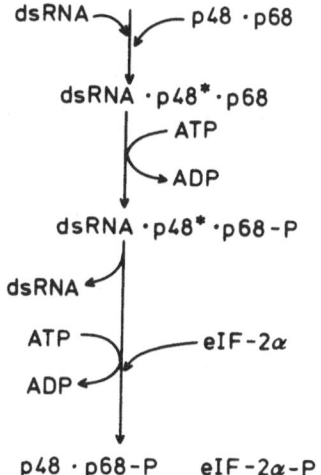

Figure 2.   dsRNA-Activated Protein Kinase Complex

The above model was supported by four independent lines of evidence. 1) The extent of 68 K protein phosphorylation roughly paralleled its ability to phosphorylate histone. 2) Regardless of the level to which the 68 K protein was phosphorylated dsRNA had no effect on its kinase activity in the absence of 48 K protein. 3) Both proteins contain ATP binding sites as shown by their ability to bind $8-azido-\alpha-[^{32}P]-ATP$. 4) The two proteins co-chromatograph on a sizing column.[14] The results, though not conclusive, support the notion of a phosphorylation cascade. This is an exciting conclusion, since even though activation of one kinase by a second kinase has been an attractive possibility, we know of only one other proven example, phosphorylase kinase (see L. Heilmeyer, this volume) and possibly a second, the S6 kinase (see G. Thomas, L. Ballou, P. Jenö and G. Thomas, this volume).

CONCLUDING REMARKS

As pointed out in the Introduction it was not possible to cover the great amount of work being carried out in the field, thus this talk should have been viewed as an introduction to the field. Indeed, for those who would wish to read further, at least two additional studies should be mentioned. The first is a series of studies by Baglioni and colleagues[15,16] demonstrating that along with blocking recycling of eIF-2, the dsRNA-activated kinase also interferes with the binding of mRNA to 80S initiation complexes. The second is a series of studies demonstrating that an adenovirus small RNA (VAI), synthesized late in infection, prevents phosphorylation of eIF-2α and thus the shut-off of protein synthesis at later stages of the infection process.[4,5,17,18]

ACKNOWLEDGEMENTS

I would like to thank L. Ballou, I. Cohen and A. Olivier for their critical reading of the manuscript. In addition I am indebted to Prof. Hans Trachsel who guided me through the vast amount of work carried out in the field.

REFERENCES

1. P. Lengyel, Biochemistry of interferons and their actions, in: "Ann. Rev. Biochem." 51:251 (1982).
2. D. L. Nuss, H. Opperman, and G. Koch, Selective blockage of initiation of host protein synthesis in RNA virus infected cells, Proc. Natl. Acad. Sci. U.S.A. 72:1258 (1975).
3. J. Marvaldi, M. J. Sekellick, P. I. Marcus, and J. Lucus-Lenard, Inhibition of mouse L cell protein synthesis by ultraviolet-irradiated VSV requires transcription, Virology 84:127 (1977).
4. B. Thimmappaya, C. Weinberger, R. J. Schneider, and T. Schenk, Adenovirus VAI RNA is required for efficient translation of viral mRNAs at late times after infection, Cell 31:543 (1982).
5. R. J. Schneider, C. Weinberger, and T. Schenk, Adenovirus VAI RNA facilitates the initiation of translation in virus-infected cells, Cell 37:291 (1984).
6. H. F. Lodish and M. Parker, Translational control of protein synthesis after infection by vesicular stomatitis virus, J. Virol. 36:719 (1980).
7. A. P. Rice, R. Duncan, J. W. B. Hershey, and I. M. Kerr, Double-stranded RNA-dependent protein kinase and 2-5A system are both activated in interferon-treated, end phalomyocarditis virus-infected Hela cells, J. Virol. 54:894 (1985).
8. K. Moldave, eukaryotic protein synthesis, in: "Ann. Rev. Biochem." 54:1109 (1985).

9. V. M. Pain, Initiation of protein synthesis in mammalian cells, Biochem. J. 235:625 (1986).

10. A. Kimchi, A. Zilberstein, A. Schmidt, L. Shulman, and M. Revel, The interferon-induced protein kinase PK-i from mouse L cells, J. Biol. Chem. 254:9846 (1979).

11. B. Safer, 2b or not 2B: regulation of the catalytic utilization of eIF-2, Cell 33:7 (1983).

12. C. E. Samuel, R. Duncan, G. S. Knutson, and J. W. B. Hershey, Mechanism of interferon action, J. Biol. Chem. 259:13451 (1984).

13. A. G. L aurent, B. Krust, J. Galabru, J. Svab, and A. G. Hovanessian, Monoclonal antibodies to interferon-induced 68,000-Mr protein and their use for detection of double-stranded RNA dependent protein kinase in human cells, Proc. Natl. Acad. Sci. U.S.A. 82:4341 (1985).

14. J. Galabru and A. G. Hovanessian, Two interferon-induced proteins are involved in the protein kinase complex dependent on double-stranded RNA, Cell 43:685 (1985).

15. C. Baglioni, Interferon-induced enzymatic activities and their role in the antiviral state, Cell 17:255 (1979).

16. A. de Benedetti and C. Baglioni, Inhibition of mRNA binding to ribosomes by localized activation of dsRNA-dependent protein kinase, Nature 311:79 (1984).

17. P. A. Reichel, W. C. Merrick, J. Siekierka, and M. B. Mathews, Regulation of a protein synthesis initiation factor by adenovirus-associated RNA, Nature 313:196 (1985).

18. R. J. Schneider, B. Safer, S. M. Munemitsus, C. E. Samuel, and T. Schenk, Adenovirus VAI RNA prevents phosphorylation of the eukaryotic initiation factor 2α subunit subsequent to infection, Proc. Natl. Acad. Sci. U.S.A. 82:4321 (1985).

GROWTH FACTOR-MEDIATED ACTIVATION OF S6 PHOSPHORYLATION AND PROTEIN SYNTHESIS

George Thomas
Friedrich Miescher-Institut
Postfach 2543
CH-4002 Basel
Switzerland

INTRODUCTION

In numerous biological systems activation of cell growth is closely coupled to the multiple phosphorylation of 40S ribosomal protein S6. This event appears to be a prerequisite for an associated increase in the overall rate of protein synthesis and to a number of specific alterations in the pattern of translation.[1-4] The purpose of this lecture is three fold. First, to present the background and rationale for studying S6 phosphorylation during the mitogenic response. Second, to review a number of key experiments concerning the possible role of this event in the control of protein synthesis and how it may be regulated. Finally, to outline the types of questions which are now being asked and in what direction the field must advance in the near future.

BACKGROUND AND RATIONALE

The activation of cell growth in the whole animal and in tissue culture leads to the stimulation of a number of seemingly unrelated metabolic events.[5] Since in tissue culture these events were expressed in a coordinate fashion it led Tomkins and coworkers to collectively refer to them as the "Pleiotypic Response".[5] In the original model which they proposed, the stimulation of protein synthesis was classified as one of the mitogen-induced pleiotypic responses. The increase in protein synthesis can be manifested in a number of different ways but it is usually observed as a large shift of inactive 80S ribosomes into actively translating polysomes. Also accompanying this shift is the movement of newly transcribed mRNA and a large pool of stored mRNA into polysomes.[6,7] These alterations in mRNA usage are also associated with a number of specific changes in the pattern of translation. A number of these proteins have been identified and should prove useful for studies on translational control, as discussed below.[2,4]

As one might imagine from the shift of ribosomes and mRNA into polysomes, it was shown quite early on that the increase in protein synthesis is due to an increase in the rate of initiation of protein synthesis.[6,7] The multiple steps in the initiation of protein synthesis as well as the components involved have been thoroughly described (see G.Thomas, this volume), however, as with many of the early pleiotypic respones we know little of the biochemical processes involved in controlling translation.[8] Since these components have been structurally and funtionally well characterized and because we know that protein synthesis is essential for cell growth, we suggested this process may serve as a useful tool for understanding how a growth factor acting on the outside of the cell is able to signal the activation of an intracellular response. It was also clear that it would be difficult to attack all of these components simultaneously, therefore we initially limited ourselves to ribosomal protein phosphorylation-dephosphorylation reactions. This was for two reasons. One, because we knew ribosomal proteins could be phosphorylated under a variety of growth conditions[9] and two, because these reactions serve as common mediators of intracellular signalling systems.[10]

S6 PHOSPHORYLATION IN SWISS 3T3 CELLS

With the above goal in mind, it was found that within 30 min of the addition of serum to quiescent Swiss 3T3 cells there was a 30-fold increase in the amount of phosphate incorporated into the protein fraction of 40S ribosomes.[1] This phosphate was shown to be incorporated into a single protein of 30,000 daltons which on 2-dimensional gels migrated in the position of ribosomal protein S6.[1] The increase in S6 phosphorylation was observed to be tightly coupled to the subsequent increase in protein synthesis.[1,11] This finding was supported by 1) Studies with inhibitors of S6 phosphorylation and protein synthesis, demonstrating that increased S6 phosphorylation did not require protein synthesis but that a dose-dependent inhibition of S6 phosphorylation led to a corresponding effect on protein synthesis[11] and 2) That the two events could not be separated by dose-responsive studies with serum or growth factors.[2] If S6 phosphorylation is involved in moving ribosomes or mRNA into polysomes then it would be expected that the more highly phosphorylated forms of S6 should be found preferentially in polysomes during the time ribosomes are shifting into polysomes. Indeed, this is what was observed, providing one of the strongest arguments to date that the phosphorylation of S6 facilitates some step in the initiation of protein synthesis.[3,12]

POSSIBLE ROLE OF S6 PHOSPHORYLATION IN TRANSLATION

At what level of protein synthesis does S6 phosphorylation exert its effect? From a number of approaches including immune electron microscopy,[13] chemical crosslinking [14-17] and chemical modification[18] it is known that S6 is localized in the small head region of the 40S ribosome, juxtaposed to the 60S subunit in a site involved in the binding of mRNA and tRNA. Furthermore, 40S subunits containing phosphorylated S6 are more efficient in binding and utilizing synthetic mRNA to synthesize polyphenylalanine than unphosphorylated 40S subunits.[19,20] This led to the suggestion that S6 phosphorylation may effect the affinity of the ribosome for mRNA. This finding is consistent with the fact that a number of changes observed in the pattern of translation following mitogenic stimulation are under translational control.[2,4] Indeed, at least two of these changes are apparently due to selective alterations in the affinity of the translational apparatus for mRNA and not due to a simple increase in overall protein synthesis.[4,21]

S6 TRYPTIC-PHOSPHOPEPTIDES

To test the above model in vitro will require a source of S6 phophorylated 40S subunits. Because of the limited amount of subunits obtainable from tissue culture cells it was decided to purify the kinase(s) responsible for carrying out this reaction. This would allow the possibility to prepare differentially phosphorylated 40S ribosomal subunits. However, before embarking on this study, it was necessary to have peptide maps derived from in vivo phosphorylated S6.[22] In addition, since S6 is multiply phosphorylated it raised the possibility that it could be phosphorylated in a random or ordered fashion. If the latter proved the case, then it would be necessary to isolate the kinase responsible for the initial site of phosphorylation. After establishing conditions for labeling S6 to shigh specific activity in vitro it was shown that the fully phosphorylated form of the protein contained 10 to 11 major tryptic phosphopeptides.[22] By carrying out a tryptic phosphopeptide analysis on each of the increasingly phosphorylated derivatives it was shown that the phosphates were added to S6 in a specific order and that the large number of phosphopeptides is most likely attributable to overdigestion with trypsin.[21] It should also be noted that individual mitogens such as epidermal growth factor (EGF), insulin and prostaglandin $F_{2\alpha}$ ($PGF_{2\alpha}$)[23] induced the phosphorylation of the same S6 phosphopeptides.

EGF-ACTIVATION OF AN S6 KINASE

Having these phosphopeptide maps in hand and knowing the dose response curve for EGF-induced S6 phosphorylation in the intact cell it was possible to search for the S6 kinase. Thus high speed cell extracts were prepared from quiescent cells or the same cells stimulated for 30 min with EGF and assayed for their ability to phosphorylate S6 in vitro.[24] The results showed that extracts from EGF-stimulated cells were approximately 10-fold more efficient in phosphorylating S6 than comparable extracts prepared from resting cells. The extent to which the kinase was activated by increasing concentrations of EGF paralleled the dose response for EGF induced S6 phosphorylation in the intact cell. Furthermore, the tryptic phosphopeptide map generated from maximally phosphorylated S6 in vitro was equivalent to the in vivo map.[24] From these results it was argued that the kinase activity followed in vitro was the activity responsible for phosphorylating S6 in vivo.

To recover full kinase activity from cell extracts required the presence of phosphatase inhibitors and metal chelators.[23] The most efficient phosphatase inhibitors were phosphotyrosine followed in order of potency by paranitrophenol phosphate, β-glycerol phosphate, phosphoserine and sodium orthovanadate.[25] In contrast to these agents sodium fluoride inhibited the basal activity of the kinase.[25] That these agents were acting as phosphatase inhibitors was demonstrated by showing that incubation in the absence of the inhibitor leads to a time-dependent loss in kinase activity.[25]

KINETICS OF EGF AND SODIUM ORTHOVANADATE KINASE ACTIVATION

The ability to recover a stably activated form of the kinase from 3T3 cells made it possible to compare its kinetics of EGF-induced activation with that of EGF-receptor binding and EGF-induced S6 phosphorylation in the intact cell. The kinase is rapidly activated reaching a maximum between 5 and 10 min and then slowly returning to basal levels by 2 hours.[25] At 10 min approximately half the EGF-receptors have disappeared from the cell surface. Another 30% slowly desensitized over the next 2 hr, with the number of remaining cell surface receptors leveling off at about 20% the initial amount. S6 phosphorylation roughly parallels the activation and desensitization of the kinase, though kineticly slightly delayed. Furthermore, S6 phosphorylation never completely returns to basal levels over the next 8 hr. The results suggest that the level to which the S6 kinase is activated and the extent of S6 phosphorylation are tightly coupled by some mechanism to the number of EGF-bound cell surface receptors.

To test the last hypothesis, a search was made for some means to circumvent the downregulation of the EGF-receptor. Others had previously shown that sodium orthovanadate alone or in the presence of EGF could stimulate cells to proliferate.[26,27] Furthermore, EGF-receptors still downregulated in the presence of sodium orthovanadate, suggesting this agent may be an ideal candidate for activating the kinase, even following downregulation of receptors.[26] The initial reasoning for employing sodium orthovanadate was because it had been shown to be a potent phosphotyrosine phosphatase inhibitor.[28] Thus, the thinking was that in the presence of sodium orthovanadate a threshold level of EGF-receptor kinase activity would build up and the cells would be stimulated to proliferate. The addition of sodium-orthovanadate to quiescent cells also led to the activation of the S6 kinase. The kinetics of activation were slower than with EGF and the kinase remained persistently at high levels for at least 2 hr. Similar results were obtained in the presence of EGF, suggesting that even though the receptors desensitized, the kinase remained activated. To test this hypothesis, cells were induced for 4 hr with EGF to desensitize their EGF-receptors and then challenged with either EGF or vanadate. EGF had no effect on the kinase or S6 phosphorylation, however, treatment with vanadate led to an immediate activation of both processes.[25] Thus in someway vanadate can circumvent the downregulation of the EGF-receptor.

CHARACTERISTICS OF THE S6 KINASE

The results above imply that the two kinases activated by EGF and vanadate are equivalent. To test this possibility extracts from vanadate- and EGF-stimulated cells were analyzed by Mono Q anion exchange chromatography. In both cases the activated enzyme elutes in the identical position, at a salt concentration slightly higher than that of quiescent cells.[29]

This result is consistent with the argument that phosphorylation is responsible for activation of the kinase. Along with EGF a number of other growth factors (PDGF, insulin, $PGF_{2\alpha}$) as well as tumor promoters and the oncogenes v-src and H-ras are known to activate S6 phosphorylation. All these agents induce the activation of this same kinase. However, the extent to which they induce the enzyme is quite distinct suggesting multiple pathways may be involved in controlling the activity level of the kinase. This is consistent with the fact that a number of these agents are also thought to use distinct signalling pathways. The enzyme has an apparent molecular weight of 75,000 daltons and an estimated pI of 5.5. It appears to be distinct from C kinase, protease-activated kinase and cyclic AMP-dependent protein kinase.[29] This fact is also supported by the finding that the S6 kinase, unlike the enzymes listed above, is apparently regulated by a phosphorylation-dephosphorylation mechanism (see Ballou, Jenö and Thomas, this volume). Though the activation of kinases through phosphorylation cascades has been an attractive possibility, only one other kinase, phosphorylase kinase [30], and possibly a second, the dsRNA-activated kinase,[31] have been shown to be activated by phosphorylation. This is an especially significant finding in the field of growth control where the search for such cascades is being activelly pursued.

FUTURE PROSPECTIVE

The extraction procedure which we first applied to 3T3 cells to recover full S6 kinase activity has been recently applied to different systems by a number of other laboratories.[32-35] In all cases a very similar kinase to the one described here has been observed. To establish whether all these enzymes are identical will require that they be purified to homogeneity. This may prove a difficult task since we have purified the S6 kinase from high speed extracts 100,000 to 500,000 fold, suggesting that this protein is present in small amounts. Along with the kinase a search must also begin for at least three other regulatory enzymes. These include the presumptive kinase responsible for activating the S6 kinase, the phosphatase which presumably leads to inactivation during desensitization (see Ballou, Jenö and Thomas, this volume) and the S6 phosphatase which appears to be a distinct molecule. It will also be necessary to prepare antibodies to these molecules to discover how they are regulated during the mitogenic response.

ACKNOWLEDGEMENTS

I would like to thank Dr. J. Knesel, Dr. A. Nair and A. Ziegler for their critical reading of the manuscript.

REFERENCES

1. G. Thomas, M. Siegmann, and J. Gordon, Multiple phosphorylation of ribosomal protein S6 during transition of quiescent 3T3 cells into early G_1, and cellular compartmentalization of the phosphate donor, Proc. Natl. Acad. Sci. U.S.A. 76:3952 (1979).

2. G. Thomas, G. Thomas, and H. Luther, Transcriptional and translational control of cytoplasmic proteins following serum stimulation of quiescent Swiss 3T3 cells, Proc. Natl. Acad. Sci. U.S.A. 78:5712 (1981).
3. G. Thomas, J. Martin-Pérez, M. Siegmann, and A. Otto, The effect of serum, EGF, $PGF_{2\alpha}$ and insulin on S6 phosphorylation and the initiation of protein and DNA synthesis, Cell 30:235 (1982).
4. G. Thomas and G. Thomas, Translational control of mRNA expression during the early mitogenic response in Swiss mouse 3T3 cells: identification of specific proteins, J. Cell Biol., in press.
5. A. Hershko, P. Mamont, R. Shields, and G.Tomkins, Pleiotypic response, Nature New Biol. 232:206 (1971).
6. C. P. Stanners and H. Becker, Control of macromolecular synthesis in proliferating and resting Syrian hamster cells in monoclonal culture, J. Cell. Physiol. 77:31 (1971).
7. P. S. Rudland, Control of translation in cultured cells: continued synthesis and accumulation of messenger RNA in nondividing cultures, Proc. Natl. Acad. Sci. U.S.A. 71:750 (1974).
8. E. Rozengurt, Early signals in the mitogenic response, Science 234:161 (1986).
9. D. P. Leader, The control of phosphorylation of ribosomal proteins, in "Recently discovered systems of enzyme regulation by reversible phosphorylation", P. Cohen, ed., Elsevier, Amsterdam (1980).
10. E. G. Krebs and J. A. Beavo, Phosphorylation-dephosphorylation of enzymes, Ann. Rev. Biochem. 48:923 (1979).
11. G. Thomas, M. Siegmann, A. M. Kubler, J. Gordon, and L. Jimenez de Asua, Regulation of 40S ribosomal protein S6 phosphorylation in Swiss mouse 3T3 cells, Cell 19:1015 (1980).
12. R. Duncan and E. McConkey, Preferential utilization of phosphorylated 40S ribosomal subunits during initiation complex formation, Eur. J. Biochem.123:535 (1982).
13. U.-A. Bommer, F. Noll, G. Lutsch, and H. Bielka, Immunochemical detection of proteins in the small subunit of rat liver ribosomes involved in binding of the ternary initiation complex, FEBS Lett. 111:171 (1980).
14. K. Tesao and K. Ogata, Crosslines betweenpoly(U) and ribosomal proteins in 40S subunits induced by UV irradiation, J. Biochem. 86:605 (1979).
15. D. R. Tolan and R. R. Traut, Protein topography of the 40S ribosomal subunit from rabbit reticulocytes shown by crosslinking with 2-iminothiolane, J. Biol. Chem. 256:10129 (1981).
16. O. Nygard and H. Nika, Identification by RNA-protein crosslinking of ribosomal proteins located at the interface between the small and large subunits of mammalian ribosomes, EMBO J. 1:357 (1982).
17. K. Terao and K. Ogata, Effects of preincubation of poly(U) with 40S subunits on the interactions of 40S subunit proteins with aurintricarboxylic acid and with N,N'-D-phenylenedimaleimide, J. Biochem. 86:597 (1979).
18. R. Kisilevsky, M. A. Treloar, and L. Weiler, Ribosome conformational changes associated withprotein S6 phosphorylation, J. Biol. Chem. 259:1351 (1984).
19. A. M. Gressner and E. van de Lour, Interaction of synthetic polynucleotides with small rat liver ribosomal

subunits possessing low and highly phosphorylated protein S6, Biochem. Biophys. Acta 608:459 (1980).

20. S. J. Burkhard and J. A. Traugh, Changes in ribosome function by cAMP-dependent and cAMP-independent phosphorylation of ribosomal protein S6, J. Biol. Chem. 258:14003 (1983).

21. H. F. Lodish, Translational control of protein synthesis, Ann. Rev. Biochem. 45:39 (1976).

22. J. Martin-Pérez and G. Thomas, Ordered phosphorylation of 40S ribosomal protein S6 after serum stimulation of quiescent 3T3 cells, Proc. Natl. Acad. Sci. U.S.A. 80:926 (1983).

23. J. Martin-Pérez, M. Siegmann, and G. Thomas, EGF, $PGF_{2\alpha}$ and insulin induce the phosphorylation ofidentical S6 peptides in Swiss mouse 3T3 cells: effectof cAMP on early sites of phosphorylation, Cell 36:387 (1984).

24. I. Novak-Hofer and G. Thomas, An Activated S6 kinase in extracts from serum and epidermal growth factor-stimulated Swiss 3T3 cells, J. Biol. Chem. 259:5995 (1984).

25. I. Novak-Hofer and G. Thomas, Epidermalgrowth factor-mediated activation of an S6 kinase in Swiss mouse 3T3 cells, J. Biol. Chem. 260:10314 (1985).

26. G. Carpenter, Vanadate, epidermal growth factor and the stimulation of DNA synthesis, Biochem. Biophys. Res. Comm. 102:1115 (1981).

27. J. F. Smith, Vanadium ions stimulate DNA synthesis in Swiss mouse 3T3 and 3T6 cells, Proc. Natl. Acad. Sci. U.S.A. 80:6162 (1983).

28. G. Swarp, S. Cohen, and D. Garbers, Inhibition of membrane phosphotyrosyl-protein phosphatase activity by vanadate, Biochem. Biophys. Res. Comm. 107:1104 (1982).

29. I. Novak-Hofer, A. R. Olivier, H. Luther, M.Siegmann, B. Friis, and G. Thomas, Growth factors and inhibitors of 40S ribosomal proteins S6 phosphorylation act differentially through a common s& kinase: reactivation following desensitization, Submitted.

30. E. H. Fischer and E. G. Krebs, Conversion ofphosphorylase b to phosphorylase a in a muscle extracts, J. Biol. Chem. 216:121 (1955).

31. J. Galabru and A. G. Hovanessian, Two interferon-induced proteins are involved in protein kinase complex dependent on double-stranded TNA, Cell 43:685 (1985).

32. R. Nemenoff, J. R. Gunsalus, and J. Avruch, An insulin-stimulated (Ribosomal S6) protein kinase from soluble extracts of H4·Hepatoma cells, Arch. Biochem. Biophys. 245:196 (1986).

33. J. Blenis and R. Erikson, Regulation of a ribosomal protein S6 kinase activity by the Rous sarcoma virus transforming protein, serum, or phorbol ester, Proc. Natl. Acad. Sci. U.S.A. 82:7621 (1985).

34. D. Tabarini, J. Heinrich, and O. Rosen, Activation of S6 kinase activity in 3T3-Ll cells by insulin and phorbol ester, Proc. Natl. Acad. Sci. U.S.A. 82:4369 (1985).

35. M. H. Cobb, An insulin-stimulated ribosomal protein S6 kinase in 3T3-Ll cells, J. Biol. Chem. 261:12994 (1986).

INACTIVATION OF AN S6 KINASE BY PROTEIN PHOSPHATASES

Lisa M. Ballou, Paul Jenö, and George Thomas

Friedrich Miescher-Institut
Postfach 2543
CH-4002 Basel, Switzerland

INTRODUCTION

Stimulation of quiescent cells with agents such as serum,[1,2] various growth factors and hormones,[3-5] oncogene products,[2] and tumor promoters[2,5] leads to the activation of a kinase that phosphorylates ribosomal protein S6. Several enzymes, including cAMP-dependent protein kinase,[6] protease-activated kinase II,[7] and protein kinase C[8] phosphorylate S6 in vitro, but it appears that the mitogen-stimulated kinase is distinct from these. Earlier we described a novel S6 kinase from Swiss mouse 3T3 cells that was stimulated up to 25-fold by serum, epidermal growth factor (EGF), or sodium orthovanadate.[1,3] EGF caused a rapid activation of the enzyme; maximal activity was expressed within 5-15 min and then there was a slow return to basal levels after 2-3 hours. Addition of orthovanadate at this time, but not EGF, led to a reactivation of the kinase. In order to recover the enzyme in an active state, phosphatase inhibitors such as p-nitrophenyl phosphate (pNPP) or β-glycerophosphate (β-GP) had to be present in the extraction buffer, otherwise a time-dependent loss of kinase activity occured. These results suggested that the S6 kinase might be regulated by phosphorylation-dephosphorylation. In this study we have examined this possibility by separating the cellular phosphatases from the S6 kinase and testing them for the ability to inactivate the enzyme. Inactivation by purified protein phosphatases is also discussed.

RESULTS AND DISCUSSION

Separation of Phosphatases from the S6 Kinase

The extreme instability of the S6 kinase in crude systems initially prevented us from developing an assay for the inactivating agent in fractionated extracts. Therefore, it was first necessary to obtain a stable preparation of the kinase. Since we thought that phosphatases were causing these losses in enzyme activity, separation of the kinase from phosphatases should yield a stable enzyme. It was previously shown that various phosphatase inhibitors could protect the kinase acti-

Table 1.   Separation of S6 Kinase and
           pNPPase Activities

| | % Total Units ||
Step	S6 Kinase	pNPPase
Extract	100	100
50% Ammonium sulfate	61	100
Fast Flow Q	17	23
Mono S	8	<0.2

vity in homogenates.[1,3] We observed that reactions containing
pNPP developed a yellow color characteristic of p-nitrophenol,
suggesting that the inhibitor might protect the kinase by acting as a competitive substrate for phosphatases that would
otherwise inactivate the enzyme. If such were the case, one
should be able to follow the putative S6 kinase phosphatase by
measuring its pNPP phosphatase (pNPPase) activity. Thus, the
kinase from orthovanadate-stimulated cells was subjected to
three purification steps and the fractions pooled at each step
were assayed for pNPPase and S6 kinase activities. All of the

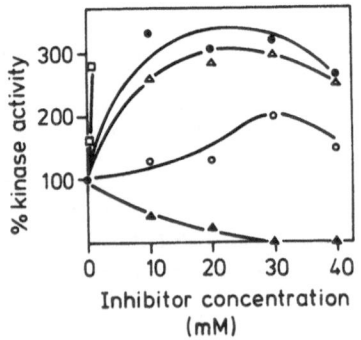 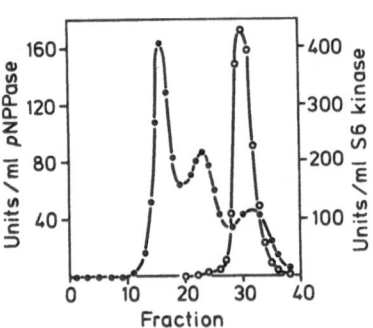

Fig. 1 (left). Effect of phosphatase inhibitors on S6
       kinase activity. Crude homogenates from EGF-
       stimulated cells were assayed in the presence of
       pNPP (●), phosphotyrosine (Δ), orthovanadate
       (□), pyrophosphate (▲), or NaCl (O).
Fig. 2 (right). Elution of kinase and phosphatase activ-
       ities from Fast Flow Q Sepharose. The dissolved
       ammonium sulfate precipitate was subjected to
       anion exchange chromatography and fractions were
       assayed for pNPPase (●) and S6 kinase (O) ac-
       tivities.

Table 2. Time-dependent Inactivation of S6 Kinase
         by Various Phosphatases

| | % S6 Kinase Activity | |
Addition	0 min	15 min
Experiment 1		
Kinase alone	100	79
Kinase + peak A pNPPase	95	68
Kinase + peak B pNPPase	84	30
Kinase + peak C pNPPase	79	1
Kinase + alkaline phosphatase	108	81
Experiment 2		
Kinase alone	100	90
Kinase + phosphatase 1_C	83	67
Kinase + phosphatase $2A_C$	93	39

In Experiment 1 the peak tubes of pNPPase peaks A,
B, and C were dialyzed against buffer with no phosphatase inhibitors and then preheated at 37°C to inactivate any endogenous kinase. The inactivation
reactions contained S6 kinase purified through the
Mono S step and 5.5 units/ml of the various pNPPases.
After 0 or 15 min at 37°C, aliquots were immediately
assayed for kinase activity. The inactivation reactions in Experiment 2 were done in a similar manner
but contained 2.1 units/ml of phosphorylase phosphatase type 1_C or $2A_C$.

buffers used contained β-GP and phosphatase inhibitors were
added to the S6 kinase assays to prevent inactivation during
the phosphorylation reactions. Kinase activities measured in
the presence of 10-40 mM pNPP or phosphotyrosine were up to
3-fold higher than those measured in their absence (Fig. 1).
Orthovanadate (1 mM) also protected the kinase, but pyrophos-
phate was inhibitory.

Table 1 shows that this purification scheme succeeded in
removing more than 99% of the pNPPase from the S6 kinase. All
of the phosphatase activity was precipitated by 50% ammonium
sulfate, and 23% was still associated with the kinase after
anion exchange chromatography on Fast Flow Q Sepharose. The
orthovanadate-stimulated S6 kinase eluted from this column in
one peak (Fig. 2). pNPPase activity was present in three
peaks termed A, B, and C, in order of their elution from the
column; peak C comigrated with the kinase. The relative heights
of these peaks varied from preparation to preparation. The
phosphatase and kinase that were pooled together at the Fast
Flow Q step were then separated from one another by chromatography on Mono S (Table 1). Since the kinase preparation was
now essentially devoid of pNPPase activity, one would expect

that the enzyme should be relatively stable. Indeed, the kinase in the Mono S pool lost only 10-20% of its activity after a 15 min incubation at 37°C (Table 2).

Inactivation of S6 Kinase by pNPPases

Using this stable form of the kinase, we were able to determine which of the pNPPases resolved on the Fast Flow Q column was responsible for inactivating the enzyme. An equal amount of pNPPase activity from each of the three peaks was added to the kinase and the mixtures were incubated at 37°C for 15 min (Table 2, Experiment 1). The kinase alone lost 21% of its activity during the incubation. Addition of any of the pNPPases induced a further time-dependent loss of activity; peak C was the most effective, followed by peak B and then peak A. By contrast, an equivalent amount of alkaline phosphatase, a potent pNPP phosphatase, did not bring about any inactivation. A drop in kinase activity was also observed in control reactions (0 min) that were not exposed to high temperature; this was most likely due to inactivation of the enzyme by phosphatases during the S6 kinase assay, a process that is not completely prevented by 10 mM pNPP.

The specificity of the peak C pNPPase was even more striking when its effect at different concentrations was compared with the other pNPPases (Fig. 3). The peak A material did not bring about any appreciable inactivation over the concentration range tested; although these results were somewhat

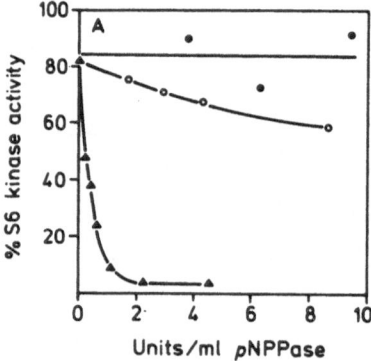

Fig. 3 Concentration dependence of S6 kinase inactivation by various phosphatases. The peak tubes of pNPPase peaks A (●), B (O), and C (▲) from Fig. 2 were dialyzed and then preheated to inactivate the endogenous S6 kinase. The inactivation reactions contained S6 kinase purified through the Mono S step and pNPPase as indicated. The mixtures were incubated for 15 min at 37°C and then assayed. One hundred percent is the value obtained for the kinase alone and no incubation.

variable, in every experiment it was observed that the peak A
pNPPase was much less potent than the other two enzymes.
pNPPase peak B brought about a 50% loss of activity at approximately 13 units/ml, while the peak C enzyme inactivated half
of the kinase at about 0.2 units/ml. The apparent inactivation
of the kinase was not an artifact due to dephosphorylation of
S6 during the assay. Under the conditions specified in Table 2,
the release of ^{32}P from S6 by pNPPase peaks A and B was undetectable, and only about 5% of the counts were released by the
peak C enzyme.

Inactivation by Protein Phosphatases

pNPP can be dephosphorylated by most protein phosphatases
under certain conditions[9] and it also serves as a substrate
for acid and alkaline phosphatases, which act very poorly on
phosphate residues in proteins. Thus, it was important to determine whether a protein phosphatase was associated with the
peak C pNPPase. Fractions from the kinase purification were
therefore assayed for phosphatase activity using phosphorylase
a as a substrate. Like the peak C pNPPase, a fraction of this
enzyme copurified with the kinase on the Fast Flow Q column but
was almost completely removed by chromatography on Mono S. This
phosphorylase phosphatase was inhibited only 20% by inhibitor-2,
a heat-stable inhibitor specific for type 1 phosphatases, indicating that it consisted of a mixture of 20% type 1 and 80%
type 2 protein phosphatases. A lower level (about 25%) of protein phosphatase activity was associated with the peak B
pNPPase, and this was mainly the type 1 enzyme. Virtually no
phosphorylase phosphatase was detected in the first pNPPase
peak (data not shown).

These data suggested that both type 1 and type 2 protein
phosphatases would be able to inactivate the kinase, but that
the type 2 enzymes would be more effective. As a test of this
hypothesis, the kinase was incubated with the purified catalytic subunits of phosphatase 1 and 2A (1_C and $2A_C$, respectively). As expected, both phosphatases inactivated the kinase
but $2A_C$ was more potent than 1_C (Table 2, Experiment 2). The
concentration of phosphatase $2A_C$ required for half-maximal
inactivation of the kinase was approximately 3-4 times less
than that required for phosphatase 1_C.[10]

Taken together, these results strongly suggest that the
S6 kinase is regulated by reversible phosphorylation. Furthermore, since phosphatases 1_C and $2A_C$ are specific for phospho-
seryl and phosphothreonyl residues, it would appear that the
phosphorylation takes place on serines or threonines rather
than tyrosines. Isolation of the kinase and its interconverting enzymes should provide important information about the
mechanisms by which mitogens trigger the proliferative response.

REFERENCES

1. I. Novak-Hofer and G. Thomas, An activated S6 kinase in
    extracts from serum- and epidermal growth factor-stimu-
    lated Swiss 3T3 cells, J. Biol. Chem. 259:5995 (1984).
2. J. Blenis and R. L. Erikson, Regulation of a ribosomal
    protein S6 kinase activity by the Rous sarcoma virus
    transforming protein, serum, or phorbol ester, Proc.

Natl. Acad. Sci. USA 82:7621 (1985).

3. I. Novak-Hofer, and G. Thomas, Epidermal growth factor-
   mediated activation of an S6 kinase in Swiss mouse 3T3
   cells, J. Biol. Chem. 260:10314 (1985).

4. E. Erikson, and J.L. Maller, A protein kinase from Xeno-
   pus eggs specific for ribosomal protein S6, Proc. Natl.
   Acad. Sci. USA 82:742 (1985).

5. D. Tabarini, J. Heinrich, and O.M. Rosen, Activation of
   S6 kinase activity in 3T3-L1 cells by insulin and phorbol ester, Proc. Natl. Acad. Sci. USA 82:4369 (1985).

6. R. W. Del Grande, and J.A. Traugh, Phosphorylation of 40S
   ribosomal subunits by cAMP-dependent, cGMP-dependent
   and protease-activated kinases, Eur. J. Biochem. 123:
   421 (1982).

7. O. Perisic, and J.A. Traugh, Protease-activated kinase II
   as the potential mediator of insulin-stimulated phosphorylation of ribosomal protein S6, J. Biol. Chem.
   258:9589 (1983).

8. C. J. le Peuch, R. Ballester, and O.M. Rosen, Purified
   rat brain calcium- and phospholipid-dependent protein
   kinase phosphorylates ribosomal protein S6, Proc.
   Natl. Acad. Sci. USA 80:6858 (1983).

9. L. M. Ballou, and E.H. Fischer, The phosphoprotein phosphatases, in: "The Enzymes", P.D. Boyer and E.G. Krebs,
   eds. Academic Press (1986), Vol. 17, pp. 311-361.

10. L. M. Ballou, P. Jenö, and G. Thomas, Protein phospha-
    tases inactivate the mitogen-stimulated ribosomal pro-
    tein S6 kinase, submitted.